ETHIOPIA,

AN ANCIENT LAND

ETHIOPIA, AN ANCIENT LAND

AGRICULTURE, HISTORY, AND POLITICS

YEBIO WOLDEMARIAM

THE RED SEA PRESS
TRENTON | LONDON | NEW DELHI | CAPE TOWN | NAIROBI | ADDIS ABABA | ASMARA | IBADAN

THE RED SEA PRESS
541 West Ingham Avenue | Suite B
Trenton, New Jersey 08638

Book and cover design: Lemlem Taddese

Cataloging-in-Publication Data may be obtained from the Library of Congress.

ISBN: 978-1-56902-426-3

DEDICATION

I dedicate this book to the peoples of Ethiopia and Haiti who despite their physical separation were historically linked in their common resolve to defend their freedom and dignity against massive external onslaught.

TABLE OF CONTENTS

ABREVATIONS

Agriculture Development Led Industrialization (ADLI)
Agricultural Implement Research Center (AIRIC)
Agriculture Marketing Corporation (AMC)
All-Ethiopian Peasant Association (AEPA)
Arba Minch Water Technology Institute (AWTI)
Centro Internacional de Mejoramiento de Maiz y Trig (CIMMYT)
Chillalo Agriculture Development Unit (CADU)
Central Statistics Organization (CSO)
Development Through Cooperation Campaign (DTCC)
Diammonum Phosphate (DAP)
Democratic Republic of Germany (GDR)
Desert Locust Control Organization (DLCO-EA)
Ethiopian calendar-Gregorian (EC)
Ethiopian Economic Association (EEA)
Eritrean Popular Liberation Front (EPLF)
Ethiopian Birr (ETB)
Ethiopian Institute Agricultural Research (EIAR)
Ethiopian People's Revolutionary Party (EPRP)
Extension Program Implementation Department (EPID)
Farmer Training Centers (FTC)
Fourth Livestock Development Project (FoLDP)
Food and Agriculture Organization (FAO)
First Livestock Development Project (FLDP)
Five Years Growth and Transformation Plan (FYGTP)
Grand Ethiopian Renaissance Dam (GERD)
Gross Domestic product (GDP)
Gross National Product (GNP)
Humera Development Project (HDP)
Inter Tropical Convergence Zone (ITCZ)
Integrate Pest Management (IPM)

Jigjiga Range Development Unit (JIRDU)
Meter Above sea Level (m.a.s.l)
Meter Below Sea Level (m.b.s.l)
Mela Yethiopia Socialist Nqnaqie (Meison)
Melka Werer Research Center (MWRC)
Minimum Package Project (MPP I)
Ministry of Agriculture (MoA)
Ministry of Community Development (MCD)
Ministry of Land Reform & Administration (MLRA)
Ministry of State Farm Development (MSFD
North East Rangeland Development Unit (NERDU)
National Democratic Revolution (NDR)
Nazareth Tractor Assembly Plant (NATP)
Non-Commissioned Officers (NCOs)
Non-Governmental Organizations (NGO)
Oromo Liberation Front (OLF)
Participatory Demonstration and Training Extension System, (PADETES)
Peasant Associations (PA)
Peasant Agriculture Development Project (PADEP)
People's Revolutionary Democratic Front (EPRDF)
Provisional Military Government (PMG)
Relief and Rehabilitation Commission (RRC)
Second Livestock Development Project (SLDP)
Seed Producers Cooperatives (SPC)
Southern Rangelands Development Unit (SORDU)
South Eastern Rangelands Project (SERP)
Third Five Year Plan (TFYP)
Third Livestock Development Project (TLDP)
Tigray Peoples Liberation Front (TPLF)
United Nations (UN)
United Nations Development Program (UNDP)
Union Socialist Soviet Republic (USSR)
United States Agency for International Development (USAID)
United States Dollar (USD)
Water Resources Development Authority (WRDA)
Wellamo Agricultural Development Unit (WADU)

GLOSSARY

Asrat	Tithe
Akniabat	Early settler
Bega	Dry period
Belg	Short rain before the big Kiremt rain
Corvee	Unpaid labor imposed on the peasant by the feudal lord
Dega	The high plateau, normally above 1750 m.a.s.l
Dejatch	Commander of the gate, title below Ras
Dergo	Supply of provisions to visitors or officials passing through the area
Desa	Land rotated between corporate members of a village every 7 years in parts of Eritrea
Enda	A clan in a village made up of few clans in Tigrai and Eritrea
Ensete	False banana with eatable root system
Gasha	40 hectare
Gebar	peasant/serf/a tax payer
Gmasharash	Fifty percent entitlement
Guilt	Non-hereditary right to collect tribute from lands bestowed to the nobility and the clergy
Kebelle	The lowest administrative unit
Ketemoch	Early towns established in the conquered areas
Kibre Negasit	Glory of kings
Kiremt	Rainy season
Kola	lowlands (hot areas)
Mar	Honey
Lij	'son' title conferred on the sons of royal family and the nobility
Maderia land	Land given temporally to people in lieu of their services
Medri Bahri	Coastal areas of historic Abyssinia (part of Eritrea)
Mekuanint	Title of nobility earned by service to the crown
Melkegna	Representative of the landlord

Mengst	Government- state-
Neftegna	The vanguards- conquerors/settlers in the new incorporated land
Ras,	'head' the highest traditional politico- military title below the Emperor
Rist-guilt	Non-hereditary land right converted into hereditary one
Semon	Church land
Sisoarash	one-third entitlement
Tabot	Replica of the Ark of covenant
Tslmi	Inheritable land used by corporate members of a village and not subject to rotation as *Desa*
Wediabat	An offspring from the original settler of the land
Weynadegan	Mid-Altitude
Woyanie	The 1942 Rebel groups in Tigray/ the current coalition members of EPRDF
Weredas	An administrative category, the Zone
Zemecha	Campaign/raid
Zemene mesafint	Era of Princess (the civil war of the mid-18th to mid-19th century)

ACKNOWLEDGMENT

When I began searching documents relevant to the subject matter almost a decade ago, the encouragement I received from my family was phenomenal. They truly inspired me to write down my life experience as related to the Ethiopian agriculture in which I am formally trained and worked for most of my adult life. In this regard I thank my wife Ethiopis and my children Aman, Rahwa and Imnet for their support. I am also grateful to Dr. Tadelle Ghebreselassie who took his time to review the technical aspect of the book. His critique on few issues helped me rethink about some of my assumptions. I am also thankful to Fentahun Tiruneh of the Library of Congress who came to my help when I needed it. The same is true for Dr Stefan Darcon who was helpful in making some crucial documents available to me when I needed them. Dr Girma Awgichew has been on my side providing editorial assistance during the final stage of the book to which I am deeply grateful. Finally my appreciation goes to Professional Staff Congress of the City University of New York for providing me with fund toward editing and indexing of the book.

PREFACE

I have spent considerable time of my career working in agriculture related fields mostly in Ethiopia. My experience in the field ranges from studying the agricultural systems of the highly developed countries of the US and Canada to Egypt's irrigated culture dominated by few industrial crops to the subsistence farming in Africa and the Caribbean. During my tenure in the Caribbean and West Africa I made it habit to interact with the subsistence farmer to understand his problem. For example in Nigeria's northeast I began to wonder if the primary bottleneck of production was soil fertility exacerbated by narrow based cropping pattern. Guyana another Caribbean nation located in main land South America had similar problems with Sugar production dwarfing food crop cultivation. Rice has been the major food crop cultivated in the Guyanese polder for as long as the first slave landed there. This, notwithstanding, the country continues importing most of its foodstuff from abroad. Generally speaking agriculture in many developing countries has similar characteristics in that it suffers from lack of inputs including moisture, plant protection measures and improved seeds not to mention technical services vital to agricultural vibrancy. This coupled with the unarticulated policy pursued toward the sector made the fight for food self-sufficiency an uphill battle.

In Ethiopia where I had spent two decades of my working life first as research worker and later as private consultant, I had the fortune to observe closely the ailing agriculture system in post-World War Ethiopia. It should be noted here that before the 1974 revolution, powerful feudal lords controlled much of the land to the determent of the peasant. The irony is that after the complete nationalization of land, the Ethiopian peasant did not fare well either. The reason lays in many interlocked factors that the book is trying to shade light on. However, it would be an exercise in futility to understand the current food predicament without probing deep into the agriculture system of medieval and pre-modern Ethiopia. I believe the past has a strong bearing on the agriculture performances of the country and also in its political civility and human right values. Nevertheless, Ethiopia has been and still is a food deficit country partially sustained by outside support. However, the story of modern day Ethiopian agriculture is the story of many countries inhabiting the Southern Hemisphere condemned to the strictly enforced division of labor rules set by the North. Like almost all African countries, Ethiopia as well exports primary produces with no value added to them. With some modesty, therefore, I could claim to be student of Ethiopian agriculture.

There are many compelling reasons that prompt me write the book. The foremost of all was to find answers to the tantalizing question; what went wrong to the country once marveled by earlier travelers for its agricultural endowment. In the case of Ethiopia the sporadic famine and the scramble to contain its destructive effect on the population is too obvious to hide. Thus, judging by the food insecurity that Ethiopia experience today, one cannot help but visit the Portuguese embassy account few centuries ago that states 'agriculturally speaking Ethiopia is within a measurable distance of Europe'. Thus, it is pertinent to examine the period between where the country was an envy of Europe few centuries ago and today. To accomplish that, one has to look into the socio-political condition of Ethiopia and its effect on agriculture development beginning the sixteenth century up to the modern times. The irony is that Ethiopia prides itself, and rightly so for being the oldest free nation on earth that Europe had failed to mess up with it. I found it relevant, therefore, to look into the correlations that exist between historical and political fact on one side and the agriculture performance on the other.

Although, it was long in the making, the Ethiopian food predicament came to the surface in early 1970s with blast. It is here that drought and famine begun to conspire against Ethiopia and the Sahelian States hurling them into center stage of Global affairs. Drought and famine became household news in the developed world to the extent that innocent school children volunteered to chip in part of their snack money to mitigate the tragedy. Since the viewing of ghastly pictures of emaciated children in 1973, food aid to Ethiopia and few other countries of the southern hemisphere become routine affair. Together with the civil war that raged throughout the 1970s and 1980s, the country once recognized as a power in the ancient world begun to acquire a status of poster child of famine, helplessness and despair, hence for writing the book to address some innocent queries posed by people living in the privileged world.

It should be borne in mind that Ethiopia under EPRDF has made some strides with respect to agriculture, infrastructure, institution building and maintaining relative peace within itself. Having said that, the change celebrated today should only be viewed in quantitative term not qualitative one. Despite The hopeful change on the ground, notwithstanding the nation have yet to shun food aid or made job available to its citizens at a required rate. On the other hand, if the widely acclaimed change is less convincing to some it is because Ethiopia was and is a bottom less barrel that requires a gigantic effort to address the massive problems inherited from its unique past.

I

THE ANCIENT LAND: BIRD'S EYE VIEW

Ethiopia shall soon stretch her hands unto God.
—Psalm 68:31

The Beginning

Ethiopia had for centuries captured the imagination of Europe and the Old World in profound ways, primarily, because it is one of the frequently mentioned lands in the bible. According to one Ethiopian website the country is reportedly mentioned 37 times in references to a variety of events, circumstance and situations.[1] It is also mentioned several times in the Holy Koran and in Arabic texts such as 'Sirat Rasul Allah', by Ibn Ishaq (704 – 767 AD). Moreover, the expeditionary force sent to Yemen in 523 AD to avenge Christian persecution by the Jewish King Dhu Nuwas put Axum at the center of the known world politics. The Byzantine Emperor Justin I, believed to have been friendly to Axum instigated the campaign to restore Christian domination in the Arabian Peninsula. The Ethiopian *Kibre Negasit* goes further to show that Ethiopia and Rome were united in the spirit of Christ to defend the faith by splitting the world into two domains, one influenced by Ethiopia and the other by Rome.[2] As early as the time of the birth of Christ,

---

1 http://www.ethiopiafirst.info/news/.

2 Miguel F. Brooks, Kebra Nagast (ክብረ ነገሥት). (Trenton: Red Sea Press, 1995), 172.

the country was mentioned in biblical scriptures where one of the three wise men that came from the east to pay homage to the infant child in Bethlehem was alleged to be an Ethiopian prince. In the middle ages too, it arouse the curiosity of European kings and princes as the land of the Prester John, a devote Christian king 'whose palace was built according to the designs of St. Thomas, the apostle of the *indes,* and contained a marvelous mirror in which he could see all parts of his dominions'.

Axum was second to Egypt and Carthage in the African world to embrace Christianity in the year 332 AD. In other words it accepted the new religion of the east nearly five decades before Christianity became the official religion of the Roman Empire by Emperor Theodosius. In truth, thanks to its unique alphabet, the Geez, Ethiopia became the repository of early Christian writings that are lost in other Christian dominions such as *the Book of Enoch* and *the Book of Jubilee.* According to Isaac (2012) the discovery of the two important Christian books by James Bruce in the late 18th century had an impact in the scholarly world similar to 'the excitement generated by the discovery of the Dead Scrolls in our century'.[3] Mariam Gishen, an ancient monastery located in the northern plateau is highly regarded by Ethiopians as a sanctuary to a piece of the crucifixion wood brought from Jerusalem to the country through the good offices of Alexandria the sit of the Holy See of saint Mark the apostle during the rule of Emperor Zera Yakob in 1446. It is also believed to be the home of the Arch of Covenant allegedly lost during the destruction of the Second Temple in Jerusalem.

The importance of Ethiopia to the rise of one of the major world religions, Islam is also underrated by modern historians. The reason is that it played a crucial role in the preservation of Islam. In this case one cannot help but hypothesize that had the conspiracy of the early enemies of Islam in the Arabian Peninsula succeeded in extinguishing the movement, the only chance for it to spread beyond its center of origin would have been Ethiopia. It is to be recalled that the first *Muslim hijra* was to Axum where the king in 615 AD promised sanctuaries to the persecuted Muslims of Mecca. Among the migrant was the third Caliph and the son-in-law of the Prophet Mohammed, Uthman-ibn-Afan who was credited for spreading the new religion far beyond its borders and who was also known to have instituted many economic and political reforms during his twelve years reign.

One could surmise that the interest of Europe toward Ethiopia might have grown deeper by the beginning of the second millennium for various reasons, among which, was the need to forge an alliance against Moslem occupation of Jerusalem. Salh El Din the founder of the Ayyubdid dynasty, the indefatigable ruler of Jerusalem and the greater part of the Moslem world has

---

3 Ephraim Isaac, The Ethiopian Orthodox Tawahido Church. (Trenton: Red Sea Press, 2012), 127.

to be dealt with decisively even if it takes resurrecting long forgotten old alliance. Late in the middle ages, Europe itself was in peril by the might of the Ottoman Empire that was spreading wildly along its eastern frontiers and the Mediterranean Sea. Thus the desire to dislodge the Moslems from Jerusalem and protect Europe from the invading Ottoman forces must have necessitated forging an alliance among European kings, hence, for Europe's' infatuation with the ancient African nation known to be ruled by a 'Christian kings'. Precisely, for Europe it was a period full of apprehension from Islamic threat but at the same time the ushering of new era of enlightenment (renaissance) and the rising of intellectual curiosity within its borders that stimulated the urge of reviving forgotten old relations nurtured by the Roman Emperor Justin in early six century AD. By the time the Ottoman Turks established a foothold in Eastern Europe and the near east in 16th century, the Red Sea littoral was already in the firm hand of the Turkish conquerors. Unfortunately, at this juncture of its history, Ethiopia was not in a position to engage with the nations outside its now ebbing and at other times flowing borders in any meaningful way let alone enter an alliance with Europe for the common cause.

The rise of Suleiman the Magnificent in 1520 and the historical epoch that soon followed kept the mountainous nation as insular as ever. More significantly, the intra Muslim-Christian war of the late 1520s wreaked havoc in the plateau to further contribute to the country's isolation from the outside world. The Christian rulers of the country unable to withstand an onslaught from within have to rely on their Portuguese allies to defeat the ferocious Muslim lowland fighters who were unstoppable for more than 15 years. However, as time goes by, Europe had successfully contained the Turkish menace and its preoccupation was shifted from religious zealousness to exploration, conquest and exploitation of resources in far off lands.

Throughout the five centuries following the Renaissance, the country like the rest of the continent had visitors from far off lands some traveling overland through Egypt and Sudan while many penetrating the hinterland via the Red Sea port of Massawa and Zeila an ancient seaport in the Indian Ocean. Luckily, many have left their impressions on the country to help us postulate what the past might have looked like, albeit, some may have been subjective when describing the country and its people while few relating their experience with racial overtones like the famous and most important travelers to the region James Bruce and Harris.[4] Coming from their dualistic belief where work ethics (secular) is regarded as supreme, the Ethiopians on the other hand uphold the

4 1/ James Bruce, Travels to Discover the Source of the Nile, in the years 1768, 1769, 1770, 1771, 1772 and 1773, 53 Vol. iii; (Edinburgh: J. Ruthven for C.G.J. and J. Robinson 1790), 2/ Cornwallis Harris, The Highlands of Ethiopia. (London: Longman, Brown, Green 2nd edition, Vol. 1. 1844) 177. 'There is not as Barbarous or ignorant people as the Abyssinians' wrote Bruce. Their dependency on amulet, witchcraft and other device of futurity testifies the fact'

theological position which defines neither faith nor is good work by itself (working the soil or giving alms) sufficient enough for salvation. They both have to complement for the wholeness of the society to the extent the devotee refraining to do any physical work during the sacred days, which in the Ethiopian Christian calendar could amount to more than half of the year.[5] Understandably, this has completely eluded the imagination of early and late travelers to harshly judge Ethiopians using myopic prism to validate their preconceived notion on natives existing outside Europe's ideal world.

Overall, their observation was useful in that it provided us with the opportunity to peek at the country's level of agricultural development at different stages of its historical evolvement. The bountiful source left by the travelers also described the politico-military situations as well as the type of governance that existed then. Here our reference is limited to the past five centuries where traveler's account is in abundance and the local manuscripts are not in short supply either. Undoubtedly, the past has a direct bearing on the country's performance and its ability to face the formidable challenges of producing enough agricultural commodities to feed, clothe and house its citizens. In other words, Ethiopia cannot escape the judgment of its distant past when facing the future.

Given Ethiopia's frequent mention by travelers in the past, it is not implausible to suggest that the country during the Middle Ages had remained somewhat obscure and left to itself. This might have prompted the English historian Gibbons to write '*Encompassed on all sides by the enemies of their religion, the Ethiopians slept near a thousand years, forgetful of the world by whom they were forgotten*'.[6] Hints from historical narratives suggest that medieval Ethiopian courts knew little about lands and peoples beyond their immediate environs and never ventured to explore it aggressively either as a matter of curiosity or with imperial motive. Since the decline of Axum in the eighth century, which by all account is known to have been a maritime power on its own right, there is little record to show that the sovereigns who came after had the same ambition as their forefathers to safeguard an outlet to the sea.[7] It is known that Axumite vessels used to sail beyond the Red Sea to Sir Lanka and by extension India.[8] There are numerous reasons for the decline of Axum as regional power. The

5 Ephraim Isaac, *The Ethiopian Orthodox Tewahido Church*. (Trenton: Red Sea Press, 2012), 55.

6 Edward Gibbon, *The History of the Decline and fall of the Roman Empire*. (London: Ecclesiastical Discord Chapter XLVII: Part V. 1776–1788)

7 G.W. Bowersock, The THORNE OF ADULIS. Red Sea wars on the eve of Islam, (Oxford: Oxford University Press, 2013), 97. Here the book describes how Kaleb, king of Axum, organizes a maritime expedition composed of an armada of 70 ships and about 120,000 men to conquer Yemen and replace unjustly ousted Christian king by the Jew, Yusuf Asar Athar also known by king Dhu Nwas.

8 Wolska-Conus, *Christian Topography III 65*. 1968-73 I: 502-5.

most salient one suspects is the environmental degradation, which triggered gradual migration to the south where rain was abundant and the ecosystem friendly. Given the current desolate state of the northern plateau one could argue with certainty that this circumstance must have contributed to the country's semi-seclusion leading to infrequent interaction with peoples inhabiting adjacent to it and with lands beyond the seas. The exception being the trickle of pilgrims making it to the Holy Land overland, which I presume would have been eager to give and receive information at the same time. Moreover, defeat of Byzantine an ally of Axum by the Persians in 570 AD and the rise of new power, Islam in the 620s must have sealed the fate of Axum not to pretend as a regional contender anymore.[9] Historians give other compelling reason for its isolation. The destruction of the important port of Adulis by Arab pirates who frequently sacked it and made it desolate for normal seas ferry to occur is one of the reason. As implausible as it may appear such traumatic events must have compelled the Axumites to turn southward giving their back to the sea and going inland. This notwithstanding, they have as a matter of fact, continued developing the culture, the art and the architect that they have initiated at Axum in places south of present day Tigrai and Wello. The many rock cave churches and the magnificent rock hewn church at Lalibela are testimony to life after Axum. Through time these seafaring people of the ancient world settled for expansion overland more than the need to have an outlet to the sea. Perhaps the desire to once again assume the role of being maritime power enjoyed by Axum in earlier centuries was seriously entertained by the rulers in the central highland with Zera Yakob/Dawit and later Libne Dingil/Glewdeious of the 15th and 16th century respectively. Unfortunately, given the relative weakness of the empire at the time it was a task difficult to accomplish. Moreover, by then, the Ottoman Turks got hold of the coast effectively blocking the Abyssinians with less freedom to maneuver, hence for their isolation and '*forgetful of the world*'. The overland route west of the plateau was also hazardous and perceived unfriendly by the Abyssinians. Because of their unique religion as well as of their cereal-livestock based mode of production they were treated differently en route to Cairo, the Holy Land and even Europe. The complete overrun of Christian Nubia by Arab nomads during the Ayybuid dynasty in the 12th century further sealed Abyssinia (Ethiopia) from the outside world. There is little doubt that the plateau further south with its amenable climate for humans, and an excellent environment for rearing livestock must have provided the Abyssinians a niche to stay off radar '*forgotten*'.

It is assumed that their seclusion from the world they knew for so long might have also affected their diplomatic prowess to influence events for their

9 Maxime Rodinson, Mohammed, (New York: Vintage Books 1974) 267.

own good. A noted Ethiopian author in his critical analysis of the court of Emperor Libnedigil, who lived in the early part of the 16th century observed that the Emperor at whose reign the Muslim-Christian armed confrontation was on the surge, could have saved his own humiliation at the hand of the Muslim warlords who rose up against him and devastated the plateau for over a decade, had his diplomatic dealings with the Portuguese Embassy in-waiting was somewhat quick and proper. The Emperor's fatal diplomatic mistake according to the author was to delay an audience requested by the Portuguese mission for months and when the meeting took place to engage the foreigners in futile theocratic debate instead of the issue at hand which was the Muslim threat or challenge to his Empire.[10] According to Bruce, the reception of the Embassy by the court was less gracious to encourage cordial discussion to continue. He concludes that it was bad omen for the embassy to be received by Ras Hadug, whom he claimed to be the commander of the asses not the chamberlain. The rift between the Emperor and the Europeans further went from lukewarm to cold because of diplomatic insensibility exercised by the chief of the embassy.[11]

In that case, we could only surmise that the delay by the court to arrange safe passage for the embassy to return home was not only because of the diplomatic hiccup but the occupation of Massawa, the seaport by Muslim Turks must have complicated matters to the worst. Be that as it may, there was no bold initiative taken to send diplomatic mission to the European courts except for one Tsega Zeab who accompanied Don Rodgrguez de Lima the head of

10 TekleTsadek Mekruia, Ye Gragn Ahmed Werera (**የግራኝ አህመድ ወረራ**). (Addis Ababa: Brhanena Selam Printing Press, 1960).

11 James Bruce, Travels to Discover the Source of the Nile, in the years 1768, 1769, 1770, 1771, 1772 and 1773 vol iii (Edinburgh: Printed by J. Ruthven for C.G.J. and J. Robinson 1790), 178. "According to James Bruce when the Ethiopian king asked present from the Portuguese king of 'sheet lead' to cover the roof of his church, Don Roderigo de Lima contemptuously remarked that this is 'trifling' to the king of Portugal who have the resource to cover all churches in Abyssinia with 'sheet lead'. Libne Dingel then added 'That since they were now upon the subject of presents, he could not help letting the king of Portugal know, that, if he ever sent an ambassador again into the country, he should take care to accompany him with presents of value'. Don Rodriguez replied " That it was very far from being the custom of the king of Portugal to send presents to any king upon earth; that, having no superior, it was usual for him only to receive them from others" He added " to accept them or not, according to his royal pleasure". We assume this must have annoyed the king further to delay their departure from Ethiopia. The ambassador's lament that he has been 'mistreated and need to leave his country, got only' injuries and wrongs' in his kingdom 'robbed of everything but the cloth, except his cloths upon his back', further added to the deterioration of relations between the bunch of Portuguese delegates and the sovereign."

the Portuguese mission to Ethiopia in 1520 to Europe.[12] Seven years earlier the queen mother Elleni is known to have sent an Armenian as envoy to Portugal to explain the Egyptian support of Muslim invasion to her territory. Prior to the near complete seal or isolation of the country from the world beyond the sea, few sovereigns among them Wedem Arad and Zara Yakob are known to have sent embassies to Europe to discuss matters concerning religious , military and technical support.[13] Contrary to the early period, however, during the 19th century there was a steady flow of European diplomats, adventurers, missionaries and gun runners handling the affairs of many of the powerful *Rases* in the north and north central highlands up to the emergence of Emperor Menelik II at the end of the nineteenth century, the standard bearer of the Shoan dynasty.

The northern nobility who were at the hub of the Abyssinian Empire for many centuries were too myopic to have an imperial ambition beyond the Nile basin to the west and the lands laying between the Rivers of the Awash and the Wabe Shebelle in the east and southeast. Despite the presence of powerful emperors who had tried their utmost best to enforce a façade of unity among the peoples of historic Abyssinia, the northern principalities remained fearlessly independent of each other. This in part contributed to the inward looking character of the state to pursue any meaningful diplomatic and trade contact with the outside world. More significantly there seemed to be no compelling reason to undertake self-motivated adventures to expose the country and its people to novelties enjoyed by other nations. There is little doubt that regional integrity was more an overwhelming concern among the principalities than a unitary system anchored in a unified objective. This is not however, to imply that there was no common ground for the provinces to meet on mutual interest under the loose confederacy of the Christian Orthodox faith. For example when a new edict was issued to bar Europeans who profess non monophysite beliefs from entering Ethiopia, it was universally adhered by every provincial

---

12 B.W.W. Dombrowski, & F.A. Dombrowski, Frumentius/Abba Salama: zu den Nachrichten über die Anfänge des Christentums in Äthiopien. *Oriens Christianus* (1984), 68: 114-169.

13 In the year 1306 a delegation composed of 30 meet with Pope Clement V and the King of Spain to discuss military alliance while in 1441 two Ethiopian embassies took part in the council of Florence where they received a church and monastery in Italy for their keep parallel to the one in Jerusalem, the Dar El sultan Church. Franz Amadeus Dombrowski in *Ethiopia Access to The sea,* 12-13. Source: 1/Crawford, *itineraries,* 214. F. Jacopo Flippo Foresti da Bergamo-suplementum chronicum data of Giovani Carignano (1291-1329) 2/pastor, *papste* Vol. 1 (1955), 332.

lord worthy of being an Abyssinian for close to three quarter of a century.[14] The first European to be admitted after a long period of pause was Charles Poncet and James Bruce who claimed to have found the source of Nile. In an effort to measure the annual rainfall precipitation of Gonder the imperial city, Bruce assisted by local craftsmen constructed a rain gauge the first modern scientific tool ever in Ethiopia. This is as far as the technical exchange apart from the steady flow of guns to the country went with Europe in the 18th and 19th century.

Despite the steady trek of Europeans to the interior, the seclusion of Ethiopia and its ability to interact with the countries beyond the sea remained unchanged throughout the end of the 19th century. In a way situation such as this has facilitated an intervention by outside forces in grand scale and weakened its position to defend itself. Historians may argue that the attempt by the Turks and their vassals who ruled the Red sea rim basin to unfold Ethiopia by alluring the lowland arid dweller to their cause using Islam as a ploy could not have materialized had the opportunity to acquire weapons was not drastically reduced by the siege. In a nutshell the prolonged civil strife between the Christian and Moslem in the second quarter of the 16th century could be characterized as power rivalry at best.

Moreover, the attempt by Europe to reduce one of the sacred institutions of the Abyssinians, the Orthodox Church, to mere appendage of the Holy See in the same period failed miserably causing dissension among believers and further increased the isolation of the country. Be that as it may, the upheaval that engulfed historic Abyssinia pitting the stalwart of the Orthodox Church and their followers with the new or perceived converts to Catholicism lasted for over two decades. The civil war subsided after the presumed catholic convert Emperor Susenyos abdicated in favor of his son but not before allegedly consuming an estimated 100,000 lives.[15] The accession of Fasil to the throne brought stability and ushered a new era of peace.[16] The new empire

---

14 The Ethiopian Christians adhere to the Alexandrian doctrine, which profess the one nature of Christ (**ተዋህዶ**), anathema to the Byzantine Church, which characterized the nature of Christ as divine and human.

15 Tekeste Negash, Distressed yet Confident Ethiopia Entering the Third Millennium. International Journal of Ethiopian Studies. (Addis Ababa: Tsehai Publishing, 2008), 13.

16 Once left alone in possession of the religion of their ancestors describes Bermudez, a short song was composed to express their joy.

"*Sweet freedom, now, the Ethiop-lamb enjoy*
Snatched from the jaws of the fell western wolves
By the holly Cyril, and the apostle Mark
Exult with joy, and Hallelujah sing."

TRAVELS OF THE JESUITS into the various parts of the world. Relation of the Expedition of the Portuguese into Abyssinia under Don Christopher de Gama, by

based in the vibrant city of Gondar had all the characteristic of post Medieval Europe town with markets flourishing and proprietor class emerging.[17] It is plausible, therefore, to argue that the isolationist path that the country was forced to adopt for so long had a negative impact in the overall transformation of the economy with agriculture as a base.

Agriculture footprint in Ethiopia extends to 1000 BC. No wonder the country is recognized as one of the centers of origin for some important crops and the center of cultivation for many others.[18] The crop species acknowledged to be of Ethiopian origin are Teff (*Eragrostic abyssinica*), Enset (*Ensete verticosum*), Niger seed (*Guizotia abyssinica*), Chat (*Catha edulis*), Coffee (*Coffea arabica*), Kosso (*Hagenia abyssinica*) and Ethiopian Mustard (*Brassica carinata*) It is also asserted that domesticated plants possibly from the Near East appeared longer than originally thought and those crops grown in Egypt and Mesopotamia has been cultivated in the Gash Delta which gave birth to Daamat and eventually Axum in the high plateau of Northern Ethiopia.[19] It is also known to host the fifth largest flora diversity, which is estimated to be between 6,500 to 7,000 species of higher plants in tropical Africa.[20] It has also been in the receiving end where it benefited from adopting exotic crops brought to its shores by European travelers in the late nineteen century. Maize the second staple crop in the country is known to have been introduced from Europe by Portuguese travelers in the country sometime in the 17th century and found growing modestly in Tigray by 1620s.[21] The Europeans were also responsible for the introduction of chili pepper (*Capsicum annuum*), an essential ingredient in the Ethiopian dish today.[22] Another tuber with tremendous potential of growing under diverse climatic condition the Irish potatoes (*Solanum tuberosum*), is a late

---

Bermudez the Patriarch; extracted from Purchass Pilgrim. Ed. J. Lockman 2nd edition. Vol 1. Printed for T.Piety, at Rose and Crown in Palermaster – row London England. 1762, P. 334.

17 Richard Pankhurst, History of Ethiopian Towns: From the Middle Ages to the Early Nineteenth Century Vol 1. (Michigan Il: Publisher Steiner University of Michigan, 1982) 259.

18 1/N.I Vavilov, Studies on the Origins of Cultivated Plants. *Bulletin of Applied Botany and Plant Breeding* v 16: (1926) 1-245; 2/ J.R. Harlan, Ethiopia: A Center of Diversity. *Economic Botany* v 23 (1969) 309-314.

19 Terttia Barnett, The Emergence of Food Production in Ethiopia. (Cambridge Monographs in African Archeology 45. Ed. John Alexander BAR International Series, Published Basingstoke Press, 1999), 763.

20 A.J Sayer, S.C. Harcourt, and M.N. Collins, The Conservation Atlas of Tropical Forest Africa. (Cambridge: IUCN, 1992)

21 James McCain, Maize and Grace: History, Corn and Africa's New Landscape, 1500-1999. Comparative studies in Society and History, 43, (Boston: Boston University Press, 2001), 246-272.

22 Tewolde Berhan Gebre Egziabher, Essay Some Important New World Plants in Ethiopia, Capsicum Pepper. (Addis Ababa: 1984), 192.

introduction to the country[23]. The tuber, which is widely cultivated in the central plateau of Eritrea, came to the scene with the advent of Swedish Missionaries in late 19th century.[24]

Historical evidence also show that steel plowshare has been in use for centuries in the plateau. A plow with exact replica of the Abyssinians was found in the archeological site of Tel Beit Mirsim dated to King Saul. The name of the plow *mahresa* or *mahareset* in early Isrealite tongue sounds similar to the Amharic and Tigrigna words for the plow Maresha/Mahresha.[25] Certainly one would argue, this must have provided an edge to the Abyssinian peasant over those using wooden hoe pegs to loosen the soil and plant seeds. The question is despite the obvious advantages of cultivating the land using iron based tools such as ploughs, sickles and forks, the Abyssinian farmer made no headway in his quest to make agricultural undertakings a sustainable enterprise.

Does the stream of modest number of European adventurers and diplomats throughout the past four centuries positively influenced Ethiopia's trajectory of development as it did to other states or communities, example Japan?[26] Moreover, one is tempted to ask, had not fate intervened to limit the country's interaction with the outside world, would the impact on the instrument and character of labor turned out differently than what it had been in the preceding centuries? Grant, intercourse with far and near neighbors could have been pivotal in sharpening the inner most desire to advance but the fact remains it is the internal dynamics that are determinant factors in propelling societies to advance from one plane to the other. Unfortunately, the internal dynamics conspired to allow Ethiopia advance steadily on the road to progress. One thing is certain, Ethiopia until the turn of the 20th century was standing alone, poor, isolated, and fearful of itself. Mildly put, since the fall of Axum its national psyche had remained buried deep in its majestic mountains and protected by horrid desert plains far well beyond the era of European colonization of the world. In fact when the 20th century dawned, Ethiopia was

---

23 Kidanemariam H.M., Project Proposal for the Development of an Ethiopian Potatoe Program. (Addis Ababa: Manuscript. 1980)

24 Personal communication with Professor Bereket Habteselassie. Although, Gebremedhin Woldegiorgis, senior potato researcher at the Ethiopian Institute of Agricultural Research (EIAR) asserts that potato was introduced to Ethiopia in 1858. *Potatoes are nuggets to save Ethiopia as Ireland shares years of expertise.* THE IRISH TIMES. April 15, 2015.

25 W.F. Albright, 1943b 78 -79, pl.61, 1-7, 14-15 The excavation of tel Beit Mirsim. AASOR 21-22 1943b) 48 and 51.

26 In 1543 Tokugawa, the Shogun of Japan at the time was credited for instructing his blacksmith to build a replica of shotguns brought by Portuguese adventurers. The Shogun was so impressed when the three Portuguese adventurers shot birds from a distant. Despite the success of Japanese artisans to replicate same shotgun, its use was frowned in Japan until the late 19th century.

pretty much ancient adorned with old drab garment struggling to assert its independence from European attempt of domination.

The lurking of Europe on its shores, the undying ambition of Mohamed Ali the Great and his descendants to control the headwaters of the Nile and the vengeful blockade of the sea ports for three centuries by the Ottoman Turks had its toll on the country. In short, when the Italians set foot on the Red Sea shores in late 19th century the country was not significantly different than it was five centuries earlier. Weakened by internal strife and unable to challenge the onslaught of the vagaries of nature, the country was wide open for direct invasion accompanied by sinister conspiracies to keep it in check.

The attempt to redirect the focus of the country toward embracing the outside world in the first quarter of the 20th century is rightfully credited to the late Emperor Haile-Selassie. Amidst protestation by close feudal trustees and objections by recalcitrant landed aristocracy, the Emperor set out to do where no other ruler before him has done. He led a delegation of dignitaries to a European tour including Alexandria and Jerusalem in the spring of 1924. Haile-Selassie, in his mind, must have viewed the royal visit as panacea to the country's ills and unlike no sovereign before him, he pursued the path relentlessly. The outcome of this interaction is hard to assess but it is symbolic for an Ethiopian sovereign to tour Europe and the Middle East on an official capacity when the entire colored world is either under direct colonial rule or an appendage to the power to be.

It is important to note here that, there were neither technical nor entrepreneur members included in the imperial entourage to photocopy ideas and innovation for use upon their return home. This in contrast to Askia Mohammed of the Songhai Empire who travelled to Mecca in early 16th century via North Africa and Cairo accompanied by men of science and jurisprudence with the specific aim of exchanging knowledge and wisdom with those who came in contact.[27] In any case the productive forces of the country were so low and the society exclusively agrarian to expect that a casual encounter such as this would have an impact in eroding all the regressive elements that kept Ethiopia underdeveloped. Simply put, the internal dynamics were not helpful enough to encourage the development of social and material forces to fundamentally change the course. Thus, it was clear by the beginning of the first half of the twentieth century agriculture begun to stagnate and food shortages appeared in the horizon. All this played to the social explosion that rocked the feudal system from its foundation half century later. In a point of fact, until the eruption of the February Revolution in 1974, the Ethiopian system of government was inherently impregnable to assail, immune to criticize and repugnantly archaic to permit any change.

27 John Hope Franklin, From Slavery to Freedom, History of African-Americans. (New York: McGraw Hill, 2011), 16.

In other words, the system in place was incapable of transforming the badly needed means of production or was unprepared to bring forth institutions compatible with the new reality to meet the challenge of the time. Finally, the feudal system relinquished power involuntarily and relegated itself to the dustbin of history by the conclusion of the 20th century. Unfortunately, reform that would have benefited the citizens of the country indirectly much less directly was not even in the figment of the feudal imagination. The hopeful dreams harbored by many with regard to the agriculture potential of the country in the early 1950s and 1960s were dashed to the ground when famine and starvation become Ethiopia's hallmark. The magnitude of the human catastrophe witnessed in Ethiopia in the third quarter of the 20th century further dampen earlier impressions by medieval and modern travelers to the country that it had the agricultural potential to feeding itself and others as well.[28] Long before drought and famine became the fixture sign of Ethiopia on the mind of many observers, Greenfield whimsies at the voices that elevated the potential of the country to "the bread basket of Africa" and "the granary of the Middle East" with monotonous regularity.[29]

When one embarks upon the task of studying Ethiopia's agricultural past, it is with the assumption that the level of agricultural production in medieval Ethiopia was more or less equal if not higher than those of its contemporaries, the European principalities. According to Alvarez, member of the Portuguese embassy to Ethiopia five century ago, agriculturally speaking Ethiopia was within a measurable distance of Europe at the beginning of the 16th century. He also relates that grains were exported from the interior to Arabia, the Mecca, Zebid, Juda (Jedda), and Toro "*And they carry the provisions to the sea to sell them*".[30] Similar account scribed by a 16th century Venetian writer details the type of crops including cereals, fruits and vegetables grown in the high country.[31] Evidence that "Abesh (Abyssinia) abounded in all necessities of life" is found

28 1/ Francisco Alvares, The Prester John of the Indies; A true relation of the lands of the Prester John, being the narrative of the Portuguese embassy to Ethiopia 1520. The Translations of Lord Stanley of Alderley (1881). Additional material by G. F. Beckingham and G. W. B. Huntingford. (London: Published for the Hakluyt Society, University Press, 1961) 2/ Alberto Sbacchi, Legacy of Bitterness Ethiopia and Fascist Italy 1935 – 1941. (Trenton: Red Sea Press, 1997) 104 quoting A. Lessona's, *Memorie* (Florence: 1958), 356 -357.

29 Richard Greenfield, *Ethiopia: A New Political History*. (London: Pall Mall Press, 1965) 321.

30 Francisco Alvares, The Prester John of the Indies; A true relation of the lands of the Prester John, being the narrative of the Portuguese embassy to Ethiopia 1520. The Translation of Lord Stanley of Alderley (1881). Additional material by G. F. Beckingham and G. W. B. Huntingford. (London: Published for the Hakluyt Society, University Press, 1961) 117-118.

31 O.G.S. Crawford, Ethiopian itineraries ca. 1400-1524. (London: 1958)

in literature based on traveler's account of Ethiopia's agricultural prosperity in the past.[32] Another traveler, Jesuit Jerome Lobo in 1626 stated,

> [t]he climate is so temperate, that at the same time I saw in some places plowing and sowing, and in others the wheat already sprouting, while in others it was full-grown and mature, in others reaping, threshing, gathering, and again sowing, the land never tiring of continual production of its fruits or failing in this readiness to produce them.[33]

Today Ethiopia's wheat yield per hectare is equivalent to what the yield was in Northwestern Europe in the turn of the 19th century.[34]

It is important to note that the country had adapted the oxen-plow culture invented by ancient Mesopotamia long since the heyday of the Axumite Empire. By the beginning of 19th century almost 2/3 of the highland Ethiopia was converted into oxen-plow culture. Although, not on the same scale as that of ecologically fragile Egypt, whose complete dependency on irrigation water is real, the Ethiopians of the plateau once upon a time are known to have laid an elaborate system of irrigation waterways feeding the table lands as well as ravines and valleys. In most cases, the irrigated culture adopted by early Ethiopians was purposefully employed as supplementary to rainfall. In other cases, irrigation water was used as a sole source of moisture to raise a variety of cereal, oil, and horticultural crop species.[35]

A cursory look at the technical, social and political reality in medieval Ethiopia may be useful to shade some light on the roots of the current agricultural predicament and could also be a path to understanding the problem of food insecurity in Ethiopia today. It will also answer the question what made

---

32 THE TRAVELS OF MARCO POLO Compiled in 1298 (Barnes and Nobles First Edition 1921, 2005) " *Abesh* Live upon Wheat, Rice, Flesh and Milk and extract oil from Sesame seed. Have an abundance of all provisions."

33 Jeronimo Lobo, *The Itineration of Jeronimo Lobo*, trans. Donald Lockhart, (London: Hakluyt Society, 1984) 245.

34 R.C. Allen, Ethiopian agriculture in a world-historical context. Background note prepared for a study on Agriculture and Growth in Ethiopia. (Oxford: 2008) 2.

35 1/ William Harris, The highlands of Aethiopia: The highlands of Aethiopia described, during eighteen months' residence of a British embassy at the Christian court of Shoa, vol. 3. (London: Longman, Brown, Green, and Longmans, 1844) 269. 2/ Walter Chicele Plowden, Travels in Abyssinia and The Galla country with an account of the mission to Ras Ali,. (London: Longmans green and Co. 1868) 5 and 136. 3/ George Visconte Valentine, Analysis of New Works of Voyages and Travels; Voyages and travels to India, Ceylon, the Red Sea, Abyssinia and Egypt in 1802, 1803, 1804, 1805 and 1806. (London: Printed for Richard Phillise, 1810) 274. 4/ A.B. Wylde. Modern Abyssinia. (Westport: Negro University Press. 1901). Wylde admired "the natives for their ingenuity and the hard work that has to be done every year to keep small water courses in order"

agriculture based human settlement possible and to what extent did agriculture developed. There is an ample evidence to show that agriculture was practiced in the Ethiopian region since 800 B.C.[36] In that case, what effect does human footprint have on Ethiopia to negatively affect its agriculture and the environment? Perhaps other socio-political factors played a negative role to retard agriculture growth contrary to what observers in the past have surmised. Could these factors implicitly answer to the many pertinent questions such as what went wrong in the recent past for the Ethiopian people to be intimately associated with hunger, misery and ignorance? In that case, it may also make sense to argue the deleterious effect–that the three-millennium long human footprint may have had on agriculture and the environment of the country.

The thrust of the book and its focus would be on the following points:

a) review of the agricultural history of the country with specific aim of establishing, defining and interpreting the link that may exist with the present;
b) the anatomical description of the feudal system and its impact on the relations of production;
c) social stability as a major factor for development;
d) critical analysis of the agricultural policies adopted and measures taken to redeem the Ethiopian agricultural from its past woes in the post Haile Selassie era;
e) the technical imperatives necessary to make Ethiopia and by extension the Horn of African countries food self-sufficient with some reference to Eritrea is in order;
f) What cumulative effect if any had European presence in Ethiopia during the 19th and early 20th century?

Physical and Natural Profile of Ethiopia

The Ethiopians pride themselves as having thirteen months of sunshine in their calendar. This, in a sense is true as it describes the geo-climatic features of the country where rain down pours in one section while a step away one enjoys the perfect tropical sun but of temperate nature. It is a country known for it is high rugged mountains and flat tablelands in near 3000-meter altitude. For almost two thousand years agriculture was practiced in the table lands. Its deep gorges and ravines also played as a conduit to cascading rainwaters from the mountaintops, which in turn forms network of rivers and streams crisscrossing the undulating and flat lowlands. The temperature of the low laying lands is above 30°c sometimes souring to over 40°c in the depression.

Ethiopia has a different soil types to allow the cultivation of variety of

36 1/Peter R. Schmidt, Matthew T. Curtis, and Zelalem Teka, The Archaeology of Ancient Eritrea. (Trenton: The Red Sea Press, 2008) 207-210 2/ Martin Bernal, The Afro-Asiatic Roots of Classic civilization, The Linguistics Evidence Vol III. (London: Free Associations Books 2006) 61.

food and industrial crops. The major ones are the Aridisols, which comprises almost half of the country's soil type followed by Entisols and the Vertisols.[37] The heavy Vertisols that makes 10% of the total soil taxonomy of Ethiopia masquerades as an impassable swamp during the rainy season while turning into a network of v shaped cracked land when the rain suddenly stops. It also has high water retention capacity, a desirable trait in soils. Likewise, the thin Aridsol soils of the majority of the lowlands pose clear danger of salinity if enough rain is not forthcoming to wash the elements responsible for salinity.

The ethnic composition of the Ethiopian people is as diverse as the areas they occupy. Their variation is manifested in the different languages they speak, in their attires, their costume and their religion. Over eighty languages are spoken in Ethiopia today. It is a country where the Afro-asiatic languages are believed to have developed and the home of many Proto-Semitic languages spoken in the north, the east and the South. It is also the last bastion of the Omotic, a major branch of the Afro-asiatic language not spoken anywhere else in the world. As of today there are over 16 proto-Semitic languages spoken in Ethiopia against few in the Arabian Peninsula and the Levant.[38] Although crop cultivation dominates as a primary mainstay for millions in the plateau and mid-altitude of the country there are a sizeable number of pastoralists and agro-pastoralist inhabiting the lowland which by comparison is double the size of the plateau. The country is also the home of the Mejenger, an ethnic group solely dependent on hunting and gathering for existence. Ethiopia is one of the ancient worlds to embrace Christianity in early 4th century and Islam soon after its revelation through the prophet Mohammed in Arabia. Strangely as it may sound for a country that embraced the three Abrahamic religions earlier than many of the nations nearby including the old world it also has people professing ancestral religions unmindful of the three dominant Abrahamic religions.

Country Relief

The surface area of Ethiopia is put at 112 million hectares or 1,120,000 km^2 excluding Eritrea which is in the range of 101,000 km^2. Out of this total land mass an estimated 34 million hectares are believed to be potentially suitable for cultivated agriculture. Currently a fraction of this arable land is in use, that is, under present circumstance the land considered to be productive for people to make a living is about 12 million hectares. An additional 1.5 million hectares of land are cultivated during the *belg* season, which begins after the *meher* or main season is over. Except for a miniscule portion where it is under irrigation almost all the cultivated land is strictly under rainfed regime.

37 Roy Donahue, Ethiopia: Taxonomy, Cartography and Ecology of Soils. (East Lansing: Michigan State University, 1972) 25.

38 Girma A. Demeke, The Origin of Amharic. (Addis Ababa: Centre Francais d'Etudes Ethiopiennes, Etudes Ethiopiennes' 5, 2009) 143.

Ethiopia is characterized by a unique physiography that lends itself to numerous valleys and gorges, mountains and hills, gentle escarpments and flat tablelands, plains and ravines. Unlike the lateral climatic bands so characteristics of the western and northern portion of the continent with its distinct layers of rainforest, the savannah, the desert and the winter rainfall in the narrow belt of the Maghreb, the East African landmass which Ethiopia is part of is formed by faulting and volcanic activities that occurred between 23 to 5 million years ago to give rise the Great African Rift Valley and the highland ridges. The lowland plains mostly of semi-arid nature are the beneficiaries of the waters drained out of the uplands by which they are surrounded. For centuries the tablelands in the plateau and the fertile lands down in the ravine and the wide-open valleys have been under terrace irrigation system. Earlier in time the plateau where crop farming has been the mainstay of the population, rainfall was believed to have been more dependable than it is today and was undoubtedly, the primary source of moisture for crop production. Based on the geographical attributes the country is classified into three distinctly recognizable physical features namely, the high plateau soaring to the sky of up to 4500 meters and above, while the low altitude Torrid Zone some of it in the range between negative 125 meters below the sea level (m.b.s.l) to about 1000 meters above sea level m.a.s.l. The mid-altitude areas lay between the upper limit of the low laying lands and the lower limit where the highland agro-ecologic zone begins. Most of the cultivated areas of the country lay between 1000 to 2300 m.a.s.l.

The highlands known as *Dega* are inhabited by sedentary population and supporting above 65 percent of the country's 110 million head of cattle, sheep, goat and donkey. This rugged plateau is mercilessly dissected into numerous gorges and deep valleys with tablelands at the top to make it suitable for crop cultivation for over two millennia. The plateau, commonly referred to as the Abyssinian highlands, is made up of a series range of mountain chains stretching from the Eritrean Sahel Mountains in the north to the most gorgeous mountains of the southwest. These almost uninterrupted chains of mountains dissect the Ethiopian landscape vertically with the Rift valley on its Eastern flank and the Abbay/Tekeze/GashSetit basin in its western flank.

Separated from these and almost running parallel to the Abyssinian highlands are the Hararghe highlands stretching from around Gelemso in the west to the Kondudo mountain group in the east. They include the Chercher highlands (or Ahmar Mountain Ranges), the Kondudo Mountain group and the high plateau between Gara Muletta and Kondudo. These mountain ranges constitute the eastern and southeastern flank of the Rift valley. The Rift valley, except for its extremely narrow body around the fertile plains of the Middle Awash basin, is an open plain in both its northern and southern approaches at times reaching over 1000 kilometers wide in the coastal area of the Red Sea. These mountains are also important watersheds to many of the important rivers that flow about the length and the breadth of the Rift valley basin and to the opposite direction, the southeastern plains. The most notable of these rivers are the Awash, the Genale, the Weib and the Wabe Shebele, which are responsible of irrigating over vast tracks of land in both Ethiopia and the Republic of Somalia.

The low altitude zones or the *Kola* are characterized by hot sometimes-inhospitable weather for most part of the year. The land gradient is flat to gentle rolling landscape crisscrossed by seasonal rivers and streams drained from the high country. Here, there is no significant rainfall to support cultivated agriculture. As a result the main stay of the population continued to be based on transhumant herding. It is in these vast natural grazing grounds of Southeast, Northeast, and Southwest plains that most of the goat, some of the cattle and practically all the camel population are found. These low laying lands are almost completely flat lands with tremendous potential for irrigated agriculture for both pasture and forage as well as for industrial commodities. The upper limit of those areas classified as lowlands is about 750 meters.

The *Weynadega* with their upper limit of about 1500 meters from sea level has an agreeable climate for human habitation. They are also well suited for rearing Livestock and draft animals. Moreover, moisture permitting many crops not commonly adaptable at either high or low altitude areas can grow there successfully. More significantly, they acted as a buffer zone between the malaria infested and extremely hot lowlands and the cool mixed agriculture based highlands. Until recently the lower limit of the mid-altitude areas remained shunned by the highland dweller because of their occasional malaria outbreaks, relative heat, and remoteness from their ancestral lands, the plateau. It is important to note that before their complete overrun by the land hungry drifting highlanders just five to six decades ago this particular zone served as vital strategic reserve for livestock grazing. Unfortunately, this phenomenon is widespread in Eritrea and can be the source of contention in the future between the highland and lowland dwellers.[39]

The Danakil depression, a unique part of the Ethiopian physiography is found located below the sea level (b.s.l.). It is estimated that about 1% of the total area of Ethiopia is found locked below sea level, the lowest point being 125 meters. Recent political map drawn after the separation of Eritrea put some parts of the Danakil depression within the political boundary of the Eritrean state.

The Abbay (Blue Nile) and the Tekezze rivers with limited economic use

39 Berhanu Debele, From institutional constraints and ethnic-related conflicts to better governance of natural resources in the Horn of Africa. *In:* Hurni H, Wiesmann U, editors; with an international group of co-editors. *Global Change and Sustainable Development: A Synthesis of Regional Experiences from Research Partnerships.* (Bern, Switzerland: Geographica Bernensia,University of Bern, Vol. 5. 2010)151–161.

to Ethiopia at the moment starts their long journey from the western side of these ranges of mountains. At its peak, the water discharge of the Abaay is estimated to be 2349 M^3/sec. In total the Abbay makes 85 percent the entire Nile water or adds about 55 Million M^3 annually to the flow. Generally, there are 13 Major river basins that could be useful in generating hydropower and irrigation. Among the most important basin not yet tapped to their fullest extent are the Omo, Abbay, Tekeze, Wabe Shebele, the Awash and the Genale.

Soils Characteristics

The soils of the Ethiopian highland have unique characteristics stemming from the way they were initially formed, hence for their variance when compared with the soils of the humid tropical lowlands. Generally, the climate of the highlands is characterized by summer-long monsoon seasons followed by a prolonged drought, a phenomenon conducive to the formation of Vertisols. It is estimated that Vertisols comprise about 24% of all cropped highland soils of Ethiopia.[40] Moreover, it constitutes about 70% of the highland soils in the traditionally cereal, oil and legume growing areas of the country.[41] Volcanic activity in the region has resulted in the formation of Andisols, although Vertisols are also found associated with volcanic ash as a parent material. The Aridisol soils are found confined in the eastern and to an extent in the southern part of the country. This type of soil unless managed carefully is prone to be saline affecting crop growth seriously.

However, similar to other agricultural soils of the tropics, the plateau soils of Ethiopia are generally deficient in Phosphorus to support optimum crop yields, except for the Andisols derived from volcanic ash. Murphy H. F (1963) using all the major soil categories in Ethiopia as a base for his findings proved that phosphorus, an essential element could be a production limiting factor.[42] There is no dispute that most Ethiopian soils lack phosphorous. Under the prevailing conditions of high fertilizer costs and low agriculture-based incomes, correcting Phosphorous deficiencies by fertilizer application must take into

---

40 S. Jutzi and I. Haque, Management of deep clay soils (Vertisols) in sub-Saharan Africa. ILCA/ICRISAT research, training and outreach project (Addis Ababa: ILCA (International Livestock Centre for Africa, 1985) 23.

41 Berhanu Debele, The Vertisols of Ethiopia: their properties, classification and management. In: *Fifth Meeting of the Eastern African Sub-Committee for Soil Correlation and Land Evaluation. Wad Medani, Sudan, 5-10 December 1983. World Soil Resources Reports No.* 56. Rome: FAO , 1985) 31-54.

42 H.F. Murphy, Fertility and Other Data on Some Ethiopian Soils. (Dire Dawa, Ethiopia: Experimental Station Bull. No. 11, College of Agriculture and Mechanical Arts, 1963), 511.

account its fixation by these types of soils to be of any use to crops.[43] Generally many of the Ethiopian soils in particular the Vertisol types are known to be deficient in Nitrogen. Nevertheless, it is uncommon to find soils deficient in Potassium.

The Climate

Rainfall

Generally, the climatic pattern of Ethiopia is influenced by the annual movements of the Inter Tropical Convergence Zone (ITCZ). The ITCZ is defined as a narrow region of weather zone imaginatively located between the hot, dry continental air of anticyclone origin arriving from the north (the Sahara) and the cooler wetter maritime air from the south (Indian and Atlantic ocean). The big rains get their moisture from the westerly, which is connected, with the north-south movement of the ITCZ. In turn the high pressure system from the Indian Ocean in February brings the small rains to the south, east and central parts of Ethiopia. Thus, the central and most of the eastern half of the country have two rainy periods and one dry season. The two rainy periods commonly known as *Kiremt* (June to September) is normally long while the *Belg* (March to May), is short supporting early maturing crop species. The annual rainfall distribution over this region shows two peaks corresponding to the two rainy seasons, separated by a relatively short "dry" period. The dry period, which covers the rest of the year, i.e., October to January, is known as the *Bega.*

It is apparent; therefore, that rainfall in Ethiopia has a bimodal character with the big rains falling during the *Kiremt* season heralding the main season while the small rains signify the start of the *Belg* season. The *belg* normally arrives in February/March and ends in April depending on the generosity of the season. The main season, the *kiremt* begins in mid-June and progresses toward the end of September. The *Kiremt* is variable in character in that it may extend up to November in the west and southwestern part of the country while exiting early in September in the north especially in Eritrea and Tigrai. The *Kiremt* rain accounts for over 60% of the annual rainfall. Although *Belg* dependent areas of Ethiopia are in the verge of physical shrinking due to poor distribution of rain or the absent of it, nevertheless, in typically normal season, the *Belg's* contribution to the overall food requirement of the country is between 5 -10%. The significant of the *Belg* rain lays more in its role as a facilitating agent for the resurgence of the fauna and flora as well as to bacterial activity in the soil. It also moistens the ground, thus easing the agro technical operations required in preparation to the *Kiremt* season. Given Ethiopia's low level of agricultural

43 D. Miressa and WP RoBarge Soil characteristics and management effects on phosphorus sorption by high land plateau soils of Ethiopia. (1999) (Madison: Soil Sci Soc Am J, 1999), 63:1455–1462.

mechanization, the Belg rains are too much of a blessing not only to the peasant farmer but also the oxen and other draft animals perpetually condemned to pulverize the soil when called upon.

The eastern and the southeastern parts of Ethiopia have two distinct climatic characteristics. The temporal distribution of rainfall over these regions shows two distinct peaks separated by a well-marked dry period. The two distinct dry periods usually fall between December to February and September to November. Although precipitation amounts between 100 – 300 mm in each of the rainy season, still it is very crucial to livestock maintenance barely staggering out of the dry period where grass and water is critically short. The inhabitants of this region are predominately cattle herders.

The western part of Ethiopia has one rainfall peak during the year. In this region as one travels northward the length of the rainy period progressively decreases, resulting from the meridional migration of the ITCZ. Typical to northern rainfall pattern, in Eritrea rainfall stops abruptly in mid-August and rarely continues to the end of it. Despite the abrupt ending of the main rainy season, the narrow green belt in the coastal plains and the escarpment facing the Red Sea receives sufficient rain in November to raise short duration crops mainly sorghum and at times wheat. This productive narrow belt in the Eritrean eastern escarpment plays a crucial role in the lives of the pastoralist in the region as well as the peasants in nearby plateau by providing a safety net during lean times.

Generally, Ethiopia's microclimate is influenced by its physiography. In other wards the occurrence, distribution and intensity of rainfall is partly conditional to the behavior of wind and atmospheric pressure, which in turn is also influenced by relief. Thus, the high plateau known as the *Dega* receives more rainfall than the low laying lands known as *Kola* of the northwest, southeast and northeast plains. The annual rainfall of the *Kola* rarely exceeds 300 mm with the exception of Gambella a region normally classified as part of the southwestern low laying areas. By comparison this area receives ample rainfall to sustain cultivated agriculture without supplementary water from irrigation. The *Weinadega* is an intermediate zone situated between the *Kola* and *Dega* areas. Relative to the *Kola* it receives more precipitation but in a customary erratic fashion and therefore not dependable. The average annual rainfall in these areas is in the range of 500 to 700 mm.

In general, the northern massif is known to receive less rainfall than the southwestern highland whose annual precipitation is over 1000mm. One can say with certainty that rainfall is more dependent and predictable here than anywhere else in Ethiopia. Here the rainy season is unusually longer extending up to 5 months. With each passing time, however, the perimeter of the drought is expanding into the one-time high rainfall areas of the country. In those areas and indeed in most of the western and central plateau rain distribution is rather becoming erratic to affect crop production and postponed the regeneration of natural pasture after a long dry spell.

Temperature

Temperature invariably varies with relief. The higher the point of reference the colder it is and vice versa. However, during certain season of the year some areas turn hotter or colder than others. Thus, in the highlands summer can be slightly cooler while in the lowlands intolerably hot. In contrast, the breeze in the lowland during autumn turns into somewhat colder days and at times very cold nights. In all the cases the mid altitude areas have almost mild air temperature throughout the year. This being the case, in some instances where the plateau abruptly cease to give way to steep escarpment like that of the eastern Eritrean plateau, the Semien highlands below the Tukil Dingai, or the steep eastern landscape of the central highlands a sudden but contrasting climatic change is encountered. In a radius of 10 -15 kilometers the air temperature drastically changes to reflect elevation difference. The degree to which temperature level drops or elevates instantly in response to relief so as to induce variability in the eco system remarkably places Ethiopia into a unique position housing distinct animal and plant life unknown elsewhere.

Therefore, it is not entirely farfetched to assume that the highland farmer was highly influenced in his cultivation practices as well as in his crop selection techniques by climatic patterns. For example, in the high plateau crops requiring longer time to mature are planted near the end of the *belg* rains, that is, before the onset of the big rains. Here the temperature level is at its highest. From the physiological standpoint, this assists in the vegetative development of the crop, that is, in the accumulation of plant biomass, before the relatively colder season of the *kiremt* sets in. Conversely, the gradual drop in temperature during the *kiremt* season allows slow dry matter accumulation in favor of a prolonged seed or fruit development resulting in higher seed yield or total bio mass. If on the other hand, the rains fail to appear in early April the farmers abandoned these long maturing crops such as corn or specific varieties of sorghum and instead prepare for the big rains to raise crops of intermediate and short duration nature such as wheat, barley, teff (*Eragrostis abyssinica*), pulses etc. The crops with relatively shorter life cycle are either planted at the beginning of the *Kiremt* rains or sometimes in the middle of it. In this case both the early and late planted crops are exposed to warmer temperatures and sunny weather conditions existing near the end of the rainy season assisting in normal seed filling and the maturation process. Some specific crops such as Chickpea (*Cicer arietinum*) are planted on residual moisture that is when the rains are formally to have ended and the temperature especially the night temperature turns chilly.

Peasants in the mid altitude region take advantage of the early rains to plant crops specific to this agro-ecologic zone namely legumes such as haricot bean, cowpea, peanuts as well as sesame, safflower, maize, sorghum, wheat and even teff (Eragrostic abyssinica). Because of the relatively high temperature found in this zone the later four crops yield less than otherwise would have been the case if planted in the colder high plateau.

Vegetation

Vegetation in Ethiopia is broadly classified into the broadleaf and hard wood wet forests and the Savannah woodland. In the high country the *Junipers procera* together with the Cypressus and Pinus are found dominating the landscape. The latter two are also found in the upper scale of mid altitude zones of the country and are believed to be an early introduction to Ethiopia or perhaps indigenous to the land. Other trees whose origin or center of cultivation is believed to be the Ethiopian region are *Dombeya torrid, Acacia abyssinica, Erythrina brucei,* and *Pittosporum viridi*. The *Podocarpus gracilior, Podocarpus falcatus,* and *Acacia umbelliferous* sps are found across ranges of altitudes except at the extreme highlands (3200 – 4000 m.a.s.l) where afro-alpine vegetation such as *Lobelia rhynchopetalun* is found growing.

The wild olive, the *Olea chrysophylla*, the *Olea africana*, the *Hagenia abyssinica* and the *Eric arborea* are well adapted to the highlands and are found distributed up to 3700 m.a.s.l. zone. In terms of their specific adaptation, the Podocarpus are found covering the Western plateau and the western edges of the Eastern highlands namely the Chercher and Arsi chain of mountains. The upper zone of adaptability for the Podocarpus is in the range of 2600 to 2700 m.a.s.l. Lately, quiet a number of Cypresses and Pinus sp. were planted in the western edge of these mountains as a result of a concerted campaign to revegetate the barren slopes. The central highland around Shoa contains the remaining *Junipers procera* woods while the Broadleaf forest are found in the south and southwest of the country, namely in areas around Keffa, Illubabor and Wellega. Yet the the recent introduced tree, *Eucalyptus species*, is the most widely cultivated with multi-purpose use. It is found growing from near sea level to around 3000 meters altitude. It is used for firewood, construction, and agricultural tools and medicinal as well. This is an indelible legacy of Emperor Menelik II who brought the tree seeds from Australia in the turn of the 20th century.[44]

Although the thorny thin leafed *Acacia species* are widespread in the mid altitude zones and mostly the lowland habitats they are also found scattered here and there in the northern and central highlands. In the past the *grar* or *momona* from the family of acacia occupied a sizable part of the landscape in the

---

44 (Notice of megabit 12, 1905 EC (March 20,1913 AD. Addis Ababa: Ministry of Agriculture.) According to earlier documents, Eucalyptus sps. was viewed with suspicion when introduced first. That is because of its fast growing characteristics at the expense of plants growing near it and because of the 'the earth which it eats'. The ministry of agriculture, therefore, passed an ordinance urging people to 'pull up 2/3 of it and let 1/3 remain'.

plateau. Be that as it may, the difference between the acacia sps found in the different altitudes of the country lays in their specific economic use and relative size. Those dominating the lowlands are bushy and some are economically important for their raisins and gums while those found in the plateau are large in size and had in the past served as an inexhaustible source of fuel wood and construction materials. The acacias are last of the little vegetation still found growing in the lowlands while they are almost extinct in the highlands.

It is difficult to state with complete certainty the extent of vegetation cover in Ethiopia at different period of its historical existence. The impression of few experts in the field suggests that at one time half of the country's one million kilometer square landmass was covered by forest. This hypothetical figure was based or deduced from the broad observation made on the climatic and soil conditions of the country.[45] Centuries later the forest cover was reduced to mere 91,000 Km^2, a fraction of the original size exposing the country's irredeemable resource, the soil, to all sorts of degradation.[46] It was not before the satellite imagery techniques become available that the world came to realize that the greener portion of Ethiopia is significantly smaller to its relative size. Based on the available information, therefore, Ethiopia is one of the countries that at one time were impressively forested land but now denude most of its vegetative cover. An eighteen century traveler in his account relates the dense vegetation of *Sycamore*, *Nabaca*, *Cedar* and *Candelabrum* that he observed in the highlands of Eritrea and areas around the ancient settlement of Yeha.[47] The northern section of the country namely Tigrai, Wello, Gonder, North Shoa and the state of Eritrea once the envy of arid Arabian Peninsula across the sea are gradually turning into dryer areas with shallow soil depth. The north including the state of Eritrea accounts barely 4% of the total forested land in Ethiopia. While the province of Shoa together with Hararghe make 3% and 0.6% respectively. Whatever is left of the forest, it is found confined to the southwest corner of the country with the broadleaf vegetation making 65% of the total

---

45 H.P. Hufnagle, Agriculture in Ethiopia, (Rome: FAO of the United Nations, 1961), 405-6.

46 Yemastaweqia Minster (Ministry of Information), Agriculture in Ethiopia. (Addis Ababa: 1964) 3.

47 James Bruce, Travels to Discover the Source of the Nile, in the years 1768, 1769, 1770, 1771, 1772 and 1773, vol. iii. (Edinburgh: Printed by J. Ruthven for C.G.J. and J. Robinson 1790), 270-271.

whole.[48] The woodland/savannah dominated by acacia and shrub trees is believed to be less than 8% down from 30% few decades ago.[49]

To add insult to injury, this particular agro ecologic zone has been the target of an extensive development schemes sponsored by successive governments throughout the decades. It was at the expense of this natural vegetation that the expansion of state run agricultural farms and state supported human settlements has been executed during the 1980s. Even today the regions with relatively abundant broadleaf woods and acacia trees are the target of foreign corporate sponsored development schemes. At present there are few minuscule protected forest of moist and dry evergreen in the central highlands and southern regions of the country good for their bio diversity interest than for their commercial exploitation. Today the church grounds are considered as the repository to many of the ancient and modern wood species otherwise extinct. The total forest cover which, in the turn of the 21-century was estimated at 3% is in constant danger resulting from reckless deforestation actions.[50]

Water Resources

At the time when human advancement in science made it promising to convert saline water into readily available commodity, Ethiopia is still in the initial stage of harnessing its surface water potential for good use. The country has 12 major river basins and an equal amount of inland lakes and water bodies. It is not by accident, therefore, that the country is sometimes referred to as the water tower of East Africa. Water, as much as 123 Billion M^2 is discharged annually from the high plateau draining toward the eastern and western basins, thus, supporting life in Ethiopia and beyond.[51] During the peak flow, that is, from mid-July to mid-August, the Abay (Blue Nile), the Omo, the Tekeze and the Baro rivers alone discharge about 15 Billion cubic meter of water.[52] Life among the pastoral population would have been unbearable without those rivers which from time to time overflow their banks to create lash green land in what is sometimes categorized as horrid climatic zone. But by far the greatest

48 Daniel Gamachu, Environment and Development in Ethiopia. In Beyond the Famine: An Examination of the issues Behind Famine in Ethiopia. Ed. By A. Penrose. (Geneva: International Institute for Relief and Development and Food for the Hungry International, 1988), 54-96.

49 UNDP Emergency Unit for Ethiopia. 1995. Country Information Brief. (Rome: FAO of the United Nations, 1995)

50 TheINDEPENDENT. (AFP 18 July 2010). http://www.independent.co.uk/environment/ethiopias-forest-cover-triples-ministry-2029508.html)

51 Gebr Tsadik Degefu, The NILE: Historical, Legal and Developmental Perspective. (Victoria Canada: TRAFFORD Publishing, 2003)

52 Sileshi Bekele, Challenges and opportunities in Capacity Building for Water Resources, R&D.(Addis Ababa: MoWR/EARO/IWMI/ILRI/ Workshop 2002)

beneficiary of this water resource is Egypt and the Sudan, which receives close to 85% of their waters from the Abay River, which also constitutes the larger portion of the Nile River, combined. The catchments area of the Abay is over 300,000 KM^2 assuring a continuous supply of water to Egypt and the Sudan.

The Base Resources

Agriculture together with livestock constitutes the primary base resource of Ethiopia. In 2013, agriculture accounted for 47% of Gross Domestic product (GDP), 60% of exports, and 85% of total employment.[53] In addition, minerals in the form of gold and fast developing light industry geared for domestic consumption have been playing a modest role in the overall development of the country since the early 1950s. Rock salt is mined in the western part of the country and the Danakil depression in the northeast which until the turn of the past century has been the primary medium of exchange for people in the market place. Despite, the wide scale use of salt extraction from the sea beginning the 1900s in the Red sea coast, rock salt has lingered as a primary cooking commodity in rural Ethiopia to this day. As far as Eritrea is concerned the primary base resource of the country is no different than that of Ethiopia except that there is insignificant amount of gold mined in Eritrea today and fish and marine salt can be considered as an exclusive to Eritrea since its separation from Ethiopia proper 1991. Although not of recent phenomenon, since the coming to power by the Ethiopian People's Revolutionary Democratic Front (EPRDF) the industrial base using local materials has expanded impressively.

There are four agricultural systems recognized in Ethiopia today. By far the most extensive and dominant one is the Seed-Farming complex. It is perhaps the oldest and widely adapted system with grain cultivation practices at the top augmented by plow culture. Roughly two-third of the country's agriculture is made up of grain-plow complex. The second most important agriculture complex is pastoral, practiced by the Somalis in the east and the Afar population of the Northeast. In reality though, the livestock-herding zone makes uninterrupted belt around the high plateau practically relegating the small ruminants, draft animals including camel in the low country. In this case the Boran and the Guji occupy the southern lowlands, while the Tigre speaking ethnic group herd their livestock in the northern plains west of the Eritrean highlands. The Nuer whose numbers are by far greater in the Republic of South Sudan also inhabit the Southwestern region of the country. The third and most significant agriculture system is the Enset planting complex widely adapted in the southern portion of the country. Depending on land size, climatic conditions and population density, Enset could either stand by itself as a

53 Ethiopia Economy 2015, CIA World Factbook www.theodora.com/wfbcurrent/ethiopia/ethiopia_economy.html.

primary crop with cereals, vegetables and coffee as complimentary or secondary to other crops. The last but insignificant of the agriculture complex is shifting cultivation practiced by few notably the Gumuz in the Nile basin.

Despite agriculture is the mainstay of the population, until recently it did not receive much attention by successive governments. By contrast, the EPRDF led government has tried to address the chronic agricultural problems facing the country by focusing mainly on the peasant farmer but also exploring other possibilities to accelerate agricultural growth. Some of those possibilities currently under review include leasing millions of hectares to foreign corporations as partners in development. The merit of such strategic choice to enhance agriculture development could be untenable in the long run especially when viewed within the context of Ethiopia's current stage of development. This coupled with uncontrollable population growth dominated by largely rural economy may conspire to unravel the foreign operated high tech agriculture scheme taunted by the government. While the Imperial government gave lip service to the development of this peasant sector, the Military led government (*Derg*) that came after it attempted to overhaul it with little benefit to the peasant. Its dogmatic approach to solving the land tenure system and the impractical method it used to boast production has left the Ethiopian peasant under the mercy of foreign aid for most of its tenure in office.

The Ethiopian agriculture has been primitively structured for so long to absorb any shock, natural or manmade. Primarily because the subsistence nature of the system did not encourage surplus production accompanied by elaborate storage facilities and at the same time create market infrastructure that in turn serves as stimulus for more production. In other words, this important sector was and still is not mechanized, deprived of modern agricultural inputs and lack the necessary technical support to uplift the archaic system into modern one. Ordinarily, peasants in Ethiopia and in particular those in Tigrai, Eritrea, Wello and perhaps Shoa and Begemdir rarely produce enough grains to carry them from one planting season to the other. Nevertheless, throughout the times the subsistence mode of livelihood has hold but tenuously on this purely agrarian society. This condition existed until the mid of the 20th century when it begun to stagnate or break down to meet the minimum standard of food requirement or more poignantly when population growth begun to outstrip food production. Normally, the shortfall in production was met by importation of grain and mostly in the form of grant and emergency aid.

A closer look at Ethiopia's annual cereal production of the past few decades clearly illustrates the degree of the country's agriculture stagnation. For example production of cereal crops in 1960 was estimated at 4,351,000 MT enough to support a population of 22 million at the time. Two decades later, that is, 1980 production increased by mere 1,068,700 MT while the population grew by 60%. Nearly forty years later that is by the year 1997 food production increased by about 23% and the population exploded to astronomic figure of close 140%. In a nutshell the growth performance of the agriculture sector was reported to be 2.1% per annum between the years 1965 and 1973, following the revolution and up to 1980 it decreased to mere 0.6%. Overall the annual agricultural growth rate in this period was much lower than the growth rate immediately before the overthrow of the Imperial government in 1974. By the year 1990 the growth rate showed negative performance of 2.5%. However, a decade after the downfall of the *Derg*, that is, by 2000 the agriculture log growth went up by 2.1% annually but only to the level of the pre-revolutionary period.[54] However, agriculture growth is reported to have accelerated 3 to 4 folds during the decade of the new millennium.[55] Currently a total of 12,086,604 hectares are under peasant cultivation producing grains including cereals, pulses and oil crops amounting to 21,857,043 MT of which 86 percent is cereal.[56] According to government statistics, a year later production went up by an additional 3.3 Million MT.[57]

The country's main agricultural export commodity is unprocessed coffee beans followed by oil crops, and raw hides. Coffee remains the primary foreign revenue earner since the 1920s and contributes nearly 80% to the GDP. In recent years, because of the extraordinary attention given by the government, flower production has grown exponentially to encourage both foreign and domestic growers to invest handsomely.

One of the most abundant and practically untapped resources of the country is the livestock sector. By all accounts, the country stands tenth in rank among countries that own huge number of livestock and certainly the first in Africa. According to the latest statistics released, the total number of livestock including cattle, draft animals, small ruminants and camel is about 110 million.[58] In the 1980s Livestock constituted 33% of the gross value of agricultural production and 15% of the GDP. Indeed live animal ranked the second to coffee export when only 12% of that was exported legally while the rest was smuggled out to the lucrative Arabian market illegally. The bulk of the animal population is found in the plateau while most of the export is from the lowlands. It is understandable that the pastoralists have a huge stake in the development of the livestock sector because most of the exported animals come from the lowlands. The pastoral population in Ethiopia is estimated to

---

54 William Easterly, Growth in Ethiopia: Retrospect and Prospect. (New York University Center of Global Development, 2006), 9. https://williameasterly.files.wordpress.com/2010/08/12_easterly_growthinethiopia_bc.pdf

55 Accelerating Ethiopian Agriculture Development for Growth, Food Security, and Equity. Report by IFPRI, ILRI and IWMI. Bill & Melinda Gates Foundation. (July 2010), 1.

56 CSA, 2012.

57 CSA, 2013.

58 FAOSTAT, (2010).

be a quarter of the total population, if not more. For long, little of the animal resources were gainfully exported either as live or in processed form. There are many reasons for that but the most significant of all is shortage of animal feed and lack of proper health care for animals. Most of the livestock in Ethiopia are in the state of not marketable condition because of the relatively poor market infrastructure and lack of adequate information on the sector. Ethiopia's main export from this important base resource is animal hides in raw and semi-processed forms.

Traditionally fish and fish products are not widely consumed by the population except for sections of people inhabiting the Red Sea coast. This is despite the abundance of it in the salted sea and fresh water lakes. The fish catch thus remains very low. In 1997 Eritrea harvested less than 4,000 tons from its vast domain of the Red Sea while Ethiopia made a catch amounting to only 10,000 tons from its fresh water lakes.[59] Almost a decade later Ethiopia's catch was in the neighborhood of 18 thousand tons.[60]

Population

In the late 18th century when Malthus came up with his essay on population he affirmed that man's ability to reproduce and hence multiply exceeds his ability to produce food so as to be able maintain with his subsistence mode of life. He based his theory on the fact that population grows in geometric mode while food and other basic necessities of life just increase in arithmetic fashion. He went on to argue that the more assured is in his subsistence livelihood the more man continues to grow in number. This, however, is not found to be true as one observes the negative population growth seen in the industrialized nations of the world where livelihood is almost assured and famine and pestilence is rarity, example Quebec in Canada in the 1990s. Again the assumption that natural and man-made calamity acts as check for population growth may have little relevance in today's world where easily transferable modern novelties such as antibiotics, vaccines, and other miracle drugs from the developed to the developing world make their way to make some difference.[61] Although in measured quantity, the drugs protected and even prolonged life as never before. One could also include in the mix, a new phenomenon hitherto unknown in international relations, that is, food aid helped prevent catastrophe of biblical proportion. Obviously, these two factors have pushed the population of many of the developing countries to a dangerous level. If one accepts Malthus theory,

59 FAOSTAT, (1997).

60 FishSTAT, (Rome: FAO of the United Nations, August 24, 2012) http://www.fao.org/fishery/ statistics/software/fishstat/en

61 Unlike the Black Death in the Middle Ages and the Influenza of the early twentieth century that claimed tens of millions of lives, the specter of the horrible Ebola epidemics was minimized by internationalizing the issue once it appeared.

as is, then the reality of Ethiopia and the Horn region will prove him wrong. Nowhere in any part of the world does one observe perfect storm develop time and time again than in the Horn of Africa and yet population level grew undiminished. Thus, despite the periodic famines that plagued the region and the continuous wars that killed, maimed and dislocated millions of people during the past 50 years the population of Ethiopia, Eritrea and Somalia in a span of 60 years went from about 23 million to 110 million today.

Generally, Population serves as pool for productive force to emerge and play an important role in societal development. Population is also essential in creating a dependable home market to increase national wealth etc. In this regard population in equilibrium with actual resources available is blessing and certainly desirable. On the other hand, for dependent nations still struggling to fulfill the minimum basic necessities of life for their population at a period where national sovereignty is but subservient to global market, where nation building process is still in the trying stage and where resources are constantly diminishing, unregulated population growth is a liability more than asset. With the income derived from sale of primary commodities falling faster than the shooting star, many of the developing states are incapable of providing basic social amenities let alone invest on essentials, a prerequisite to building vibrant society enjoying education and health care benefits. In the case of Ethiopia and the nations surrounding it, the situation is further compounded by the unending wars fueled by uninhibited flow of arms to the warring sides.

Although Ethiopia's ills have their roots somewhere beyond the new phenomenon of population explosion, the fact remains that unchecked population growth in the 20th century did not help in breaking the vicious cycle of underdevelopment. However, it is important to state that population as such cannot be regarded as solely responsible for the abysmally low performance of the agriculture sector in Ethiopia. Nevertheless, beginning the 1960 the unchecked growth of both human and livestock population together with the lack of progress in the agriculture sector accelerated the downward spiral of the economy. However this may be, it is safe to argue that food production might have been somewhat in equilibrium with population growth in the years before the 1960s not to warrant massive food aid as was the case in the 1970s and the decades that followed. This again is an assumption based on empirical deduction rather than on statistical facts.[62] Be that as it may, chronic famine

62 Data compiled from various source to illustrate that food production was not catching up in required amount with the geometrically increasing population since 1960s. The consumption/head is appeared skewed in 1984 and 2003, because it was the worst drought period in Ethiopia.

YEAR	POPULATION	CEREAL(MT)	POP. INCREASE%	CONSUMPTION / HEAD
1960	23,844,300	4,351,000		0.18

and frequent droughts covering about a third and sometimes half of Ethiopia's farmlands is a phenomenon that was reported after the 1960s. Evidence indicate that food aid was sent to Ethiopia for the first time in 1959/60 under the United States Agency for International Development (USAID) program called PL 480 Title II to two districts of Eritrea.[63]

According to the US Bureau of the Census the population of Ethiopia is growing at a rate of 3.0% annually. The population, which was estimated at 20 million in 1950, has passed 60 million marks in the year 2000, that is, an increase of 200% in half a century. A decade later the population is speculated to have jumped to 90 million. Out of the total population of 90 million about 55,050,000 are in the productive stage of life, that is, between 15 years and 60 years of age. An equal number or 42.8% of the population are below 15 years of age described as not yet in their reproductive age. This segment of the population is regarded as the primary contributors to the population explosion in Ethiopia for this century. In contrast, the population above 50 years of age makes one-sixth of the total population. It is important to note that in a span of quarter of century since 1984, Population has doubled while food production failed to keep up with the pace of population growth.[64] Ethiopia ranks number 14 in the scale of world population ahead of Egypt.[65]

The plateau, where the ecology is long neglected and the physical environment highly degraded, the habitat is increasingly becoming inhospitable for people to pursue a predictable life. The rough estimate of those living in the plateau is put at approximately 60 to 65 million. Although overpopulation is universal in the highlands some tend to be better off than others. For example, in South central Ethiopia the land to man ratio is alarmingly narrow that people cannot depend on farming alone and thus support themselves as migrant labor working for the sugar estates and cotton plantations located outside their immediate surroundings. This is not an off-farm employment opportunity where subsistent farmers throughout the world practice before returning home in time to cultivate their fields, but something permanent. In addition, a substantial number of the population is found engaged in the tertiary sector of the economy. In the districts of Sidama, Hadia and Kambatta the

1976	33,717,000	4,458,600	41	0.13
*1984	39,868,572	4,730,950	67	0.12
1997	58,412,000	9,472,807	144	0.16
*2003	70,000,000	9,006,237	193	0.13
2013	90,000,000	21,583,523	277	0.24

63 Margaret Bonner, USAID/ Ethiopia Program from FY 1951 to FY 1978. (Addis Ababa Mission, 1978).

64 US Census Bureau. International data Base, (2010).

65 CIA FACT Book, (2010).

concentration of rural people per square kilometer has been in the range of 480, 344 and 484 respectively.[66] Theoretically, each able bodied peasant in these districts would receive as his/her share a parcel of land ranging between 2066 to 2907 M^2. From this argument it is not difficult to guess that within the past quarter of century the situation has gone from bad to worst. The picture if not worst is the same in the northern plateau especially in Wello, Gonder, Tigrai and few others.

---

66 CSA, Population Projection of Ethiopia for All Regions at Wereda Level from 2014 – 2017, (Addis Ababa: 2013)

II

FOOD AND FAMINE

"ቆጭቋጭና አራሚ የሌለው አታክልት አይረባም። ንጉስና መምህር የሌላት አገር እነደዚሁ ናት"

—**Debtera Zenebe**,
Church scholar of the mid nineteen century.[1]

Global Perspective

For the first time in human history scholars came to agree that food shortages is not a remote possibility but a reality that the world has to face in unison to overcome it. True, after the Second World War death caused by famine was regarded as a passing phenomenon that once nature finds its balance things will return to normal. The chronic famine that India has to endure for decades was treated as such. In 1943, close to four million people were known to have perished in Bengal alone. Similarly, 45 to 50 million of famine-induced deaths were reported in China during the Great leap Forward experiment of the late1950s.[2] The most tragic of all was the pre-Second World War Ukrainian famine that claimed the lives of 6 – 8 million people. This and many others were however rationalized as isolated regional problems that need not draw universal attention or require global solution.

---

1 "A garden without a gardener is useless (overrun by weeds). The same is true of a country without (an effective) king and a teacher." Mashafa c'awata sigawi 1857, (Addis Ababa: Goh Tsibah Press, 1924).

2 Frank Dikötter, *Mao's Great Famine: The History of China's Most Devastating Catastrophe, 1958-62.* (New York: Walker Publishing Company, 2010), 333.

However, beginning the early 1970s and throughout the 1980s the issue of resource depletion and the fair allocation of resources together with the protection for the environment become a matter of international concern. It was argued that a global Climatic change was not hoax but a reality and if not addressed in time it could have dire consequences on agriculture threatening the very survival of humans in the planet. Many of the untested assumptions that climatic change could bring a detrimental effect to agriculture production alarmed many governments. As a result numerous initiatives were taken to address this crucial issue that seem to concern poor and rich, developed and developing countries alike. In light of the greenhouse effect that seemed to affect world climate and in turn disrupt food production the world over, the Kyoto Protocol was initiated in 1997 and adopted in 2005 for the sole purpose of curbing carbon emission responsible for global warming.[3] It is also fair to assume that the hysteria was prompted in part by grain production shortfalls in many countries during the early 1970s. The huge grain purchase by the Soviet Union in late 1970, the cruel drought in California in the mid-1970s and the persistent drought in sub Saharan Africa throughout that period caused legitimate concern. Moreover, in1986 a severe drought hit the west and southwest of the USA destroying crops worth 2 billion dollars. Despite the ominous sign of climate change from 1950 to the mid-1980s, total world grain production expanded 2.6 fold. This may be attributed primarily to the improved food production technology in the developed countries and in part by the Green Revolution.

Amidst the hysteria that global climatic change may trigger shortfalls in grain production worldwide, came the catastrophic drought of the Sahelian drought followed by the much-publicized drought, famine and revolution in Ethiopia. During the summer of 1973 an estimated 200,000 peasants perished in Wello and unknown number of drought victims in Tigrai and the middle section of the Afarland. It is assumed that many more must have perished silently in the remote corners of the country including the nomads in the southwest rangelands and the Ogaden of eastern Ethiopia.

Record show that the events that occurred in the 1970s was momentous for the world organization and regional groupings to attempt mediate trade regulations, provide forum for dialogue between the countries of the South and North and extend aid and support to those in need. The United Nations (UN) without completely shifting its focus as being the political safe valve for hot bed issues became an active body in coordinating technical supports and facilitating multilateral and bilateral aids to the countries of the Southern hemisphere. Ethiopia as a victim of natural calamities and as the founding member of the organization was a beneficiary of many of the largess that the

3 Kyoto Protocol to the United Nations Framework Convention on Climate Change (New York: United Nations 1998)

UN has to offer. Thanks to US major contributions to the world body, the UN through its branches continued to be involved in technical but mostly temporary aid relief to Ethiopia

Almost at the same period many international organizations supported by public donations matched by government grant funds mushroomed overnight to the extent of overshadowing many of the activities initiated by many governments in Africa. The Non-Governmental Organizations (NGO) as they came to be known played a vital role in famine relief. In some cases besides handling emergency crisis at particular place and time, they also stepped their bounds in areas where the local institutions appeared to be weak and ill prepared to handle the crisis. Among many of their interventions is the one that they helped pave the way for the formation of an early warning system to forestall impending disasters. For the system to function at the expected rate, the collection of information on the ground and the rapid processing of the data become vital. True enough, since the consolidation of the information system television viewers of the developed world were spared of the ghastly sight of skeletal beings very common in their screens during the 1970s and 80s. In mid-1990 there were 250 big and small, white and colored, politically motivated and humanity driven NGOs in Ethiopia.[4]

The History of Famine in Ethiopia

Today more than ever, Ethiopia including the states surrounding it lives under the perpetual shadow of famine resulting from food shortages. People requiring food aid has been increasing exponentially since the 1960s. A glimpse at statistical data on famine in Ethiopia shows that in the 1970s, some one million people were affected by drought. A decade later areas hitherto unaffected by the drought was included in the list of aid recipients bringing the total food aid dependent population into six million. Earlier in 2000 a distress call was issued to fill the gap resulting from production slump in the neighborhood of 830,000 MT. In early 2002 an estimated 14 million of people were categorized as needing food aid. By the time the *meher* (kiremt) and the *belg* season were formally declared absent, the status of some 10 million Ethiopians were upgraded to dangerously vulnerable position and therefore, in dire need of food

4 No doubt the NGOs have played crucial roles in saving lives by distributing food aid in the drought and famine affected areas. However, the negative effect of food aid on the agricultural sector should not be underestimated. As subjective as it may seem, the negative impact of food aid on the overall agricultural performance of the country was hinted by agriculture workers in the field. The distribution of food aid has also distorted market prices for certain commodities like wheat and oil and thus discouraging farmers in high potential areas to improve productivity and increase production.

for mere survival.[5] To make up production shortfalls, the government as always apportioned funds from the national treasury to finance the purchase of grain worth 43 million dollars. This scenario was repeated in 2003 where 4.5 million people in addition to 3.4 million people already in the food aid program were threatened with specter of famine. Finally in 2009, after three successive years of short rains, the government came with an estimate figure of 6.2 million people needing immediate food aid on top of the 7.5-million chronically food-insecure Ethiopians receiving assistance through a largely U.S.-funded program known as the Productive Safety Net.[6] The benevolence of the *meher* season notwithstanding, the plea for more food aid is louder as ever. For example a year after the bountiful harvest of 1996/97, the country was reported to have needed over a million ton of cereal and non-cereal food aid. During this revelation, production shortfall for cereal and pulse was estimated to have been at 28 and 19 percent respectively.[7]

In alarmingly repeated fashion, the people leading a precarious life in the Horn of Africa are increasing in number by each passing year. There are numerous underlining factors responsible for the deepening crisis that the region is currently facing. For now suffice is to pinpoint that by the end of the1950s the subsistence production system in Ethiopia begun to sway precariously into the danger zone where hunger and starvation begun to come out on the open. In almost a decade after that it left a large segment of the population vulnerable to famine. This was amplified by the successive drought and locust outbreak that devastated the extreme northern corners of the empire in particular the highlands of Eritrea and Tigrai throughout the third quarter of the last century. In hindsight, one can argue that these events laid the corner stone of the country's dependency on food aid for decades to come. Significantly enough, these events also internationalized famine and hunger on world scale to the extent that it became the concern of civil societies in the industrialized world. But to be fair, the events of the past half-century are the symptoms of the long existing structural problems that had festered Ethiopia for over three centuries.

Throughout its long history, Ethiopia has been faced with many devastating famines some of biblical proportion. Between 1500 and 1900 there were 30 major outbreaks of famine often with localized effect but at times covering a wide range area.[8] Half of the famine incidence occurred between

---

5 1/Ethiopian Herald, (Addis Ababa Jan 20, 2000) 2/UN IRIN (March, 2000)

6 Peter Heinlein, Ethiopia's Meles Accuses Donors of Exaggerating Food Crisis. (VOA, October14, 2009)

7 FAO/WFP, Special Report: Crop and Food Supply Assessment Mission to Ethiopia. Economic and Social Development Department (Rome: FAO 21 Dec 1998)

8 Girma Kebede, State and Development in Ethiopia, (Humanity Books, 1992), 157-8.

1540 and 1812.[9] In the last century alone Ethiopia witnessed 7 major famines of greater scale.

However, the most celebrated famine was the one that occurred in the late 19th century. The Great Famine of 1888 – 1892 known as "yekfu qen" in Amharic and "akahida" in Tigrigna ravaged the entire country leaving fields uncultivated, forcing people to migrate aimlessly and exposing the once impregnable country to European invasion. The great famine and its lingering effect were felt all over the Ethiopian region. Villages were abandoned and residents helplessly roaming from one part of the district to the other in search of food. A European who lived in Ethiopia for a long time describes the famine as dreadful that claimed untold lives. He writes in *Les Missions Catholiques* that during twenty years of his residence in Ethiopia he had seen all the natural and manmade calamities but had 'never been witness to so frightful a spectacle' as that of the famine that befall on the land of the Ethiopians.[10] In the most salubrious and agriculturally speaking the most productive district of Eritrea, Seraie over half of the 275 known villages and hamlets were completely abandoned without a single soul left to vigil during the slow death of these left behind.[11] Roughly, one third of the population and 80% of the cattle were known to have perished during the Great Famine.[12] Most of the recorded famine is believed to have a profound effect in the highlands of Abyssinia where settled agriculture was the rule more than the exception. The extent of the cattle plague and drought was felt all throughout the lowlands and not sparing someareas in the Great African Rift Valley.

Grant, the cause of the famine could be generally attributed to the injudicious interaction of man with his environment. This phenomenon of man to nature imbalance was glaringly noticeable in the past 80 to 100 years where the human appetite for raw material reached the stage of insatiability. In this case the equilibrium that existed between man and nature has been upset more dramatically in recent years than at any time in history causing among other things, recurrent drought. To ordinary Ethiopians, the role that nature played in exacerbating shortage of food followed by societal dislocation needs no explanation. But there is more to it than simply blaming nature and justifying the death of the countless victims of famine in the past century as divine

9 Mengustu Woube, The Geography of Hunger: some aspects of the causes and impact of hunger, (University of Upsala, Department of Social Economy S-751 22 Uppsala research Report #95, 1987)

10 Richard Pankhurst, The history of Famine and Epidemics in Ethiopia prior to the Twentieth century. (Addis Ababa: RRC Publisher, 1985), 230. Pankhrust quoting original source of Father Picar from his note *Les Missions Catholiques.*

11 Ruffilo Perini, Da qua a Mereb **(መረብ ምላሽ).** (Chiaramonte Roberto Edition, 1905), 86-93. (Tigrigna)

12 Girma Kebede, State and Development in Ethiopia, (Humanity Books, 1992), 158.

scourge. Instead, by way of diagnosing the maladies that afflicted Ethiopia for several centuries it is possible to provide a sympathetic argument if not a defensible one to the chronic food crisis that the country continued to experience since the early 1960s. What are the causes of the chronic food shortages that Ethiopia continues to experience for the past half century? What historical elements if any are responsible for the wider gap that perceived to exist between potential and actual food production? Some of the factors that could be attributed as the primary contributors to the current food shortages and underdevelopment of Ethiopia are described in the following sections.

Feudalism Described

Ethiopian feudalism was perhaps one of the most important social variables of all that was responsible for the country's agrarian stagnation. As rightly argued by Cohen and Weintraub (1975) the effects of kingship, the conservative value of peasants, and the change-resistant religious institutions are but secondary issues that one should not dwell excessively when explaining the causes of Ethiopia's static agricultural conditions.[13] Indeed feudalism remained to be the most distinguishable variable more than anything else to explain Ethiopia's predicament. It provides the link necessary to explain the intricate relationship between the land tenure system and the producers, the peasant folks.

The feudal system, which has existed in Ethiopia since the beginning of social stratification has barely changed with the changing times. The social categories namely the royalty and the warrior class, the spiritualists and church functionaries and those who by providence or royal decree condemned to eat bread from 'the sweat of thy brow ' remained as rigid as possible. Above the surface, save for some peculiarities Ethiopian feudalism resembled those of other nations with similar agrarian system. Broadly speaking the closest of the land tenure system resembling that of Ethiopia is of the ancient Israelites where the land is apportioned into private (communal, kinship and village corporate), the Royal land and the priestly land.[14] What distinguishes Ethiopian feudalism from others is that it had a variety of features at various regions. In other words, it did not have universal character to impact uniformly throughout the land. The difference in modality between the northern part of the country and those of the recently incorporated regions of the south is obvious. Having noted that extraction of surplus was uniform across the cultivator.

This rather diverse and flexible production relations evolved in time at different parts of the provinces hindered the precipitation of a unified understanding of the system and on the management of land and labor. In this

13 John M. Cohen and Dov Weintraub, Land and Peasants in Imperial Ethiopia. (The Netherlands:.Van Gorcum and Comp. B.V – Assen, 1975), 18-19.

14 Oded Borowski, AGRICULTURE IRON AGE ISREAL. (Winona Lake. Indiana: Eisenbruns, 1987), 22-30.

case the provincial lords continued to extract both commodity and labor surplus at whim. Likewise, the absence of recourse for something unjust (either through local hierarchical structure or through an Imperial agent) stifled initiatives one expected coming from the producer, the trader and the artisan. Instead, the ultra-parasitic relationship instituted a millennium ago between the feudal lords and the peasant/serfs continued up to the close of the 20th century.

The Anatomy

Before the sweeping land reform of 1975, land was owned under a variety of tenure systems. Suffice is to say that land tenure system in Ethiopia was extremely complex with Wello alone believed to have had over 100 different systems of surplus extraction. Generally speaking there may have been as many as twenty varieties recognized by Ethiopian agrarian experts.[15] In a sense what seemed to be applicable to one region or a district within a region might not necessarily be relevant to the other. Technically, though land belonged to the sovereign until its supremacy over it ended in early 1974. In this regard, there is little distinguishable feature one can speak of between the successive monarchies who took the realm after Menelik II, the founder of modern Ethiopia. The constitution of 1932 and its revised 1952 constitution, unequivocally put land, the people and the law as belonging to the sovereign. Similarly, the Land Proclamation No. 31 of 1975, which in effect striped the big landowners the right to own huge lands, declared that 'all rural land shall be the collective property of the people'. Moreover, the law prohibited the private ownership of land and forbade the sale, lease, mortgage and transfer part or whole of it under any circumstance. In effect the people were given the usufructuary possession right of the land amounting to no more than 10 hectares each. The New constitution drafted by the EPRDF, the successor to *Derg*'s rule in September 1995 also put land under the ownership of the 'people and the government'. Like those before it, the law did not go beyond making land available to those who work on it with no provision for sale and mortgage attached to it.

Before the 1974 popular upsurge that propelled the army to power, the land question was such a burning issue that it became a battle cry for students determined to challenge the old order. As mute as it might have been, land issue was also a subject of debate in the Imperial parliament. For the peasant serf who lived all his life yearning to own a piece of land with no landlord hovering over him like a vulture, land reform was a matter of death and life. According to Desalegn (1984) the land tenure system in Ethiopia can be placed into two broad categories. In the north, access to land was usufructuary with some rights

15 Desalegn Rahmato, Agrarian Reform in Ethiopia, (Uppsala: Scandinavian Institute of African Studies, 1984), 17.

dictated by local laws and those edicts emanated from the crown, while in the south, land was held privately but not in the strictest sense of the word.[16] Since the time of the *ancieñ regime* and even today land was technically placed under the guardianship of the State/kings. In days gone by, the feudal lords and their agents may own the serf and his produce but not the land as such. In contrast the people pursuing agriculture of variant kind such as shift cultivation but more significantly cattle rearing in the southern and western lands before the Menelik conquest had different take on land. Their outlook and philosophy on land and its utilization was surprisingly different from the northern sedentary agriculturist who perhaps might have been required to conform to relatively rigid laws and regulations governing land. Thus, within a definite boundary of their hold, clans pursue an egalitarian system of land administration without out infringing into the right of others outside their communal grazing ground.

Unlike their northern compatriots, the Oromos of the south and the west might have had an equal and unlimited opportunity to access land, pasture and water as late as 17th century. It is known that by tradition no man may enclose for himself more than he can cover with spear or stone throw. However, distinguished warriors and very rich men could claim nominal ownership to large areas of land without prejudicing the basic tenet of communal land ownership. Thus, people held in high esteem by the society can buy favors from elders to own lands who in turn will allow relatives and friends to come and reside in their land.[17] It is assumed that peoples other than the Oromos in the south might have similar arrangement with regard to land. The rapid conquest of the late 19th century by Menelik changed everything. Below I shall try to explain briefly what the land tenure system outside the historic Abyssinia looked like.

After the conquest, land in the South was appropriated in three forms; private, the state and the church. Under private ownership there were numerous ways by which surplus was usurped from the serf, that is the *gebar*. By far the most extensive and equally onerous within this system was the *gult*. Gult can be described as an economic unit where the rightful holder is responsible for both the administrative and judicial matters within the unit. Persons entitled for *gult* lands are those closer to the Emperor, traditional nobilities and those who demonstrated valor and an outstanding service to the crown. It is important to note, however, that Menelik did not allow landholding titles over the dispensed *gult* lands but gave the right to extract tributes from peasants living in the grant lands. It was after the Italian defeat in the early 1940 that Emperor Haile Selassie converted *gult* lands to freeholds with obligation to pay land tax to the government. In later days the *gult* was transformed into *rist-*

16 Ibid 17-18.

17 G.W.B. Huntingford, The Galla of Ethiopia: The Kingdoms of Kafa and Janjero. Part II. (London: Lowe Brydone Printers Ltd., 1955), 27.

gult by which the holder was permitted to hold it permanently and even pass it to his descendants. This notwithstanding, the ultimate power rests on the sovereign who decided on how much, to what extent and for how long one can enjoy the privilege of owning the land. In the final analysis *gult* or not, land was distributed by the sovereign not as the matter of right but more as benevolence gesture by the Emperor. In many cases, extensive *gult* lands were administered indirectly by the resident representative of the land lord, the *melkegna*. By virtue of being closer to the land and the gebar, the *melkegna* is no less powerful than the feudal lord who may be absent from his fiefdom. He, in turn extract surplus commodities and labor from the serfs at will.

In another category of land holding, the State normally referred to *mengist* own millions of hectares from which it dispense favors. At one point land was distributed to ex-service men or civil servants in lieu of their services. Before the decree to establish the Ministry of Pension in mid-1950s was announced, many civil servants were bestowed with lands ranging from 40 hectare to 200 as a retirement benefit.

As an extension to the age-old customs practiced in the north, here too the sovereign distributed land to the churches that soon begun to emerge alongside the settler community. True to its tradition, thus, the church possessed huge tracts of lands in which the *gebar* has to toil for the benefit of the church functionaries and those associated with it. These church lands were called the *semon* and their size ranged from 10 to 12% of the total cultivated lands including those in the north and other parts of the country.[18] Although church lands were normally operated by the *gebar*, the *rim* was another land holding relationship within the *semon* system where individual priests or learned men of the church were given access to land in lieu of their services. Upon the death of the *rim* landholder his possession is taken away and given to someone active in church duty. However, in many cases the family raises one of its own to be a priest in order to inherit the land. Technically, the church lands as others are not for permanent keep as the Emperor can take them away. However, except in rare occasions where the sovereign or the overlord is in disagreement with the church or the patriarch that land is rarely taken away.[19]

In the eve of the 1974 Revolution, the state land or more appropriately the *mengst* (government) land were estimated at 46 percent in the south of the

18 Desalegn Rahmato, Agrarian Reform in Ethiopia, (Uppsala: Scandinavian Institute of African Studies, 1984), 19.

19 I have yet to come across a literature which verifies my claim, However, I suspect that lands belonging to the religious order of Dekike Estifanos in the 14th century must have been taken away from the order so as to force them submit to the Emperor's will that they completely rejected. We learn from Dr Getatchew's book 'Dekike Estifanos' that the religious order was persecuted harshly to an extent that some burned alive.

country.[20] It was indisputable that the state owned more land than any in the group. These were lands from which the Emperor derives benefit either by directly exploiting it or distributing it as he finds it fit. The *mengst* lands as they came to be known were not fully utilized except for a fraction of it. For all practical purposes, they were kept in reserve for future distribution as gifts and favors or as a tool for welding more power. All lands that were distributed to various groups of people since 1942 came from lands classified under the *mengst* holdings.

The North constitutes the core of the Abyssinian Empire. It is geographically distinct and historically unique where for over two millennium human activities had left its heavy imprints not only on its landscape but also on the mind of its people. Throughout history, land had played a major role in defining social, political and moral boundaries within the society and also shaped the psychic of the people and its rulers. Despite their omnipotence, the Emperors approach all matters concerning land cautiously. It takes a sovereign wrath to expropriate land belonging to minor chiefs or the hereditary lord. Nevertheless, land per se belonged to the emperor with minor difference that in the northern region communal or kinship ownership stands on the way. The *rist* system was based on specified kinship, with hereditary rights based on blood ties. A *rist* holder has in alienable right to cultivate the land as long as he lives. When he passed away his land will be divided equally among his offspring. In the event that the *rist* holder dies without bearing children, the land will automatically pass to the nearest kin. Interestingly enough the system gives some room to accommodate members who does not fall into the category of kinship but by virtue of their living among corporate members for a considerable time are provided with parcel of land for sustenance. A new phenomenon to the *rist* system was the one found in parts of highland Eritrea where land is kept in corporate state and made to rotate every seven years among the cultivators. Although hard to rationalize for its institution, the rotation system named the *desa* was the opposite of what was once an adapted practice in some regions of Eritrea called the *Tslmi* which is purely private and individually held. Before the revolution that made land property of all, the owner of *Tslmi* land has the full right to rent, mortgage and exchange the land.

Like the *gult* lands in the south, the northern noblemen and hereditary chieftains possessed *gult* lands with special rights to pass it to their offspring. However, to continue enjoying the privilege of having *gult* land they are required to give service to the crown and pay tribute regularly. Since large *gult* holders are potential contenders for power, slight misunderstanding with the sovereign could result in land revocation triggering rebellion that ultimately may wreak havoc among the peasantry. Like the southern *gebar,* the peasants in the north pay tribute to the overlord and the crown for the parcel of the land

---

20 CSO Statistical Digest, (Addis Ababa: 1971), 11.

that they are entitled to cultivate. They also give free labor to the *gultegna* whenever requested some times out of respect but most often due to coercion. The difference between the northern peasant and the southern serf was more of a cosmetic than substance in that, the former despite his wretched condition had a sense of false security camouflaged in an exhaustible vanity while the later lived under perpetual insecurity of eviction.

The church and the crown existed side-by-side complementing each other. Since the early times, the Emperors had recognized the church as a valuable appendage of the Realm useful in the expansion of their domains. It was also regarded as an essential tool to remind the people that the 'king's rights were divine and to be feared and respected as one fears and respects God'.[21] Thus, they generously donated lands upon which the churches and monasteries heavily depended on for their living. Among them, the Emperors, Sayfa Arad, Dawit and Yeshak who lived in the 14th century were known to have been particularly generous to Debre Bizen an ancient monastery in the eastern escarpment of Eritrea and to Aba Medhanine and Aba Debre Abbay in Seraye by providing lands for their upkeep.[22] Almost all the renowned churches and ancient monasteries in north Ethiopia had *semon* lands some covering entire districts and some amounting to few villages. Likewise the churches and convents in the South enjoyed similar privileges accorded to the ancient church traditions in the North. The relationship between the churches/monasteries and the peasants residing in those lands was no different than the fiefdoms run by the *gult* owners except for their religious shenanigans.[23]

As explained earlier tenancy rate was by far widespread in the south than in the north of the country. According to earlier released statistics the proportion of tenancy in the south ranged from 37% in Sidamo to 73% in Illubabor. The next highest rate of tenancy was found in Keffa with 59% followed by Wellega and Shoa each amounting to 54 and 51% respectively. Almost half of the lands in Harar and Arssi were under tenancy holding.[24]

21 Tirfe Mammo, The Role of Indigenous Knowledge, Traditional Practices, and Local Institutions: the Ethiopian case. (Trenton: Red Sea Press, 1999), 89.

22 Richard Pankhurst, The Ethiopian Boarderlands. Essays in Regional History from Ancient Times to the End of the 18th century. (Trenton: The Red Sea Press, 1997), 38.

23 Francisco Alvares, The Prester John of the Indies; A true relation of the lands of the Prester John, being the narrative of the Portuguese embassy to Ethiopia 1520. The Translation of Lord Stanley of Alderley (1881). Additional material by G. F. Beckingham and G. W. B. Huntingford. (London: Published for the Hakluyt Society, University Press, 1961). 92-93. "According to Alvares, the peasants (Tigrai/Medribahri) pay tax for the *semon* lands in many ways; Each village from supplying millet, rye and barley and teff it grows every year to providing one horse if not 50 cows in place of horse every three years.

24 CSO 1967 and 1968, Addis Ababa, Ethiopia.

Tenancy outside the core was new phenomenon promoted by the forbearers of the Shoan Dynasty in mid-19th century. However, Emperor Menelik with the help of the Northern chieftains and co-opted local *Balabats* from the conquered regions eventually perfected the tenancy system that it required the heavy handedness of the 1974 February Revolution to do away with it permanently.

The Effect

Based on earlier accounts from foreign travelers, we know that Ethiopia was endowed with fertile land and pleasant climate. Some of the early travelers were also well disposed to the industriousness of the peasantry who always 'positioned them for double or triple cropping'. Yet many commented that "With all these advantages of triple harvest which cost no fallowing, weeding, manure, or other expensive process, the farmer in Abyssinia is always poor and miserable." [25] The apparent contradiction between the abundant natural resources together with industrious nature of the farming population that the country was known to possess and the glaring misery that was so obvious to a casual foreign observer of the time, can only be explained by focusing on the social relationship that existed between the ruler and the ruled, the peasant and the feudal lord, the layman and the church hierarchy.

As indicated earlier, the Ethiopian society was moderately regimented into distinct classes and within classes into 'castes'. While the primary task of the peasant was to produce crops and raise cattle ostensibly for his well-being, due to the existing social relations he was obliged to maintain hordes of retinue led by the warrior class with the Emperor and his *Rases* at the helm. After all there has never been a standing army paid from the coffer of the state until recently, that is, until 1941. Moreover, the priestly class, which amounted to hundreds of thousands, was also included in the entitlement leaving a bare minimum to the producer.[26] Thus, the church hierarchy as an appendage to the ruling class has an equal share, if not more of the material proceedings, frees labor and many hidden obligations that the peasant is required to fulfill. Within the cultivating/cattle-rearing category of the society, there was an assorted layer of pottery makers, cloth weavers, and blacksmith, goldsmith, entertainers and traders very important elements in the social stratification process. Like the peasants, however, they were mercilessly exploited, socially discriminated, professionally discouraged to be seriously considered as harbingers of change

25 James Bruce, Travels to Discover the Source of the Nile, in the years 1768, 1769, 1770, 1771, 1772 and 1773, 53 Vol. iii; (Edinburgh: Printed by J. Ruthven for C.G.J. and J. Robinson, 1790), P. 315.

26 Ethiopia is known to have over 32,000 churches, 365,000 priests and 57 Bishops. World Council of Churches http://www.oikoumene.org/en/member-churches/regions/africa/ethiopia/ethiopian-orthodox-tewahedo-church.html.

or pioneers to Ethiopia's material and social progress. For rapid technical progress it is imperative that the cleavage that is bound to appear between the rural settings and urban development need to be nurtured.

So far, no one has ever described the savagery and inhumane character of Ethiopian feudalism in more graphic form than did Asbe Hailu in the first quarter of the past century. In an Amharic language article titled *sile mengist gudat*, Asbe writes:

> No sooner has the peasant unloaded the tribute due to the lord than the later 'congratulates' the peasant for having come just at the right time to be sent to the lord's measured lands somewhere beyond the Awash, from which the peasant is supposed to bring a load of teff (**small grain indigenous to Ethiopia*). The toil-worn peasant supplicates, pleads and laments: Oh, Sire, it is harvest time in our area and if I don't harvest now, before the approaching rains, Sire, I will be finished, evicted, uprooted. Oh, Sire! No heed to his pleading and lamentations. He must go to the measured estate and collect the load of teff as the lord ordered. The peasant has no choice and he submits. Cursing the day he was born, like the Biblical Job, he takes a whole week's work... [on his return] another order! Endless! The peasant now collects the whole lot of the grain, ground to flour and stores them properly. While he does this, he runs out of his provisions and in hope of keeping his belly full hangs around feast places and come back exhausted, sick, and diseased. Like a sick dog with his head resting on a heap of animal dung, the peasant passes his last torturous and agonizing days below the fence of the lord's compound. When at last he dies, the lord's household servants carry out the body on a stick and after a few scratchy digs; they 'bury' him in a ditch. A lady living nearby asks a lady of the lord's household: 'Sister, I saw a dead body leaving your household for burial today. Who could he possibly be?' 'Don't mind him, sister,' retorts the lady from the lord's household,' he was not human born; he was only a *gebbar*.'[27]

This is a telling story on how the *gebar* fared under the harsh system of feudalism until its demise in the fourth quarter of the last century.

A *gebar* is a peasant serf from whom a civilian and military official is empowered to extract labor and agricultural commodities. Based on oral agreement therefore, the serf is required to pay to the landlord one third of the produce or a quarter or half as the case might be. Technically, payment to the landlord is dependent on how much and what kind of instrument of labor and input each supplys. If all inputs including the instrument of labor is supplied by the tenant serf, the custom allows the landlord to retain a third of the share to himself while the remaining two third to the serf. In this case, the arrangement is called *sisoarash*. Similarly if all is supplied by the landlord, half of the produce will go to the serf and thus he is no more *sisoarash* but *gmasharash* (half sharecropper). In a majority of cases, the serf gets only one-third of the produce. On top of this, he is obligated to pay the tithe, *asrat* as part of revenue to the landlord and ostensibly through him to the government. The peasant's

27 Asbe Hailu. 'sile mengst gudat', (Addis Ababa: Brhanina Selam Printing Press, July 22, 29 1927) (Amharic)

obligation was not limited to paying dues what one might innocently characterize as a rent to a land owned by the feudal lord. However, as Bahru Zewde (1991) summarized it succinctly, 'The ruling class appropriated not only surplus products from the *gebar* but also surplus labor'. Among many other duties that the *gebar* is obligated to perform were the following.[28]

1. Corvee on state land or governors land, the *Hudad* (this takes 1/3 of gebbar's labor time)
2. Ground grain for the lord or *melkegna*
3. Built granaries and fences of the lord
4. Provide and transport commodities belonging to the lord
5. Acts as a custodian of a prisoner
6. Supply provisions, *dergo* to visitors or officials passing through the area
7. Collect firewood to the lord's household, twice or thrice a year
8. Pay in *mar* or honey tribute for the preparation of the lord's honey mead drink.

It is difficult to enumerate all the tributes that the peasant was forced to pay in return for his safety and that of his family. In late 1840s the British consul to the Court of Ras Ali described the fate of the peasant in the following manner:

> The impost over the peasant is numerous. They pay a certain portion in kind to the Ras or other great chief and sometimes a regular tax in money. Their immediate governor then takes his share in kind of every grain say one fifth and feeds besides a certain number of soldiers at the expense of each household: he has rights to oxen, sheep, goats, butter, honey and every other requisite for subsistence: he must be received with joy and feasting by his subjects whenever he visits them, and can demand from them contributions on fifty pretexts - he is going on campaign, or has just returned from one, he has lost a horse or married a wife, his property has been consumed by fire or he has lost his all in battle or the sacred duty of funeral banquet cannot be fulfilled without their aid.[29]

Similarly, Alvarez observed peasants pay for church lands in grains and "every three years each village nearby pays one horse if not 50 cows in place of horse." At one time the monastery at Debre Bizen received 33 horses from the *semon* lands.[30] Before Emperor Yohanness returned the land to its rightful owners in late 1870s, sizable lands in Digsa, Mai Imako, Adi Hadid, Hadish Adi in what was to be the province of Achele Guzai belonged to the Church of Axum

28 Bahru Zewde, A history of Modern Ethiopia 1855-1974. (Ohio University Press, and Addis Ababa University Press:, 1991) 87.

29 Walter Chicele Plowden, Travels in Abyssinia and the Galla Country with an Account of the Mission to Ras Ali, (London: Longmans green and Co. 1868) 137.

30 Francisco Alvares, The Prester John of the Indies; A true relation of the lands of the Prester John, being the narrative of the Portuguese embassy to Ethiopia 1520. The Translation of Lord Stanley of Alderleyu (1881), (London: Published for the Hakluyt Society, University Press, 1961), 92-93.

Tsion.[31] In another part of Achele Guzai, at Degaan 5 villages were Debre Libanos's gulti in Shoa.[32]

At the dawn of the 20th century, one safely argues, that the Ethiopian system of feudalism showed little change if at all. There were neither new laws governing land tenure system nor decrees protecting the peasant from unfair treatment by the landowner. His initiative to produce beyond subsistence was systematically stifled. Instead, land gradually begun to be consolidated in more or less centrally controlled manner, but with no apparent means of administrating it fairly. The outlying provinces were left under the ruthless feudal barons and their vassals to exploit the peasant serf at will. As time goes by, the condition of the rural population continued to worsen to the extent that many peasant serfs abandoned the land to its disastrous fate causing massive erosion, degradation and deforestation. In 1913 when paying tribute became too oppressively cumbersome to Wellega peasants their leader plead on their behalf to the in-term ruler of Ethiopia Lij Eyasu 'I pleaded to you my master, while you are here to exempt me as I am in difficulty…..you my master, have seen that the country is uncultivated, this being the result of too much tribute.'[33]

In the word of a pre-war Ethiopian intellectual, "The Ethiopian feudal lords and their lesser chiefs are living luxuriously, pre occupied with drinking and eating like 'a fattened cattle' ready for Easter slaughter.[34] All this is at the expense of the poor and the disadvantaged.

Feudalism in Ethiopia can best be described as malicious pest that did not only harm part of the plant's luxuriant foliage but also that attacked the very root system that anchors the plant firmly to the ground. As pointed out earlier, the system was notoriously incoherent and institutionally deficient to enable it function uniformly and in a symbiotic fashion. This is best illustrated by the different system of tenure that existed between and within the regions of the country. As opposed to other systems like it, European type for example, feudalism in Ethiopia failed to beget its own antithesis that would bring about transformational change to the rural setting and the country as a whole. Unfortunately, the feudal system was inherently deficient for one to expect a model evolve out of it where direct agriculture consumption, as in subsistence, is substituted by indirect one through market mechanism supported by modern means of exchange. Had this scenario existed, no doubt, progress in agricultural techniques, development in trade and commerce and the rise of market

31 Ruffilo Perini, Da qua a Mereb **(መረብ ምላሽ).** (Chiaramonte Roberto Edition, 1905), 60. (Tigrigna)

32 Ibid., p. 73.

33 Donald Donahue and Wendy James.The Southern March of Imperial Ethiopia: Essays in history and social anthropology, *Letter from Ghebregzeiabher to Lij Eyassu Megabit 2, 1905 E.C. and 11 march 1913 G.C.* (Cambridge University Press. 1986), 59.

34 Afwerk G.J. , Guide du voyageur en Abyssinie (Rome: 1908), 43, 182.

economy would have been the logical outcome. Instead, Feudalism in Ethiopia led a parasitic relationship that deprived the producer the right to enjoy the fruit of his labor and hampered his ability to set aside a reasonable amount of surplus for the purpose of transforming his surroundings.

On the other hand, the Ethiopian peasant sedated by Christian demagoguery propagated by the willful partner of feudalism, the church, never had the initiative to consciously improve his miserable lot for many centuries. Moreover, because of the stifling effect of the environment that he was condemned to operate, the peasant farmer lacked an elementary means of circumventing the onerous system to utilize any surplus to his advantage and by extension to the country.

The general picture of the landholding size at the time when the land reform decree came into effect appeared as follows. Before the revolution of 1974, some 31.1% of the peasantry had an average land holding size of 0.5 hectare. However, 2/3 of peasants owned between 0.5 to 1.0 hectare of land before or at the time the land reform decree was promulgated.[35] Despite the modest increase of cultivated land three decades later the ratio of man to cultivated land was as wide as ever. The reason is that the unchecked population growth in the past few decades shrunk land-holding size further reducing to mere 25.7% those operating between 0.5ha and 1.0ha while peasants cultivating below half a hectare remained practically unchanged.[36] Other studies show that at least 63% of the household has less or equal to 1.0 hectare at his/her disposal.[37]

It is also interesting to notice that at this period peasants owning a pair of oxen constituted a mere 29% while equal or slightly higher percentage own none. Similarly those owning a single ox amounted to 33%.[38] Despite the country's phenomenal number of livestock, the inaccessibility of farm animals for the majority of the peasants in this Ox-Plow culture continues to complicate agriculture production. Survey conducted 1984 on four contrasting districts show that in the well-endowed area about a quarter of the peasant household own no ox while in the areas designed as marginal over half of the peasants own not even one.[39] On the average 35% of the districts' resident have to rely on their own physical power to work on the land or rent oxen. This figure

---

35 MoA Sample Survey. 1974-75. Vol. 1 (Addis Ababa: July 1975), 54-61.

36 CSA, Ethiopian Agricultural Sample Enumeration 2001/02. Country level Part I and II. (Addis Ababa: 2003).

37 Brhanu Nega, Berhanu Adenew and Samuel Gebreselassie, Land Tenure and Agricultural Development in Ethiopia. Ethiopian Economic Association/ Ethiopian Economic Policy research, Workshop. (Addis Ababa: November 13, 2002)

38 MOA. Vol.1, (Addis Ababa: 1975), 50.

39 Desalegn, Agrarian Reform in Ethiopia. (Uppsula: Scandinavian Institute of African Studies, 1984), 5.

agrees with another study conducted almost at the same time.[40] Nearly two decades later the ratio did not change one iota.[41]

Tractors and their accessories were novelty to Ethiopia with few concentrated in the Setit Humera, Cheffe and perhaps the Awash valley in the early 1960s. During this period the estimate is that there were no more than 200 tractors and perhaps 20 combiners in the whole country. By Mid-1970 the tractor number increased to 2,500.[42] The dramatic increase of tractor power at this period can be attributed to the establishment of commercial farms mostly owned by local entrepreneurs. More than a third of the tractors were employed in the medium to large-scale farms in the sesame-sorghum belt of the Seti-Humera plains of Northwest while the rest in the irrigated Awash valley and the Shashemenie area of central Ethiopia.[43] Tragically, the introduction of the tractors to hitherto unknown areas, according to experts of the Chillalo Agriculture Development Unit (CADU) in Arssi Administrative Region were instrumental in the decision of the absentee landlords to evict tenant farmers from the land. In a word it was the main triggering point for the landlords to take drastic action against the peasants. As a result between 1966 and 1974 some 5,000 families or approximately 20,000, souls in the Chilalo district were evicted from areas run by the CADU.[44] Thousands more become landless in the entire country displaced by tractor power. One tractor in the field is estimated to displace 20 tenant farmers.

The attempt made by the imperial government to somehow change the grim picture of the country's agricultural condition was not serious. This can be corroborated by the amount of budget allocated to the agriculture sector in comparison to other sectors. According to Assefa and Eshetu (1969) the sector received a mere 5% of the total during the decades of the 1950s and 1960s.[45] For example The Third Five Year Plan (TFYP-1968-1973), the one that appeared to be focusing more on agriculture, had benefited the miniscule

---

40 MoA and FAO, Highland Reclamation Study Working Paper 11 (Addis Ababa. 1984).

41 Tewdaj Mogues, Must drought and poverty go hand-in-hand? College of Agriculture and life science (Madison: University of Wisconsin 2004) http://news.cals.wisc.edu/departments/highlights/2004/03/19/must-drought-and-poverty-go-hand-in-hand

42 Mekonen Getu, Socialism, Participation and Agricultural Development in post-revolutionary Ethiopia: A study of Constraints. Doctoral Thesis (Almqvist and Wiksell International. c1987), 51-52.

43 Thomas Killion, Workers, Capital and the State in the Ethiopian region, 1919 – 1974, Vol. I. (Stanford University, 1985), 539.

44 Micheal Stahl, Research Contradiction in Agricultural Development: A study of three minimum package projects in Southern Ethiopia, Report # 4. (1973), 5.

45 Asseffa Bekele and Eshetu Chole, A Profile of the Ethiopian Economy (Oxford Press: 1969), 48.

commercial farms more than the predominantly peasant farms which has received an insignificant share of 1%. In contrast, mining, telecommunication, industry, power plants etc. got the lion share of 90% of the total budget.[46] This was the reality of Ethiopia and its people before the advent of the creeping turned into violent revolution, which consumed anybody within its reach and outside it. Throughout Haile Selassie's reign the peasant sector was never given serious attention not even by the feudal lords in the south and west of the country whose relatively luxurious life depends on the productivity of the serfs.

Peace and Security

The absence of peace for periods longer than normal and the lack of security to enable one plan ahead did not only interfere with the daily routine of the peasant population but prevented surplus accumulation and stifled innovations that would have otherwise helped challenge problems encountered during the process of production. Continued warfare in Ethiopia, therefore, has worked against the consolidation of power, accumulation of wealth and emergence of technical inventions that should have been within the grasp of the Ethiopians whose introduction to iron smelting, pottery, wood and stone curving and fine art was from time immemorial. It should be underlined here that war is not peculiar phenomenon and certainly not exceptional to Ethiopia. In contrast Europe might have had experienced more destructive wars than Ethiopia. The question, however, is did war played a constructive or destructive role in the development and consolidation of the state? In the case of Europe, historians contend that its rock solid unity and economic viability today is credited to the effect of the endless war waged in the past few centuries. This alludes to the fact that war helped Europe to consolidate itself from fragmented mini principalities to full-fledged nation states capable of solving internal and external problems.[47] Similarly, one could also argue that had it not for the union forces in the US triumphed decisively over the southern secessionist during the American civil war of 1861-65, the US map would have been somewhat like a checker board than a compact one with dire social and economic implications. Who knows? The difference is that in the Ethiopian case regional wars seemed to have played negative role on the national consolidation process and in the advancement of the social forces capable of transforming the society.

Regional wars has been frequent and in cyclical mode to affect the normal progression of the productive forces beginning from the first half of the last millennium. After the environmentally induced retreat of Axum toward the south in the seventh century, the central core, most often identified with

46 John M. Cohen and Dov Weintraub, Land and Peasants in Imperial Ethiopia. (The Netherlands: Van Gorcum & comp. B.V.- assen, 1975), 8.

47 Christopher Clapham, War and State Formation in Ethiopia and Eritrea. Failed States Conference. Lancaster, UK: Lancaster University)

Christian Abyssinia has for centuries posed menacingly and acts predatorily over the peoples of the peripheral regions. This is not a one-time affair but a continuous practice where the aggressor returns back more prepared to the same theatre of war to replenish what was consumed from the seemingly inexhaustible resources that the fringe territories was believed to have possessed. It is not out of order, therefore to hypothesize that absent of centrally administered system of rule, the perpetual state of war and the unending arms raids visited upon the weak did not in any way help the country to extricate itself from the state of backwardness. Its present quandary is therefore, the carryover effect of the senseless wars waged across the land starting from the Muslim revolt against the Abyssinian rulers in the early 15th century.[48] In other words the predatory nature of the feudal class with the Emperor as the head has hampered the transformation of the state into a viable unitary entity where laws are promulgated to remind the sovereign of his boundaries when dealing with his subjects while educating the citizen of his rights and obligations vs. the state.

The rivalry among the Christian rulers of the central core is also a major factor to consider. Notwithstanding, the wars waged outside it, here again the peasants in the plateau were robbed of their produce, dislocated from their place of origin and cruelly crushed to the satisfaction of raiding party. It is not uncommon that regional chiefs challenged each other on territorial issues; personal vendettas or marriage arrangement went bad.[49] At the end all the personal animosity or infringement on ones honor becomes a pretext to wage war.[50] In such circumstance it is the peasants in particular those on the losing side who carry the brunt of this periodic confrontation. In turn the victims of past violence bid their time for the table to turn and exact revenge. It is for this reason that one may assume that the subordinate Islamic principalities in the peripheries supported by the Ottoman Turks daringly invaded the plateau in

48 Record shows that beginning the reign of Emperor Amdetsion (1314-1344) war against the Muslim principalities of Yifat and other lowland peoples was common. This notwithstanding, popular legend has it that the Emperor Libne Dingel on whose reign the strong Muslim army led by Imam Ahmed challenged Abyssinian rule used to lash the ground with cane 40 times a day conjuring war.

49 Herbert W. Blundell, The Royal Chronicle of Abyssinia, 1769–1840, (Cambridge: Cambridge University Press, 1922), 321. "The steps taken by Queen Menen to marry Yohannes III, invited a wrath on the population of Gondar by troops loyal to the deposed Emperor Sahle Dingel. According to Chronicler 'the soldiers broke into all the houses ate up all that the inhabitants of the city (Gonder) had to eat."

50 Historians assert that the assassination of Emperor Iyoas in early January 1769 by the provincial lord Mikael Suhul of Tigrai herald the beginning of the long drawn civil war that lasted for almost three quarter of century. The reason for the murder of the Emperor is attributed to his closeness to the strong Yeju house in Wollo, his maternal family.

1527. It was on these relatively fertile highlands with an agreeable climate far from the oppressive heat and dry environment of the lowlands that the major battles of the 1527 – 1543 civil wars was fought leaving the entire Christian enclave destitute. It was also during this period that the vagaries of nature, instead being tempered by human intervention went amuck.

The effect of the Christian-Moslem war had lingered on throughout the 16th century and affected food production seriously. The prolonged agrarian-pastoral war that had "almost succeeded in uprooting Christianity from Ethiopia" was brutal in its consequence plunging the Christian kingdom as well as the Islamic principalities into darkness.[51] Portuguese expeditionary forces that came to the aid of their co-religionists, the highland Christians in July 20, 1541 were visibly sadden to observe the one time fertile Eritrean/Tigrai plateau turned into rubbles.[52] It is also reported that in the middle of the 1527-43 Muslim-Christian wars major outbreak of contagious disease and famine decimated the central highlands to seriously check the advance of the invading Muslim forces. It is important to note that few years before the war, the country was reported to have been a supplier of provisions to Arabia, the Mecca, Zebaid, Juda (Jedha) and Toro" across the sea.[53] As late as the 18th century a European traveler marveled 'there is no country whatever better peopled or more fertile than Ethiopia'.[54]

Following, the agrarian-pastoral war of the 16th century, succession to the throne from within the ruling class and outside it, remained contentious for nearly half a century. This by itself became the primary source of disruption in the production and supply of food for a long time. However, with the accession to the throne by Emperor Fasil in 1635, matters begun to settle in favor of law and order. Fasil, who came out triumphantly against his catholic converted father settled permanently at Gonder, the first modern city ever to have been established after the fall of Axum and the withering of Lalibella in the 13th century. According to Merid Welde aregai, viable urban centers except for Gondar was extinguished throughout the 10th or perhaps the 11th century never

51 Mohammed Hassen. *Futuh Al-Habaša*: The Conquest of Abyssinia [16th Century] by Šihab ad-Din Ahmad bin Abdal-Qader bin Salem bin Utman-Review-, (Addis Ababa: International Journal of Ethiopian Studies, Vol. 1, No. 2, 2004) 184.

52 1/ Miguel de Costhansa, The Portguess Expedition to Abyssinia in 1541 – 1543. Transl. Richard Stephen Whiteway, (1902) 2/ Emir Idole the most trusted general of The Imam was defeated and killed in what is now Southern Eritrea.

53 Francisco Alvares, The Prester John of the Indies; A true relation of the lands of the Prester John, being the narrative of the Portuguese embassy to Ethiopia 1520. The Translation of Lord Stanley of Alderley (1881). Additional material by G. F. Beckingham and G. W. B. Huntingford. (London: Published for the Hakluyt Society, University Press, 1961), 127.

54 Charles Jacques Poncet, A Voyage to Æthiopia Made in the Year [sic] 1698, 1699, and 1700. (London: reprinted by the Hakluyt Society, 1949), 130.

to appear again before Addis Ababa was made sit of the Empire by Menelik II in the turn of the 20th century.[55] For once with Gonder as a center of the empire in the 17th century, the roving imperial court assumed the formality common to sedentary governments. Thus, law and order gradually took firm hold in the realm. Trade and commerce had begun to flourish dramatically. Citizens doing business with the city and its environs were also subjected to taxation.[56] Gonder was a vibrant city where all the merchants meet in a wide spacious place to treat of their affairs. These they expose their merchandise to sale. The market lasts from morning to night. All sorts of commodities are "sold here".[57]

As described by Poncet (1762), the first European traveler to enter Abyssinia after close to three quarter of century edict banning Europeans from entering the country, the highland region was relatively peaceful and prosperous ready to trade with the outside world through its agents in Massawa of which an Armenian, Agapyri is one of them.[58] The evolvement of Gonder as a center of government heralds a new beginning to the country. More significantly the citizens of Gonder were beginning to be stratified into social class as aptly described by Pankhurst "differentiated from soldiery, and that they may have become a power to be reckoned with".[59] At this time, Gonder was a large city with close to 50-65,000 inhabitants, with rich aristocrats owning real state in the city, consuming population,' people of consequence' ' breed up about court' 'influence and riches' 'rich traders, weavers, and tailors." This period was considered relatively stable, although, not completely tranquil. After all it was at this juncture of history, that the gradual but peaceful expansion of the Oromo toward the central and northern Abyssinia begun to gather momentum.

However, the Gonderian era of peace and prosperity did not last long to enable the country meet the challenge of sporadic famine brought about by violence that begun in mid eighteen century and continued throughout the nineteen century. From a historical perspective, this period known by historians as *zemene mesafint* (era of Princesses) could be considered as the most violent, insecure and debilitating period. Because of the complete breakdown in law and order, the peasant cease to produce food and instead chose to transform himself into a ragtag retinue attached to one of the numerous warlords as a

55 Merid Wolde Aregai, Technology in Medieval Ethiopia. New York Public Library the Schomburg Black research Center. *Microfilm.*

56 Merid Welde Aregai, Gonder and Adwa the Tale of two Cities, (Addis Ababa: 8th International Conference of the Ethiopian Studies Vol.2, 1989), 62.

57 William Foster, The Red Sea and Adjacent Countries at the Close of the Seventeenth Century (London: Hakluyt Society, 1949).

58 Poncet, Charles Jacques, A Voyage to Æthiopia Made in the Year [sic] 1698, 1699, and 1700. (London: reprinted by the Hakluyt Society, 1949), 227, 264.

59 Richard Pankrhust, (1982). The history of Ethiopian Towns: From the Middle Ages to Early Nineteenth century, (Stuttgart: Franz Steiner ver Log GMBH), 170.

means of protection and/or sustenance[60]. It should be borne in mind that beginning 1750 and up to the accession of Emperor Menelik II to the throne, that is, in the late part of 19th century big and lesser warlords contending for power plagued the country.

The growth of Gonder as future metropolis of the entire Ethiopian Region was nipped from the bud when succeeding Emperors residing in the numerous medieval castles built around Gonder were turned into mere figureheads. As pauper monarchs, they lacked the resources to raise imperial army and enforce the law. Most of them ruled at the whim of powerful warlords and some unfortunate ones were subject to murder by lesser lords elevated to prominence by sheer military valor and intrigue. The despicable position of the monarchs during the period is best explained by the letter sent from the last Emperor of the Gonderian era and before the closing of the *mesafnt* era to a French resident in 1852 pleading for money 'as much as you can [so that] I can buy myself grain to eat, the telescope (gift you have sent) is of no use to me now'.[61]

Within the plateau, notwithstanding the communality of culture that is deeply rooted in the Orthodox Church and the mode of production, which is identical in character, regionalism remained the defining feature of the country until the ascendancy of Emperor Haile Selassie I to the throne in late 1920s.[62] As indicated earlier, the regional autonomy and the rivalry that engenders among the regional nobilities reached its tipping point in mid-18th century triggering a century long civil war. In this historical epoch suspicion and indifference among the regions was rife to encourage the welding of an effective alliance against intruders who challenged Ethiopia's independence. No doubt during this period Ethiopia stood vulnerable to more deadly European machination, regional armed invasion and later on the permanent presence of the colonial powers in its historically shared shores of the Red Sea and the Indian ocean. The social upheaval unprecedented in the history of the

60 Contrary to the civil war that raged in Ethiopia for about three quarter of a century, the American one had a synergic effect in stimulating growth on the modestly developing manufacturing and trade sectors existing before the brutal war of 1861 – 1865 that claimed close to 700,000 people. In that trying period the Northern half prospered, pursued bold diplomatic offense most often on the position of strengthen and finally able to rebuild the ruined Southern half in no time. Unfortunately, the civil war in Ethiopia was more depressing than uplifting.

61 Sven Rubenson, Correspondence and Treaties 1800 – 1854. (Addis Ababa: Addis University Press, 1987). 198. Correspondence of Emperor Yohaness the III to Micheal D'Abbadie July 1852 Doc 145.

62 The semblance of a unitary Ethiopian state engineered by Menelik is largely attributed to his skillful application of less and less crude violence to resolving issue and his reliance more on feudal trickery, wisdom and patience to ensure the integration of the peripheral regions and calm potential rivals to his domain.

country also brought with it high level of banditry causing disruption in internal trade as well as widespread famine throughout the land. This lead the nineteen century Ethiopian chronicler to lament that "there was a great famine and many people perished of hunger.[63]

The era of the *zemene mesafint*, was the period where cultivation of crops had for all intent and purposes interrupted if not completely ceased. In a bid to wrestle power, regional warlords fought against each other incessantly destroying all living and non-living things that they remotely suspect will help sustain their foes. Between the years of 1769 and 1800 alone the northern section of the country had seen over 46 major and minor battles. During the same period one particular Emperor, Tecle Giorgis is known to have been enthroned and dethroned five times in his lifetime. Within this short span of time, that is, between 1769 and 1852 where the chaos begun to wane after Tewodros II triumphantly secured the throne, about 23 persons all claiming to be the descendants of Emperor Fasil, who built the permanent Capital at Gonder ascended the throne, albeit some for few years.[64] This led a chronicler at the time to sadly scribe "All sons of kings that have been scattered like dust before wind. Alas I and woe in me! What avails that kingdom that was snatched away by the hands of underlings".[65] The power struggle further intensified during 19th century simply because of the European factor that begun to intervene stealthily but determinately in the guise of conducting business with various warlords eager to secure recognition and arms.

Never in the long history of the country had various Ethiopian feudal lords unashamedly posed as the rightful heir to the throne as they did during this period. With proliferation of firearms violence attained a different form unprecedented in the country's history. By some estimate there was about hundreds of thousands of guns in Ethiopia before and shortly after the Battle of Adwa in 1896.[66] Thus, importation of firearms added another dimension to the seemingly unending conflict in the country resulting to close 78 small, large and very large wars in almost all over Ethiopia up to the eve of 20th century.[67]

---

63 Herbert W. Blundell, The Royal Chronicle of Abyssinia, 1769–1840, (Cambridge: Cambridge University Press, 1922), 492.

64 Here, like Emperor Tecle Giyorgis who had the misfortune of ascending and descending from the throne 5 times, Emperors Sahle Dingel, Gigar, Demetros, and Solomon III were twice made Emperor of the land. In effect within eighty years there were 18 different emperors in the helm. In contrast the British queen Victoria ruled uninterruptedly for almost 64 years.

65 Blundell W. Herbert (ed., tr.), *The Royal Chronicle of Abyssinia, 1769–1840*, (Cambridge: Cambridge University Press, 1922), 471.

66 Richard Pankhurst, Guns in Ethiopia. Transition, No. 20, (Bloomington: Indiana University Press. 1965), 26-33.

67 1/ The author has meticulously compiled all the minor and big wars that occurred during the epoch using different historical texts as a source. It is possible there may have been small and large skirmishes between and among warlords that

This period was crucial not only from the standpoint of causing serious crisis on food production but also from the fact that it was the beginning of the complete reversal of the technical advances made earlier. In other words, centralization was sacrificed for regional interest making the country open to all kinds of European machination.

Being one of the few countries which adapted iron plowshare early in history, it is certain that Ethiopia must have had its own blacksmith who smelt, cast and mold iron ore to fit the design of their creation. The fineness of jewelry and the elaborate embroidery work still evident today must have suffered tremendously during this period. Moreover, the magnificent architectural wonder of the churches at Lalibela and similar but by far smaller rock hewn cave churches in Tigrai attests the technical competency of the people in the region. Merid azmach Abiy, the founder of the Shoan dynasty in early 19th century, was reported to have employed Ethiopians who profess Judaism, the Felashas to make axes, sickles, plowshares, hoes, and *wegel* by which to clear the dense forest of Yifat and cultivate crops. [68] He also supervised the manufacturing of gun barrel in his palace workshop and was keen in producing gunpowder by grinding rock Sulfur.

It is ironic, however, that few warlords worth of attention by European travelers or embassy at the time has to lament about the extreme level of backwardness of the country and hence a request for foreign technicians and skilled persons to help it develop.[69] In a last ditch effort to convince Menelik import skilled men from Europe so that 'knowledge will increase and our country will be saved from strife and misery', Blatta Gebre Egziabeher risked

---

escaped the search of the author 2/ Richard Pankrhust , Economic History of Ethiopia 1800-1935 (Addis Ababa: Haile Sellassie I University Press 1968) The book gives detailed account on the amount of arms available in Ethiopia before the conclusion of the *zemene mesafint* and the dawning of the 20th century.

68 Bairu Tafla, Asmegiorgis and his work: History of Gala and the kingdom of Shoa. (Stuttgart: Franz Steiner Varlag, 1987), 513.

69 (1) British and Foreign State Papers 1861 – 1862 Vol. Lil Great Britain & Abyssinia. Correspondence Respecting Abyssinia 1846 – 1860. Dej. Wube to Plowden "If you can find 5 to 3 workmen if possible at Massawa, if not by sending to your country, builders or mason, *technicians* bring them for me. If they wish for opportunities I will give land etc."

(2) Alan Morehead, The Blue Nile, (New York: Harper Collins publishers, 1962), 252. Some of Tewedros foreign correspondence with European sovereigns was aimed at obtaining skilled men. In the book *Blue Nile*, Tewedros proudly declares "I need people with skill from Europe not people who bring me Bibles " I am sure his comment was to underscore the fact that his country embraced Christianity longer than many European countries to want or yearn for Bible. Moreover, to appease Tewedros of his rage, Queen Victoria agreed to meet his request for skilled men by stating 'we gave our sanction…for workmen to be at your Majesty's service as you desired to employ in Abyssinia'.

his life by smuggling a letter addressed to the Emperor from occupied Eritrea.[70] Unfortunately, neither the sovereign nor the regional rulers pursued relentlessly to import knowledge or better still to look internally for solutions. Despite their continued fascination with Venetian and other foreign manufactured ware glasses, silk dresses from the east and other fancy commodities available at that time, the feudal lords including the church looked down on the very people who were disposed to create fine arts and develop tools of production in the country. While people engaged in iron smelting, handicraft, pottery and elaborate artifacts remained in the fringe of the society, stigmatized and unworthy of respect and encouragement, the Ethiopian court since the early 15th century was known to have been importing a variety of goods from abroad. According to historical records, the most favorite goods traded with the country for the consumption of the feudal lords was finely embroidered Indian silks, ornamentals, cotton cloths, magnificent swords and even drinks.

Raids and Dislocations

In the Ethiopian context, sudden raids as opposed to full-fledged wars had a special psychological effect on peasants. While wars need an advance preparation by both warring parties and by implication allows sufficient time to make life saving decisions; in contrast raids came when least expected and very swift as that. The effect of raids on the psychology of the peasant not to mention on his property is beyond description. The word raid as a means to settle score with rival feudal lord or simply to prey on the weak for survival has so much so entrenched itself in the Ethiopian vernacular that today any campaign involving military expedition or planting trees or cleaning refuse from neighborhood area is undertaken in the name of *zemecha* literally meaning raid.

For logistic reasons, wars and raids are carried out before the onset of the rain or after it. In both cases, the peasant is the loser. If it happened before the rain, he is deprived of the opportunity to prepare the land and plant crops. This was the fate of Hamasien farmers in the 1870s that confided their predicament to a foreigner "but the worst part of all is that he (Ras Woldemicael) passed during the rainy season expelling us and saying 'do not stay in your country".[71] Conversely, if the raid is

70 Tekeste Negash, No Medicine for the Bite of a White Snake: Notes on Nationalism and Resistance **in** Eritrea, 1890-1940. *Letter from Blatta Gebre Egziabeher to Emperor Menelik II* found in "A Memoire of Blatta Gebre Egz iabeher Gila Mariam from Tzzadda Kristian, Hamassien, 1889 EC" (1896/97 AD) A. E., Pacco 1008 (Uppsula, 1986), 58.

71 British Museum Department of Oriental, Gordon papers 12913. Words related by peasants in today's Eritrea to Gordon Pasha an English troubleshooter in the region and the Governor of Khartoum on his way to meet Ras Alula.

postponed after the rains all the same the marauding forces take away his long awaited harvest and his livestock as booty. Such was the life of the Ethiopian peasant and the cattle herder for centuries end. There have been many terrible raids recorded in the past that left the people completely destitute and therefore migrate to areas far away from their birthplaces. At times, raids were intolerably frequent that few such as the loyal Kentiba Hassen and his Habab people of the extreme north of the country were forced to seek Italian protection at the coast long before the resistance to colonial conquest was exhausted. Italians admit their ascension to the plateau in 1889 was made possible by the active cooperation of some raiding bands and dislocated refugees.

As historical records show, many raids of horrific magnitude were undertaken against the agricultural communities of the country for an imaginable period of time. At the height of his power in 1776, Ras Mikael Suhul, dubbed as the king maker by historians and his vassal Dejatch Kifle Yesus practically destroyed Qulala in Semien and burned Enzageder and Mecha to the ground. Bruce described many of the horrible scenes of which one was around the Blue Nile region that Mikael Suhul has passed through in his campaign to defeat rival Warlords.[72] It must have taken years for these districts to revive again. In 1820 Ras Woldeselassie admitted to an English adventurer that he lived pretty well on corn belonging to rebels in one mountainous country 'burned the town and marched to Aderghso to find corn fields ready for harvest. It took us 5 days to burn the fields to the ground.' At another district near the river Munnai, the finest in Abyssinia where cattle and corn is abundant it took the *Ras* one whole week to destroy everything standing.[73] Pankhurst (1968) relates story of raids and mayhem on the rural population caused by unpaid soldiery, vindictive militias and common robbers during the first half of the 19th century. To witness the cat and mouse game played between the peasants and traders on one side and the overlord and his rag tag retinue on the other is amusing but sad at the same time.[74] Upon the return of the defeated Gojami forces from the ill-fated Mahdist campaign of the Sudan in 1888, Emperor Yohaness for no apparent or may be flimsy reason took the liberty to devastate Damot the breadbasket of the Amhara land.

---

72 Bruce, James. Travels to discover the source of the Nile, in the years 1768, 1769, 1770, 1771, 1772 and 1773. (Edinburgh: Printed by J. Ruthven for C.G.J. and J. Robinson, 1790), 434-5. Bruce observes "whilst not a living creature appeared in those extensive, fruitful and once inhabited plains, an awful silence reigned everywhere around.'

73 1/ James C. McCann, People of the Plow: An Agricultural History of Ethiopia, 1800-1900, (Madison: The University of Wisconsin Press, 1995), 81. 2/ Nathenial Pearce. The Life and Adventures of Nathaniel Pearce Vol. 1; (London: Henry Colbum and Richard Bentley, 1831), 68-69.

74 Richard Pankhurst, Economic History of Ethiopia 1800 – 1935. (Addis Ababa: Haile Sellassie I University Press 1968), 564.

Despite his revolutionary zeal to lead Ethiopia toward the path of unity and modernity, in his final years Tewedros revert back to the old ways of replenishing stock by launching raids on some of the prosperous regions of Meccha and Zage. Summarizing the eyewitness account of Stern, Rassam and Massaia where the former two were English captives in Tewedros's camp, Pankhrust described the aftermath of the raids as 'horrific' were regions reduced to acute poverty, with plow oxen and cattle slaughtered, granaries despoiled, people killed and mutilated and the youth taken away to fight'.[75] Alulla's raiding campaign against the low countries of Medri Bahri and the Naib's limited armed incursion into Hamasien are but few examples of the insanity that gripped Ethiopia's feudal class before and during the European scramble to subjugate the African people. Moreover, after the dust settled, that is, when Menelik's effort toward a unified feudal system begun to take shape raids continued against the southern people by rogue chieftains. As late as 1910 Ras Woldegiorgis and his vassal Dejatch Damte plundered what is left of Maji and the Boma plateau simply because of the grudge he hold against the Viceroy Biteweded Ras Tessema who ordered him to the capital for a different assignment that he seemed not to fancy.[76]

Pankhurst (1968) in his voluminous work dealing with the Economic History of Ethiopia from 1800 to 1935, relates an eyewitness report from European travelers at the time that living on war booties was so common that peasants were enticed to enlist in the ragtag army than plow the land.[77] Giving yourself to the service of one regional warlord is 'the only and sure way to live in Abyssinia'. To avoid of becoming victim of plundering and other crimes one become a plunderer and 'live a happy life'.[78]

Armed raids were carried out on various pretexts by the most powerful including the emperor or by an aspiring feudal chief. In the 19th century alone close to 42 major unprovoked raids were believed to have taken place in

75 Richard Pankhurst, A Social History of Ethiopia. Institute of Ethiopian Studies. (Addis Ababa University: Kings Ripton, England: ELM Publications, 1990), 320.

76 Mersie Hazen Woldekirkos, Memoire on the reign of Menelik II, 1935. (Addis Ababa: Zamra Printing press reprinted 2002), 105.

77 Richard Pankhurst, The Economic History of Ethiopia 1800 – 1935. (Addis Ababa: Haile Selassie I University Press, 1968), 578. "According to eyewitnesses and as related by King Sahle Selassie himself in different occasions some 180,000 head of cattle were rustled from the peasants in the south. His grandson Menelik II also robed the southern peasants of their cattle amounting to 134,000. At 'Gore the entire flocks destroyed'. In numerous occasions' we fire few shots and the natives run away and all their grain brought to the camp'; 'formerly were the owners of head of cattle but now all has been raided'; 'they had formerly been wealthy owners of many head of cattle today they had scarcely any to speak of' and ironically one feudal lord was quoted as saying 'It requires more than seven years to reconstitute what can be ravaged in three months'.

78 Afevork G.J., Guide du voyageur en Abyssinie (Rome: 1908), 235.

Ethiopia.[79] Some of the causes that may trigger raids are territorial disputes between districts and provincial chiefs. Disloyalty to or outright rebellion against overlord could also invite blanket punishment by the powerful. Since the royal coffer is considered private and therefore not used in the service of the nation, it is not uncommon to encourage a vassal to conduct raid in lieu of his service. It is indeed a cruel way of paying or compensating an underling for his valuable services. The peasant army is rarely sanctioned for its frequent raids sometimes on the way to fighting an enemy or returning back home after epic battles. In traditional Ethiopia raids were justified to punish areas suspected of harboring a known enemy of the crown or recalcitrant feudal lord. Sometimes, lesser chief conduct raid as sign of displeasure toward his overlord or even the Emperor. But in a true sense of the word, raids were carried out as a sustenance means for the unpaid soldierly and to amass wealth for the feudal lord and the nobility. In the words of an Ethiopianist 'The peasants suffered greatly as a result of onerous traditional tax system, looting, and the frequent fighting against internal and external enemies that is certain to involve them. This in turn produced an atmosphere of insecurity, uncertainty, and arbitrariness which had an inhibiting effect on production'.[80]

At almost every junction of the Ethiopian history the road is littered with such occurrences. It is sufficed to mention one more anecdote by a 19th century chronicler who explains vividly of the peasant's predicament. The Ethiopian chronicler and no doubt a close observer of the changing times writes, how during the transition to power, that is, when Itege Menen the mother of the powerful Ras Ali who married Emperor Yohaness III, the son of Emperor Teclegiorgis who deposed Emperor Sahle Dingil wreaked havoc on the production and supply of food instantly. He writes 'the city of Gondar and its (environs) underwent grave disturbance; the soldiers broke into all the houses and ate up all (the grains *mine*) that the inhabitants of the city had to eat. There was a great famine and many people perished of hunger".[81]

One that should not pass undetected was the massive dislocation of southern people for internal slavery between the close of the 20th century and up to the coronation of Emperor Haile Selàssie in 1923. In one historical incidence alone over 40,000 people of the Gimira nationality was believed to have been transported all at once to Addis Ababa to be sold as slaves in the Abyssinian household.[82] Ethiopia is also known to have provided a substantial amount of slaves from the conquered people of the south to the Indian and

79 The author has compiled all the raids that occurred during the epoch using different historical texts as a source. It is possible there may have been small and large raids not accounted for during the compilation of the data.

80 Richard Pankhurst, The Economic History of Ethiopia 1800 – 1935. (Addis Ababa: Haile Selassie I University Press, 1968), 223.

81 Herbert W. Blundell, The Royal Chronicle of Abyssinia, 1769- 1840 (Cambridge: Cambridge University Press. 1922), 492.

82 Bahru Zewde, A history of Modern Ethiopia 1855-1974, (Ohio and Addis Ababa University Press, James Curry London, 1991)

perhaps the Ottoman Empire markets.[83] The *habshi/habsan* who later sours to prominence in south India, Gujarat and Janjira trace their root to South Ethiopia. The most notable Emirs and princes of their time were Malik Anbar, Habshi Sultan of Begal and Sidisidi. A novel by a prominent literary figure, Dejazmatch Girmachew Teklehaworiat highlights the agony of an Ethiopian slave shipped to Arabia and the Middle East during the late 19th century.[84]

Natural Calamity

Unless assuaged by technical advances to minimize its effect on the lives of people it can certainly be a drawback to national development and threat to one's existentialism. Ethiopia has equal if not more share of nature-induced cataclysm in the past. The catastrophes that befell the country at various times of its history were not limited to the physical violence that the ordinary people were subjected to for centuries but also to the natural calamity that seem to occur from time to time. Pankhurst (1968) using documents from primary sources listed over fifteen major locust invasions throughout the 1880s, one lasting for a stretch of four years uninterruptedly. The infestation of this veracious pest continued throughout the good part of the 20th century with one vicious attack after the other. The latest of such event to occur was in Tigrai and Eritrea in 1956-58 and 1963-64. Coincidentally, it is at this juncture of Ethiopian history that for the first time food aid was sent from abroad to districts in Eritrea and Tigrai and later in the decade to Wello.

In fact, locust has been a problem pest for peasants in the north and central part of the country for longer time than recorded history. As early as 1520s foreign travelers to present day Tigrai and Wello had their amazement tempered by the local population to the abundance of crop they witnessed in those areas "honored guests, don't be amazed, in the years that we harvest little we gather enough for three years plenty in the country, and if it were not for the multitude of locusts and hail, which sometimes do great damage, we should not sow the half of the wheat we sow".[85] Historical record show that in 1796 Begemder and other districts including Wegera and Gonder were reported to have been devastated by swarm of locust repeatedly. This happened few years after the gradual recovery from the excruciating famine called the *Qechinie* literary meaning "one that emaciates". The leader of an English diplomatic and trade mission to the Shoan court in

83 Harold Marcus, *History of Ethiopia.* (University of California Press 2nd Edition. 2002), 55.

84 Girmachew Teklehawariat Ali Habeshi: Ye Adenu Janderba, (Addis Ababa: Rome Printing Press, 2000 EC(2007AD) (Amharic)

85 Francisco Alvares, The Prester John of the Indies; A true relation of the lands of the Prester John, being the narrative of the Portuguese embassy to Ethiopia 1520. The Translation of Lord Stanley of Alderley (1881). Additional material by G. F. Beckingham and G. W. B. Huntingford. (London: Published for the Hakluyt Society, University Press, 1961), 189.

early 19th century describes how 'vast flights of locust which has assisted to lay the country bare, still carrying their desolation (effect)... and their destructive course toward Abyssinia. They quiet darkened the air as the moment that the caravan halted'. Locust plague was unending affair in the north and eastern part of the country that Sahle Selassie the king of Shoa implored an English visitor "locust are destroying the crops...have you no medicine to drive them away".[86] The famine of 1958 in Tigrai and Eritrea was preceded by two consecutive locust invasions. Today alongside Nairobi, Addis Ababa hosts the Regional Headquarter of Desert Locust Control Organization (DLCO-EA) for the management and control of the viscous pest.

Likewise, disease epidemics have been a common phenomenon in Ethiopia. Diseases such as smallpox, typhoid, diarrhea, and 'malaria' had inflicted serious damage on the population longer than any calamity that the country had to endure. Unfortunately, disease and pestilence appear to go hand in hand with the conditions prevailing in the country at a particular period of time. Thus, the frequency and spread of disease appear to coincide with poor harvest generally caused by drought, wars and locust infestation. When James Bruce arrived at Gonder via Massawa in 1769, he witnessed Smallpox epidemics all along the way to the interior.[87] Coincidentally, this was the beginning of the hard times where provincial sentimentalism accompanied by cruel armed confrontation was at its highest point and where significant warlords challenged openly the legitimacy of the Emperor at the center. According to historical records, there was a major outbreak of epidemics in the Shoan camp at the time of the decisive battle of Embabo against the Gojamis that took place in 1882. Incidentally, this historic battle assured the Shoan ruling house led by Menelik II to seize power and rule over Ethiopia immediately after the death of Emperor Yohannes at Metema. The partly typhoid and partly cholera was known to have decimate the entire population of the central and north central region of the country. Few Years after the incident at Embabo, major cattle plague that almost destroyed the entire livestock population in the country was accompanied by diarrhea epidemics causing heavy loss of lives. An Ethiopian octogenarian living in the second quarter of the 20th century recounts the effect of the cruel famine to his famous author son by stating that "both people and cattle became insane [and died]. The people gathered to pray 'oh lord! Have mercy upon us... the land became fallow'.[88] He further relates how in late 1850s many people from Shoa including those seeking

---

86 William Cornwallis Harris, The Highlands of Ethiopia. Vol.3 (London: Longhmans, Brown, Green), 209.

87 James Bruce, Travels to discover the source of the Nile, in the years 1768, 1769, 1770, 1771, 1772 and 1773. (Edinburgh: Printed by J. Ruthven for C.G.J. and J. Robinson, 1790)

88 Meshaf Tizita Ze Aleqa Lema Hailu, (Addis Ababa: Published by Mengistu Lema, 1959) 142-143 (Amharic)

appointment by the Emperor at Gonder fell ill to mysterious disease and vanished. In all cases widespread disease did not only herald famine and death but also was harbinger to more conflicts. Thirty years after the visit of Bruce to Ethiopia, Ras Woldeselassie of Hintalo related to Pearce that 'smallpox is worse than locust' as the northern region continued to be plagued with the disease.[89]

Cereal Dependent Culture

As in the ancient times, today people whose livelihood is built on the consumption of tuber, corm and underground stems fare better than cereal or even legume dependent societies. There are few reasons for that, one being its longevity in storage, at least for some, but also its storability in situ. Many of the tuber crops are drought tolerant and on hectare base they yield more than cereal crop. Thus, in technologically disadvantaged societies their advantage is enormous in maintaining normalcy and stability among people whose norm is to stay put in their ancestral lands.

The effect of the chain of natural calamities that befall Europe throughout the middle ages and up to the beginning of the industrial revolution would have left the old continent staggering had it not been of one historical incident that took place in the sixteenth century, namely, the Spanish conquest of the Americas. Undoubtedly, the most valuable gift that the Spanish conquistadors' gave to Europe and to an extent to the world in general was the introduction of many commodities among them the life saving energy concentrated potato crop *Solanum tuberosum* from the central Andean plateau. This strategic crop is 10 times more productive than cereal crops and can be utilized in many forms including consumed as alcohol. The argument that this tiny historical consequence might have contributed to making Europe forge ahead is pluasable. This is particularly true, in view of the recent scientific revelation known as the *Little Ice Age* that stretched between the years 1500 and 1850 where crop performance in Europe was dramatically altered to the worst and where cereal seed yields was reduced to bare minimum. Throughout this period the land base for cereal cultivation begun to shrink and production per unit area decreased substantially. The introduction of energy and protein providing crops from the new to the old world may have worked in favor of Europe and may also have laid ground for world population to explode.[90]

89 Nathenial Pearce, The Life and Adventures of Nathaniel Pearce Vol. 1, (London: Henry Colbum and Richard Bentley, 1831), 68-69.

90 A careful examination of the climate record reveals that Europe experienced a prolonged cooling period known as the *Little Ice Age* between the years 1560 and 1850 affecting seriously the cultivation of traditional crops such as wheat. Instead many countries affected by the sudden climatic change begun to adopted the newly introduced hardy tuber and corn as an alternative to the established crops. The

Dependency on cereal crop for existence may have its drawbacks when considering two factors. One is the output expected from unit area and the other is the formidable adversary that the cereal crops face during their biological progression. Furthermore, in traditional societies ridden by societal contradictions and lack of security, cereal crops with their seemingly defiant upright posture could invite untold destruction and plunder. It is therefore, plausible to suggest that the adaptation of cereal as the primary crop in the Abyssinian plateau might have contributed to the down spiral of food production in the country leaving a bitter legacy of famine and starvation that Ethiopia experienced in the second half of the 20th century.

Thus, dependency on cereal as staple food hurt the peasant in two ways. Unlike root crops with phenomenally high tuber yield, cereals produce a mere 25% of that potato crop and other tubers. Indeed, Ethiopia is a living proof that cereal based culture without the support of production enhancing inputs to raise production per unit land is untenable. Incidentally, the mean cereal yield per hectare in 1935 as it is today stand at ton and half.[91] The second drawback of cereal based cultivation system is that unlike many root crops that you can store in situ and therefore withdraw the produce from the ground when the need arise, cereal crops are harvested right away and if stored in environmentally uncontrolled places could raise the post-harvest loss incidence to almost half. More significantly, cereal fields on top of being exposed to biotic enemies they have also been an easy target of plunder to marauding feudal armies in the past as they are today to government and guerrilla armies.

In ancient Ethiopia and as it has been throughout the three quarters of the last century, the primary target for destruction was a crop standing above ground. Obviously, this action is a prelude to more plunder and death unless a political compromise is forthcoming. The methods of destruction assume many forms among them burning the crop to the ground until the rival force submits or swear unwavering allegiance to the powerful. At times cultivated fields are left to rot while the victim continues to refuse for submission. But most significantly, harvested crops are looted leaving nothing to the contender. It would be astonishingly illustrative to present a typical scenario that occurred in late 1820 to a settlement in northern Ethiopia where a powerful feudal lord relates his expedition to a visiting European." We lived pretty well on corn belonging to the rebels. It took us 5 days to burn the fields to the ground". No doubt such measures may have an entrancing effect on the powerful to destroy the defiant looking upright yellow corn foliage or the elegant grayish wheat and barley fields in

---

Little Ice Age, Ca. 1300 – 1870. Environmental History Resources. http://www.eh-resources.org/timeline/timeline_lia.html

91 Alberto Sbacchi, *Ethiopia Under Mussolini: Fascism and the Colonial Experience*. Third World books, 1986.

place of the real enemy who took flight to the hills.[92] In contrast the effort required to destroy crops stored below the ground surface wouldn't be as easy as standing corn or sorghum. Such cruel measures taken by the warlords during the *era of princesses* were repeated over and over again spilling out to the early twentieth century.

92 Pearce Nathenial, The Life and Adventures of Nathaniel Pearce Vol. 1, (London: Henry Colbum and Richard Bentley, 1831), 68-69.

III

ETHIOPIAN AGRICULTURE IN THE TWENTIETH CENTURY

Ethiopia's history has been renowned for 3,000 years but during that long period the Implement of the farmer, the businesses of the small trader and all other spheres of life has seen no changes. There have been continued ignorance and low standards of living. There has been no progress whatsoever.[1]

The Background

Ethiopia escaped European colonization of sheer luck or perhaps due to the deeply held aphorism that in time of distress 'Ethiopia stretches her hand unto God' for guidance.[2] This is not to imply that an immeasurable sacrifice was not paid to safeguard the country's independence in the face of immense adversities. More importantly, it is far from underplaying the significant victory scored at Adwa in 1896 where rag tag peasant army defeated Italy and by extension undermined European superiority complex. It is simply stated to underscore the point that at the time of the Italian attempt of colonization in late 19th century, the country's marginal resources were thinly

1 Broadcast by the 1960 coup plotters on Ethiopian Radio Station, December 13th, 1960.

2 Psalm. 68:31.

spread to enable Emperor Menelik II feed a standing army of close to one hundred and twenty thousand for a considerable length of time to drive Italy out into the sea. By not doing so, it made it clear to any student of history the condition under which the Ethiopian army challenged the foreign forces and the state of Ethiopia's agriculture at the time. It must be borne in mind that the war of 1896 was conducted at the time when the country was emerging from a staggering famine and disease epidemics that had befallen upon it shortly before the carving of a sizeable landmass by Italy from Ethiopia to establish a colony soon to be christened Eritrea. In fact, the drought situation was still lingering in Tigrai and some northern districts when the war to safeguard the country's independence was waged in 1896. Suffice is to say that food was very scarce throughout the empire including the newly incorporated regions of the south and west of the country. Consequently, shortage of food forced the victorious peasant army to disband quickly and return to the various provinces where it originally came from.

Under such circumstance, it was not hard for Italy to realize that Ethiopia was incapable of sustaining a protracted war because it lacked basic logistic capabilities together with insufficient food resources essential for a standing army. In retrospect, one cannot help but wonder that had Italy being a relentless enemy with unlimited political and material support from the home country, perhaps Ethiopian history would have read differently. In the interim, the Italian army using food as a *raison d'être* took calculated risk by invading a frontier district shortly after the consolidation of their newly acquired colony was complete. Notwithstanding, their ulterior motives, the Italian colonial army crossed the border of what is now Eritrea entered Adwa and performed war dance before their sudden retreat. An eyewitness to the 1890 event recounts on how the once vivacious town of Adwa has turned into orphan with only one fifth of its population remained holed in the famous church of Medhanie Alem fearful of their lives.[3] Before the retreat, the Italians distributed food to the hungry population, which even enticed their archenemy, the feudal chieftains to expect food aid from Italy at a later date.[4] In a letter sent to Rome, the colonial ruler of Eritrea, General Barateri informing his government on the food aid that the feudal chiefs requested from Italy "the 20,000 sacks of grain which they (the feudal rulers across the border) asked for (some) two and half months ago, I have not sent them even a single one'.[5]

---

3 Fiseha Giorgis Abyezghi (Debtera) ca. 1897 (Tigrigna), Tarik Itopia, The manuscript was published with Italian introduction by Dr. Yakob Beyene 1979. P125-126. In the meantime Hailu Habtu republished the book with his own commentary in late 1990s.

4 Erlich, Haggai, Ras Alula and the Scramble for Africa. (Trenton: The Red Sea Press, 1996), 151.

5 ibid. 173. Haggai Quoting Barateri from the Archivio Storico Ministero dell'Africa Italiana A.S.MAI. 3/6-148, Barateri to MAE, 18.8.92.

This is not the first time that the foreign powers occupying the coastal city of Massawa used food as a ploy to meet their objectives, that is, to weaken further the shaky confederacy that existed among the provinces of northern Ethiopia and the newly incorporated lands of south, east and west of the country. Indeed Egypt and later Italy regarded food as Ethiopia's major predicament and continued to use it to their own advantage during the last part of the 19th century. A case in point is that of Ras Woldemicael who wreaked havoc on the highlands of Hamasien and as a matter of habit raided the mid-altitude regions of Halhal-Bogos, in search of food and for sanctuary. His army of 7,000 men lived on food subsides guaranteed by the Egyptians and another Dejatch Debeb, the nephew of Emperor Yohaness IV, practically lived on the Italian handouts for long time. With the change of the political climate, their power ignominiously declined as soon as the puppeteers begun to pull the plug.

After the war, the situation in the north particularly in Tigrai was much more disturbing. For obvious reason, Tigrai besides being the battleground for the many wars against Italy had been passing through hard times because of various natural catastrophes. Moreover, war among the now disempowered Tigraian nobles were so rampant that it played in the hand of the Italians whose ambition remained to conquer Ethiopia.[6] On top of being stripped of their basic means of livelihood because of the loss of animal power few years earlier, disease epidemics were also rampant on the population. There is no doubt that the prolonged drought of the late 1880s and early 1990s had seriously affected food production in Tigrai and perhaps the colony of Eritrea despite the Italian subsides which appeared to be handy for some Eritreans associated with the colonial governing structure. Nevertheless, the depression that seemed to grip the northern highlands for some time was partially eased by the availability of food imports and the expansion of public works in the European section of Abyssinia. To that effect people attracted by job prospect and above all the availability of food begun to trek to Eritrea from as far as the province of Gojam. The acuteness of food shortage at the time is best appreciated when one examines the plight of the onetime powerful lords who after the war begun to live on charity and outright banditry. As habit dictated, many of the feudal lords lived expecting gifts from individuals that are more fortunate and sometimes by confiscating property of the innocent.[7] In short, this was more or less the scenario of the food situation and more broadly the agricultural condition of Ethiopia in the early part of the 20th century looked like.

---

6 Report of Governor Orero to MAE (*Ministere des Affaires Etrangeres.* ASMAI 3/5-33).

7 Erlich, Haggai, Ras Alula and the Scramble for Africa. (Trenton: The Red Sea Press, 1996), 194 Haggai quoting from Mulazzani Report in Archivio Storico Ministero dell'Africa Italiana A.S.MAI. 3/17-136, 26.7.96.

Despite, Menelk's success in incorporating the well-endowed South and West, the state of affair almost remained the same, that is, the combined national wealth was continuously misdirected or inefficiently utilized to benefit the productive forces of the country. The lack of resolve to build a more enlightened government that is inimical to the entrenched feudal system made any progress impossible. As a result, the means of production and the infrastructure that is crucial to supporting the growth of the economy remained as archaic as it has been since the middle ages. In other words, the extraction of surplus from the peasant continued unabated while accumulation and conversion of the surplus to meaningful purpose was far from being realized.

England at this period of its history not only devised the means of advancing agricultural science to benefit the farming community as a whole but also allowed farmers to be partners in the process. In contradistinction to the 18th century English gentleman who was rewarded by the agricultural societies of his country for 'choosing to spoil his hands with dirt' the Ethiopian 'gentlemen' in the word of an intellectual at that time was busy expropriating 'injera, honey, fattened ram, eggs, chicks, tej and tela' from the poor.[8] To encourage and further invite community participation 'for the good of agriculture', the various societies in England had made it a tradition to award prize money to small farmers and laborers who demonstrated proficiency in plow.[9] In the Ethiopian model of feudalism those who deserved commendations and adulations were ironically, the feudal parasites whose vision of Ethiopia was a loosely held archaic Christian confederacy living at the expense of the serf. With no provision for surplus accumulation and lacking clear intention of plowing back some of the resources for sustained development; the country was forced to languish into wilderness for a long time.

Land and the Tribute

Throughout history, land has been contentiously held in Ethiopia that it became a convenient weapon for political maneuver and feudal trickery in the Ethiopian body politics. It has been used as tool for power realignment among the highest of the high and a bait to extract services from the downtrodden. The reason is that land was acknowledged to be the only and most valuable asset upon which people built power, wealth and prestige. Remarkably, it remained so up to the dethroning of the last Emperor, Haile Selassie in 1974. In days gone by, appointment to a post or inheriting a position of power was regarded as a proper avenue for land aggrandizement and access to free labor from the peasantry. In the same manner, matrimonial relationship entered for

8 Afewerk G.J., Guide du voyageur en Abyssinie (Rome: 1908).

9 Kenneth. Maclean, 1950. AGRARIAN AGE: A Background for Wordsworth. (New Haven CT: Yale Printing Press 1950), 6.

this and this purpose alone was not viewed as sinful act to warrant confession from a priest. Despite the emperor's absolute power over land matters, the revocation of *gult* by Imperial orders from recalcitrant feudal lord had never been regarded as permanent or irreversible.

In the Northern part of the country, land entitlement goes hand in hand with eternal residency in the area ascribed by blood ties. Therefore, earlier settlers and their descendants make a corporate body. Regardless of the social status or material well-being of a person by virtue of his birth right is assured a secured membership in the corporate village and therefore access to a piece of land. By implication people with blood ties to a corporate village, *enda* or the *akni abat* have their livelihood assured. In Eritrea long time residency in one locality or village without necessarily claiming a blood relationship to it, entitled one with a piece of land. This new system was of recent phenomenon introduced by Ras Alula, the governor of the coastal lands to benefit his Tigrain soldierly who accompanied him during his sojourn to Hamasien[10].

Before the revolution of 1974 that made land the collective property of both 'the people and the State', the peasantry especially, the northern peasantry has been viscously sucked into endless land litigation to secure corporate membership when challenged. Usually, this leads the litigant to fall prey of unscrupulous government officials who mercilessly extract bribes on false pretension that they will adjudicate in his favor. The Tigrigna adage '*Ni Risti Yiwegalla Ansti*' says it all when one speaks of land in the Ethiopian context. Nominal ownership of land does not only bring to a person a semblance of a secure livelihood for him and his family but also is the source of pride having been able to trace his blood tie to the early settler. In early traditional West African societies, this notion is farther stretched to deprive a reigning king to have any say regarding land if he did not belong to the family of the first person occupying the ground in his kingdom.[11]

Before the advent of the Abyssinian conquest of late 19th century, land in the south, east and west of the empire was regarded as a collective property free from excessive abuse. The mode of production being primarily cattle rearing, the society's wisdom dictated that the resources available are shared collectively. With the introduction of rigorous agricultural practices common in the northern highlands, production relation begun to change in the south. The new conquerors together with the indigenous chieftains begun to introduce production relations based on serfdom where producers retain half

10 The local law known as 'wedi arba' meaning 'one who lived for 40 years' in a village not necessarily connected to it by blood. According to my maternal grandmother her grandfather who came to Hamasien from Tigrai as retinue of Ras Alula stayed in Tsada Kirstain, a village in the outskirt of Asmara where his descendants still live.

11 John Hope Franklin, and Alfred Moss, (McGraw Hill: From Slavery to Freedom. History of African-Americans 2000), 17.

or three fourth of their produce for themselves and the rest left to the overlord and his functionaries. Hitherto unknown levies and dues accompanied by *corvee* were also introduced with laws and standards not uniformly applied throughout the land.[12] It is to be recalled that Menelik never consented that the newly incorporated lands to the Empire be under anybody's tutelage but him. In as far as the Emperor was concerned the powerful feudal lords who made his southern and western campaign successful were mere adjudicators and administrators at best. Their power can be limitless short of claiming the land as their own. As beneficiaries of the new conquest, their usufruct right is boundless operating on immense lands with the serfs working for them. Their primary function was to collect tribute from the subject serfs and pass it to the court with reasonable portion of it for their keep. In essence, it was not different in the north where the *gult* holder or the governor extract labor from the peasant and enforce payment of tribute due to the court and partly to him. The major difference is that the northern peasantry as corporate member of a village had an opportunity to access land while people from the recently incorporated south did not have that prerogative or are prevented from exercising their right to own land. In the final analysis, it is in the perception rather than the reality that made the northern peasantry appeared to be in a better position than the southern or western serf.

Soon after the death of Empress Zewditu in 1930, the crown prince Teferi Mekonen, later Emperor Haileselassie I, ascended the throne to rule the country for close to half a century. Characteristically, he pretty much continued the same policies pursued by his predecessor with slight exceptions that he allowed his loyal governors and noblemen to amass more wealth unchecked. He even allowed a clause to be attached in the draft of the first constitution adopted in 1931 to the effect that the feudal lords could own *gult* or *rist* in their respective provinces that they could pass to their offspring. The compromise was reached as a result of ensuing court battle between the Emperor and the newly upcoming *mekaunint* and ministers of low pedigree on one side and the nobility on the other who insisted that the provinces be converted into their fiefs with the understanding that they be allowed to pass it to their offspring. However, more formal land grant and the conversion of *gult* into freehold did not come before the Italian defeat and the return of the Emperor from five-year exile in 1940.

To the peasantry nothing seemed to change but pay more tributes than he could spare. He lived in constant threat of violence where mere infraction of the feudal laws and/or rebellion could result in eviction and occasional bodily harm. A look at the plea made by a vassal chief in Wellega in 1913 clearly indicates how little things had changed to improve the wretched condition of

12 Desalegn Rahmato, Agrarian Reform in Ethiopia, (Uppsula: Scandinavian Institute of African Studies, 1984), 17.

the peasant who, few years earlier lived under dire condition brought about by the wrath of nature and the onerous extraction demanded by the feudal lord. A ruler of a small conquered kingdom in the west writes to Lij Eyassu, the king in waiting that 'concerning this overdue payment of tributes, I pleaded to you my master. Have seen that the country is uncultivated, this being the result of too much tribute, not because of any wrongdoing in my part. And now, if I order the people to give the three year overdue tribute, the country will be completely abandoned.'[13] The likelihood that *Lij* Eyassu exempted the people of Wellega from paying annual tribute is remote.

Despite variability in the type of tributes paid by tenants, there were few common and applicable rules over the empire. For example, consistent with the spirit of the first constitution of 1931, the law states that the word of the landlord takes precedence over the serf's and that the courts are not obliged to hear the case submitted by the tenant (No 5680). For any unexplainable reason should the landlord find it expedient to evict the person occupying his land, the serf has to comply without expecting any recourse from the courts (No 3272). Finally, regulation Nos 3287 and 3418 reiterate that when tenant is asked to leave the land he should do so but dismantle his house and expect no compensation from the landlord.

The Protagonist of Land

Although land and issues surrounding it has been the driving force for the Ethiopian state to exist the way it did for many centuries, not very many had raised the land question in a more revolutionary way than the student movement of the 1960s and 1970s. This tiny section in what is otherwise predominantly agrarian society identified themselves early on with the peasant serfs. They indicted the system as having major structural defect hindering agricultural growth and well-being of the peasant. However, few exceptions are mentioned in historical records as harbingers to the reformist's movement that appeared near the end of this century. It is known that in early 17th century, Emperor Zedingl expressed his wish to introducing land tax based on productivity. Apparently this unpopular move plus power rivalry within the ruling circle must have triggered anger among the feudal class, that the Emperor was dethroned after reigning two tumultuous years.[14] No one after him ever tried to regulate land use or impose tax based on-productivity to foster

---

13 Donald Donahue and Wendy James.The Southern March of Imperial Ethiopia: Essays in history and social anthropology, *Letter from Ghebregzeiabher to Lij Eyassu Megabit 2, 1905 E.C. and 11 march 1913 G.C.* (Cambridge University Press. 1986), 59.

14 Historians attribute Zedngl's downfall from power to his attempt of relegating the Sabbath to ordinary weekday contrary to the Ethiopian Christian tradition. During his short stay in the throne he was also eager to learn European laws and governance from the Portuguese in the country, which could have further infuriated the establishment.

agricultural growth. Century later, that is at the height of the *Zemene Mesafint* era, the strongest of the strong, Ras Gugsa proclaimed that henceforth land will be centrally administered and that the hereditary chiefs have no private land they claim their own. Gugsa himself an outsider asked, "What is the landowner whose property is stronger than himself? Holders of lordly estates and holders of fiefs, let me say to you that there is no hereditary right of suzerainty. *Rabbi* give it to whom he pleases he has given it to me. I alone allocate it at my pleasure". Gugsa's intention was not clear in that whether depriving the feudal lords tracts of land meant to encourage centralization based on peasant farming or simply was consolidating land to further increase the power of the king.

It is difficult to predict the outcome of this decree had it survived long enough to be an institution by itself. One thing is clear, had centralization came to Ethiopia earlier than later and at the same time a progressive land tenure system was introduced in earnest, Ethiopia's position would have been different than what it is now. In other words, the picture of the country's agriculture would perhaps have evolved into more market oriented than simply a consumption based on subsistence production. Inevitably, the producer would have ceased to be subsistence but an engine to creating markets, flourishing towns and a factor to fueling curiosity among citizens for technological invention. Had such course of historical development took place, Ethiopia would not have been a country besieged by host of social and economic ills. *Lij* Eyassu the heir apparent of Menelik II was rumored to have promised the farmers at Harar to restore their expropriated land when and if regained power again. Unfortunately, the promise was unfulfilled because of his capture and subsequent imprisonment at Fiche away from Harar.[15]

Before, the celebrated tour by the Reagent, *Ras* Teferi to Europe and the Middle East in 1924, very few Ethiopians ventured outside their country. Those who travelled abroad together with those who by sheer luck came in close contact with Europeans and other peoples of the east became outright critic of the feudal system after realizing the extreme backwardness of the country they adore. Addis Hiwot (1975) in his book, *Ethiopia: from autocracy to revolution*, describes each of the enlightened individuals dub as Japanizers based on their aspirations for Ethiopia and their theoretical outlooks on land and government.[16] Among them was the arch critic and later fascist apologist Afwerk Gebreyessus who openly called for the abolishment of feudal rights and the dissolution of provincial offices in favor of centralism. According to him the twin evils that hold back Ethiopia's development was wars and raids waged by the feudal lords and starvation caused by the combined effect of insecurity and natural calamity. Driven by an extreme hatred of the system and

15 Harold. G. Marcus, HAILE SELLASSIE I: the Formative years 1892 -1936, (Trenton: Red sea press 1995), 19.

16 Addis Hiwot, Ethiopia: from autocracy to revolution, (London: Occasional Publication No 1. Review of African Economy, 1975)

an utmost disdain to the nobility who kept his country under extreme backwardness Afwerk christened the date that Italy occupied Ethiopia as "The era of Mercy'.

Afwerk's utter disgust with feudalism was left for Gebrehiwot Baykedagn to elaborate on the system in political economic term. He argued that land should not be concentrated in few hands. This he said is a 'bad omen' for a developing country. If land is distributed among the people he rationalizes 'it creates wealth for them and strengthen the government too.' He went on further to advocate for the people's right to sell, mortgage and exchange land 'which is the only source of income they (peasant) have.'[17]

Asbe Hailu writing satirically on the feudal system and the premises upon which it was built challenged the government to introduce land taxation system in the country whereby an annual income tax of 15 Birr be imposed on approximately 4 million hectares owned by handful royal family members and hereditary *Rases*.[18]

Unfortunately, it took quarter of a century after the column first appeared in Berhanna Selam newspaper for the government to introduce land tax in 1944. Moreover it took another twenty years before the *Riste-Gult* landholders were made to pay tax on their holdings 1964.[19]

Like Gebrehiwot before him, Deressa Amante was extremely apprehensive of European machination and called for self-reliance based on agriculture development. As a Director in the Ministry of Agriculture (MoA) before the Italian invasion of 1935, he had the privilege of understanding Ethiopia's predicament to feeding its own citizens and vigorously fought against the feudal system. In his contribution to a newspaper, he equated Ethiopia with the Garden of Eden but further lamented that 'while America is feeding others, we are still poor' because of the archaic system of feudalism imposed upon the people.[20] Ironically today the source for Ethiopia's

---

17 Gebrehiwot Baykedagn, Mengstna Yehizb Astedader 'Government and Public Administration', (Addis Ababa: 1953EC)), 88. (Amharic)

18 Asbe Hailu. Sile Mengst Gudat. 'Bankrupting a Government'), Addis Ababa: Brhanina SELAM Printing press July 22, 29 1927) (Amharic).

19 The holder of the inheritable *Rist-Guilt* collects tax from the *gebar* (serf) ranging from 5 birr to 15 birr depending on the fertility status of the land and in return pays a mere 3.5 birr per *gasha* to the imperial government. Under the *Sisso* arrangement, the hereditary chief give two-third of the tax collected from the *gebar* to the government while keeping the remaining one-third for himself. The Bill of 1964 abolishes this traditional system of taxation and obligated the landholder to pay direct tax to the government ranging from 5 to 15 birr Per Gasha.

20 Deressa Amantie. Sile Giziew Yegenzeb Chigir. (On Temporary Financial Problems-, (Addis Ababa: Brahanina Salem printing press. February 11, 1932) (Amharic)

industrial investment comes from agriculture with modest foreign involvement as elsewhere in the continent of Africa.

Blatta Gebregziabher Gilamicheal of Tseadakirstian, self-educated critic of the feudal system and an enemy to the occupying power occupying his birth region *Medri Bahri*, lamented on the misery of the countryside long before many did. He prophesied that 'the tears of the poor are going to accumulate and descend on the king'. He also equated the status of Ethiopia as a nation to a slowly dying leper.[21] Gebregziabher, who suffered greatly in the prison of Naples and later at Nocra, a desolate island prison in the Red Sea was an arch critic of Menelik and the feudal system in almost unorthodox way. In his 1897 memoire, he lambasted the practice of billeting soldiers as a curse that 'spoiled and destroyed the country'. He went on to equate the parasitic nature of the unpaid soldierly as 'locusts and hail' descending upon a land.[22] Strangely, it sounds as if he read the constitution of the US drafted in 1787 which clearly forbids the quartering of soldiers in any 'house, without the consent of the owner'.[23]

However, the challenge to feudalism in the most articulated form came from the student movement that begun to gather momentum in the aftermath of the failed coup d'état of December 1960. Given the small number of the university students throughout the 1950s, it is not surprising not to have a vibrant student organization to catalyze for change.[24] The University College of Addis Ababa, the center of dissension until the demise of the monarchy, was established in 1952 with 200 students and 25 staff mostly expatriates. By the early 1960 the number of students rose to 1,500 with few scattered campus around the country. Almost a decade later the number of high school and university students reached a mere 380,000 and 2,160 respectively.[25] Thus, it is safe to argue that because of their miniscule size at the time, student activism for radical land reform was mute. It took decade for this numerically insignificant student body to gradually begun to identify with the country's problems by raising issue not popular with the ruling circle. No wonder it was

---

21 Bahru Zewdu, Pioneers of Change in Ethiopia. (Addis Ababa University Printing Press 2002), 125.

22 Tekeste Negash, No Medicine for the Bite of a White Snake: Notes on Nationalism and Resistance **in** Eritrea, 1890-1940. (Uppsula: Reprocentralen, HSC, 1986), 6 -7. From. "A Memoire of Blatta Gebre Egziabeher Gila Mariam from Tzzadda Kristian, Hamassien, 1889 Eth. C." (1896/97) p. 51, A. E., Pacco 1008.

23 Amendment III of the United States Constitution.

24 Syoum Gebregziabher. 2012. *The Symphony of my Life*, (Trenton: The Red Sea Press), 208. In his autobiography, Dr. Syoum Gebregziabher who in the early 1960s was a professor at the Haile Sellasie I University recalls that in 1960/61 Student enrollment were 381 of which 27 were students from various African countries.

25 Patterns of Progress, (Addis Ababa: Ministry of Information, 1967)

the students who came out in support of the coup when the news hit the airwave.

In early 1960 the student body called for land reform and boldly displayed banners reading 'land to the tiller'. Through time the slogan 'land to the tiller' became the battle cry of those who had much to loss than gain, college and high school students all over the country. At last, their persistence demand and uncompromising stand against feudalism paid off by triggering the historic February 1974 Revolution where few had anticipated to take place. It is however; less accurate to exclude other important factors that help galvanize the popular uprising into achieving its objectives. In other words, the existing religious and national oppression gave further impetus for the popular uprising to continue unhindered. It should be noted here, that under the ancient regime, Christianity was the undisputed religion of the state while the Amhara (and Abyssinians) ruling class supported by assorted landed aristocracy and bureaucracy from all regions dominated the political scene. Therefore, political and economic oppression combined with religious inequality that existed among the population of the Empire culminated into a revolution with far reaching consequence.

Given the spontaneity of the popular uprising and the absence of an organized body to commandeer or channel the discontent manifested by the day throughout the middle portion of 1974 in the capital city and other parts of the country, the army assumed responsibility and enacted sweeping reforms. Among the measures taken by the army committee, later known, as the *Derg* was the nationalization of rural lands in March 1975. The decree made it clear that no compensation is to be paid to the original owners meaning the landlords and if they wish to stay in the land will be allocated 10 hectare of land just like any other peasant farmer. The government also put all commercial farms under state control and forbade the hiring of wage labor on private lots.

In contradistinction to the harmless version of land reform draft tabled to parliament in 1969 by the newly established Ministry of Land Reform & Administration (MLRA) this was radical in every respect. Unwilling to act on the draft, the landed aristocracy packed house of parliament returned the draft bill to the Ministry for milder version so as not to antagonize entrenched interest groups. Tragically the second version albeit, in much watered down form was again rejected and left in limbo for four years until its fate was decisively resolved by the ensuing popular revolution. It was understood from the outset that the reform envisaged by the imperial government before its timely demise in late 1974 was not designed as such to cause an imbalance in power structure leading to a revolution from the bottom. It was soft and harmless to cause alarm for many of the land owning class.[26] It is no hard to imagine, therefore, that when the land reform decree was announced the students as champions of the 'land to the tiller' hailed the move as historical almost instantly.

Revolution and the Land

As obstinate as it has been for centuries, the imperial system finally caved in for the student demand. Indeed, it was a matter of time before land is distributed to the landless. Following, the unsuccessful coup d'état of December 1960 that left dozens of noblemen and dignitaries dead in the palace massacre, a sort of uneasiness begun to set in among the ruling class. In general, this terrible incident among others was viewed as a wakeup call for the *ancien regime* and its staunch supporters that the time has come to deal with the land issue, no matter how painfully. Thus, to address the underlining problems pertaining land, a department was established within the Ministry of Community Development (MCD) in 1964, which soon was upgraded to ministerial level. Its mandate was to draw plans on how equitable land can be distributed or allocated to landless. It is not quite clear, however, whether the plan would address the concern of the tenant farmers or change the fundamental structure whereby democratic rule and justice prevail. Obviously one would assume that the landless and the poor peasant on whose back the landowner lived for ages would be the primary beneficiaries. Since the military coup of 1960 and up to the eve of the popular uprising in 1974, the imperial court instead carried out the largest land handout to civilians residing in cities, the civil servants, retired civil and army personnel, the army loyalists and the Eritrean unionists without the slightest regard to the needs of the tenant farmers and the landless.

Based on a rough estimate, this time alone close to 4 million hectares of land were known to have been distributed to yet potential landlords, albeit most of them from humble background.[27] Meanwhile, the land reform bill worked out by the staff of the Ministry aimed at benefiting the poor and the landless failed to see the light of day leaving the serfs in a more desperate situation. It was at this historic juncture that the 1974 February revolution erupted with numerous forces claiming to be the rightful heirs to lead and cajole it to its final consummation. Later, the interest of the student lead movement was fatalistically collided with the interest of the factions of the army to affect many

26 Ras Imru a member of the royal family was an exceptional figure that disposed most of his inheritance to the serfs. No wonder when Imru died in mid-1980s the *Derg*, the archenemy of the feudal class gave him state funeral.

27 Harrison Dunning, Land reform in Ethiopia; a case study in non-development. UCLA law Review, Vol. 18, No2. (UCLA, 1970), 290.

Ethiopian lives in much horrendous way.[28] It should be clear that the revolution was sparked by what is termed as the Sector Program in education that limit the number of students going to higher education but allowing them to join vocational trade schools instead. Action by taxi drivers affected by gasoline prices added more fuel to the fire creating difficult situation for the government to control. The inferno that started in February in February 1974 was so ferocious that there was no force to extinguish it. All throughout the four month of the revolutionary fervor, there was no recognizable group or acknowledged leader to navigate the course. In the interim a well-organized formidable forces, the army came to the scene with touchy slogan "Ethiopia first" ready to challenge any one stood on its way. In retrospect, it was the weakness of the civil institutions, the absence of political parties and the lack of social means of communication that brought the army into the revolutionary platform making it an arbiter and later owner of the revolutionary process.

In as far as depriving the landed aristocracy of its ill-gotten wealth and relieving the peasantry from the onerous task of serfdom, one can argue unabashedly that the swift action on land reform by and large has achieved its symbolic target. No doubt centuries old of injustices were redressed and dreams seemed to be in the verge of being realized. One point that need to be stressed here is that, had it not been of the land issue taken as a pivotal to striking the imperial government, the student movement would not have come out to be as radical as people imagined it to be in the 1960s and 1970s. The student movement at that period was more radical than many in Africa and their tentacles spread to Europe and North America through their affiliates outside Ethiopia. Radical student movements in the Diaspora such as the Ethiopian Student Union in America and Ethiopian Student Union in Europe catalyzed the mass movement at home by supporting the major player on the ground the University Students' Union of Addis Ababa.

It is important to note here that in the eve of the revolutionary upsurge the illiteracy rate was at more than 95% and the number of the working class amounted to less than 100,000. The implication is that in the absence of home grown ideology nurtured by the wishes and desires of multitude of peoples across sectors of the society, the revolutionaries begun to embrace foreign ideologies in their totality and accepted imitations on social and economic remedies quiet irrelevant to the country. The result, as we shall see later brought the country into the brink of civil war, foster mass discontent, armed clashes, famine and death. Understandably, the land reform promulgated by the Provisional Military Government (PMG) was welcomed and supported by the

---

28 No one can give an accurate figure on how many young people died during the Red and White terror campaign launched by the radical left and the *Derg*. In the infamous 'May Day Massacre' of 1977 alone one thousand young people were reported to have died by the Special Forces.

students and the intelligentsia not to mention the peasant farmer who stands to gain more than anybody else in the society. The question arises was land reform consummated as envisioned by its earlier proponents the radical students or more appropriately by the ultimate beneficiaries, the peasant farmer? Was the land reform decreed at the height of the famine helped in addressing food insecurity? What transformation occurred if at all in the Ethiopian agriculture as a result of the radical land reform engendered by the 1974 revolution?

It is important to note here that the students and the radical section of the intelligentsia pushed for an immediate land reform based on purely socialist lines even if it means the expulsion of all categories of landlord from the land. In comparison, the interim government waged a futile struggle to explain the effect of sweeping land reform could have on production. It advocated for gradual dispossession of the landlords.[29] In the course of the revolutionary process, "limiting ownership of land based on its beneficial use by the owner" was found to be unacceptable by the provisional military government, the *Derg* and instead went further to nationalize all lands without much qualification. Thus, unlike in China where the rich peasants together with the middle peasants, who constituted the backbone of agricultural production, were spared from the shock of blanket land nationalization, in Ethiopia the opposite was true.[30]

As the political debate degenerated into mundane on whether it is in the nature of the army to lead a revolution and truly align in the struggle with the down trodden or by its very constitution is the tool of the ruling class and therefore not to be trusted. In this regard, the leading opposition group to the *Derg*, the Ethiopian People's Revolutionary Party (EPRP) expressed its unequivocal distrust of the army and demanded for its return to the barracks. In August 1975 during its founding congress, EPRP resolved to replace the *Derg* with 'provisional people's government'.[31] At the same time few within EPRP posed to evaluate the merit of the sweeping land reform decree passed by the army. In other wards after the demise of the old order this important item in which Ethiopia's future hinges upon was taken as side issue instead as central by all actors in the political arena. Similarly, section of the intelligentsia in the Mela Yethiopia Socialist Nqnaqie (Meison), a group with an intellectual capacity, political acumen and arm chair revolutionary experience withhold their critical analysis on the reform but choose to ally with the Provisional

29 Ethiopian Herald April 19, 1974.

30 John Wong, Chinese land Reform in Retrospect. Land tenure Center, (Madison: University of Wisconsin, 1974), 4.

31 Prison confession of Berhane Meskel Reda a prominent Ethiopian student leader in the 1960s and one of the founding and politburo member of EPRP. http://www.assimba.info/uploads/files/dosse/Berhanemeskel_%281%29.pdf) 14.

Military Government in the hope of weakening their political nemesis, the EPRP. In hindsight the reform in its crudest form can be equated to a dynamite ready to blow taking with it the entire edifice to the ground. The flaw in the land reform law was exposed during its implementation on the ground.[32]

Implementation of the Land Reform

The land reform in Ethiopia was overdue for two basic reasons. First, the peasantry who for centuries was the backbone of the country's economy has been shut off from accessing land freely. This prevented him from contributing meaningfully to the country's development by increasing production and raising productivity. The peasant lived under constant threat of violence, eviction leading to eventual destitution. The other concern that made the land reform such an urgent matter was that the ineptness of the Ethiopian landed aristocracy to utilize the land effectively. Unlike the huge land owning class in Latin America where they employ the land productively and become net exporters of beef, Soybean and Wheat, the Ethiopian feudal chiefs were satisfied with the traditional provisions and extractions from the serfs to contemplate pursuing modern agricultural enterprises.[33]

The decree on land reform announced by the *Derg* came as no surprise even to the serfs whose trust on the government has been at its lowest since the incorporation of the new territories by Emperor Menelik in late 19th century. The question remains would the land reform enacted in March 1975 change their lives. In hindsight, it will take more than simple declaration of the land reform law to effectively address their concern. For one the reform was not accompanied with comprehensive plan of action. It lacked technically supported land categorization format that would assist in the execution of the reform in a systematic fashion. It did not attempt to categorize the size and type of holdings permissible under peasant holding and the bourgeoning commercial farms, albeit small. It simply refused to recognize the right of citizens to own and operate commercial farms of any sort. To the contrary the government moved close to half million peasants from their ancestral land to give way for the construction of socialist enterprises, the cooperatives and the state farms.[34] Preceding the land nationalization decree, there were no seminars

32 Dagnachew Yirgu a newly appointed Minister of Agriculture in 1975 argued against drastic measure that could harm future agriculture production and productivity. The revered plant pathologist who came from a humble background was immediately suspected as feudal remnant of the past and mysteriously disappeared on his way to inspect farms in Wellega.

33 One of the major landholding feudal personality, Ras Mesfin Sileshi is reported to have owned 100,000 hectares of land at the time of the revolution.

34 Desalegn Rahmato, Agrarian Change and Agrarian Crisis: State and Peasantry in Post-revolutionary Ethiopia. Africa: Journal of the International African Institute. Vol. 63 (1) (1993), 47.

held, no workshops conducted or consultation made with the peasant farmers on how best to distribute the land. The left leaning military government and its advisers found it inconvenient to study land reform models of other countries other than the ones adopted in the socialist countries, which have already proved unworkable.

Quiet understandably, the country was at the lower stage of its agrarian development to expect the subjective forces compliment the already ripen objective conditions such as famine, mal administration, nationwide unrest and the student fervor for change. It is only through the convergence of the two forces that a meaningful change is expected. At this crucial juncture of Ethiopia's history, the social forces responsible of inducing genuine change were numerically minuscule, intellectually immature and ideologically speaking externally inspired. In short, the self-proclaimed vanguard of change, the educated class, besides being qualitatively insignificant were in tune with Marxian philosophy that may have little to do with the Ethiopian reality deeply anchored in medieval beliefs and mores. In addition to the archaic mode of production that seemed to dominate the agriculture sector, the size of modern institutions including schools of higher education was practically small or nonexistent. In retrospect one is obliged to ask the wisdom of imposing scientific socialism in such a backward agrarian society where even the working class is not qualitatively mature and quantitatively insignificant to lead and/or be an equal partner in the revolutionary process. After all, the majority of the working class at the time was nothing more than individuals super imposed in urban settings with their rural consciousness intact, to fathom the implication of forces unleashed by the revolution.

Therefore, the tactics and strategies needed to administer the technical and a judicial imperative of the reform was recklessly left to small influential groups who were closer to the *Derg* and to the student groups outside it. Meison, one of the early established and influential political groups gave its tacit support to the model expounded by the PMG with regard to the dispensation and administration of land. Until their forcible expulsion from the 'coalition' they continued supporting many of the senseless measures and bravado actions taken by the *Derg*. Immediately after the decree to nationalize the land went into effect the government launched Development Through Cooperation Campaign (DTCC) that sent off around 70,000 high school and university students in early 1975, to the countryside to 'teach' and 'learn' from the peasantry at the same time. However, the main task of the campaigners later known as *zemecha* participants was to herald the proclamation of the land reform issued by the government and assist the farming communities in organizing themselves. In a way, it is here that the seed of discord was planted resulting in a prolonged urban warfare under the banner of 'Red Terror' and 'White Terror'.

No sooner did the *zemecha* students settle in their respective duty stations scattered throughout the country, that major rift begun to develop among and within the implementers of the reform and between the local authorities. Moreover, in many areas of the country and in particular where the freehold tenure sway most, contradiction begun to surface among the peasantry and the *zemecha* folks who wanted a self-governing units established in line with the socialist doctrine. Ironically, this did not settle well with the government who had its own agenda regarding peasant radicalization, that is, a peasant who is less vocal but subservient to the state. How else would the regime face the challenge posed by the separatists, ethnic nationalists, and those radicals in the center vying for power without the support of docile peasantry who for centuries has been the cannon fodder in internal and external wars. More importantly, some sections of the peasantry were less convinced that the landlords were the mortal enemies deserving complete annihilation as advocated by the radical students.[35]

Therefore, and because of the lack of clear directions from the top, each and every group was determined to put its agenda on land reform as it saw it fit but more poignantly as its ideological inclination dictates. As a result there was no or little standardized directives to implement as the countryside become a hot bed between factions supporting the provisional government and those opposing it, between the radical students and exceptionally conscious serfs on one side and the feudal lords who resisted the reform on the other. Finally, with the support of hard core Marxists on its side the government begun to draw land utilization plans modeled after the Chinese but soon abandoned that in favor of the so-called non-capitalist model promoted by the Union Socialist Soviet Republic (USSR). Nevertheless, among the mandate given to the *zemecha* participants when executing the land reform, was to assist in the setting up of' peasant associations. It is presumed that once peasants are organized and are made free from obligation to the landlord, food production will improve.

35 Aregawi Berhe, A Political History of Tigray People's Liberation Front 1975-1991. (Tsehai Publishers and Distributors, 2009), 74. In part this can be explained by (probing into the agrarian system and the ways it evolved. Unlike the European feudal system, in Abyssinia there was no lordly estate where the serfs are completely shut off from associating with the feudal lord. To the contrary there was frequent interaction with the serfs, be it in the act of extracting tribute or in taking part in the celebration of one of the numerous religious holidays, in public feast etc. The truth is even at the outbreak of the February revolution the majority of the feudal lords lived a lifestyle not excessively different from the peasant. The peasant's woe was mostly the abusive aspect of the tribute itself and its unpredictable nature. Aregawi Berhe, one of the founding members of the TPLF a major partner in the EPRDF lead government observes that the peasant fighters did not want the traditional relationship [lord-peasant], which they considered a better one, to be eroded.

However, it did not take time for nature to impose its will on the Ethiopian landscape as it did for centuries that is, hitting the country yet with another big drought and famine in 1978.

In September 1977, a decree to establish a nationwide peasant association was proclaimed, in part thanks to the *zemecha* students who paved the way for the proclamation to be heard loud and wide. The establishment of the All-Ethiopian Peasant Association (AEPA), an umbrella of the *Kebelle*, *Wereda*, District and Regional Peasant Associations gave further credence to the land reform announced earlier. The AEPA from its lowest administrative unit, the *Kebelle* up to the highest body, the Regional office, was empowered with a wide range of responsibilities, including implementation of government land use directives; adjudication of land disputes; engage in development programs, such as water and land conservation; construction of schools, clinics, and cooperatives; and above all facilitate the collection of taxes. The mandate of the Peasant Associations (PA) extended further to include afforestation work, rural road construction, and data collection. Finally, the PA was responsible of supplying the army with the necessary manpower to augment the forces fighting the civil wars in the north and other parts of the country.

Each PA was granted 800 hectare of land divided among 200 of its members to be operated separately or independently. By late 1980 there were over 20,000 PA all over the country including Tigrai and Eritrea where the situation was volatile and murky. Side by side, a provision was also made to push as much as possible for the formation of Production Cooperatives, (PCs). This entails the submergence of individual needs and aspirations for the common good. Thus, for all intent and purposes, agriculture in the peasant sector was put under two categories, the service and PCs. The private farms, which were largely mechanized, were taken over by the government to be managed eventually by salaried bureaucrats and in some cases by the very persons whose commercial farms was expropriated under the land nationalization decree of 1975.

The SC as the name implies were entrusted to distribute agriculture inputs to members and help market their produce directly to consumers without having to deal with unscrupulous middlemen or traders. In addition they were designed to provide basic consumer services, promote local cottage activities and indoctrinate members in socialist ideology. However, the land and other means of production remained individually managed with practically no or little pressure by cadres to enroll the farmers into working together as commune. In an effort to consolidate the organizational structure of the PA and push forward with the formation of SCs, the 1975 land reform proclamation encouraged the amalgamation of 3 – 8 adjoining associations into one. Despite this and the obvious biasness of the field workers toward SCs, there were no more than 11,000 of such cooperatives all over the country in early 1980. Only those with relatively developed manpower and sufficient capital were recognized as credible Service Cooperatives (SCs) association.

However, the associations were never free from other obligations primarily the supply of human cannon fodders to fight in the civil war raging throughout the land and the obligatory supply of agricultural commodities to government grain agency at below the market prices. Obviously, they were not exempted from paying dues that range from land to health to education as well surtax in support of the war or alleviate drought. Irrespective to which category of peasant association one belongs there was no exemption from involvement in the war in one-way or another. At one time, some sections of the Eritrean peasants were asked to contribute money, albeit, voluntarily for the rehabilitation of the Ogaden region just liberated from brief occupation by the invading Somali army in 1978. Suffering from the brunt of the war many districts politely refused to be part of the campaign or refused to pay at all.[36] However, the call to aid the people in Ogaden after the brutal war of 1977 was more for symbolic reason than a substance one.

The PCs on the other hand were regarded as an ultimate vehicle to making Ethiopia self-sufficient in foodstuff. To this end the government, against all indicators, devoted considerable amount of resources for its success. True to their conviction to socialist ideology, the higher echelons of the Workers Party of Ethiopia, a party formed in the first half of the 1980s, believed wholeheartedly that the success of the communes would enhance the image and ensure the triumph of socialism in Africa. Despite the over zealousness of many in the government and party, the establishment of full-fledged PC were made to pass through three phases of developmental stages. The first in the process entails the pooling of land to the extent of obliterating individual field boundaries and sharing the means of production. The second stage requires each member to submit all materials and equipment under his/her disposal to the cooperatives including livestock. At this stage the only property under his/her command would be a garden plot of land. The third and final stage of the communal life will begin when all private properties are abolished, the farms become mechanized and the workers organized into farm brigade. This being the highest stage of socialist life, income would also be distributed based on the quantity and quality of labor invested, that is, one stage less than the socialist motto introduced after the death of Lenin, 'each according to his need'.

Although, more than 2920 cooperatives were formed between 1977 to 1990 only 255 were qualified to receive certificate of eligibility to allow them secure heftier loans to purchase machinery and other farm inputs. The ultimate goal has been to collectivize the Ethiopian farming system by 50% at least before the dawn of the new millennium. Instead at its twilight, the regime

36 Some villages around Asmara were asked to contribute to the rehabilitation effort touted by the government.

succeeded in converting 340,000 hectares under communal management with 240,000 members to its credit. At the end of the day the innovative assets of the PC did not go beyond 262 tractors, 10 combine harvesters, 2200 improved animal breeds and no more than 2870 beehives.[37]

Considering the disposition of the farms, that is, the machinery they inherited, the skilled manpower they possess and the overall modern agricultural practices that were in place before their nationalization, the State Farms one would think had all the prerequisite of becoming successful. After all, these were high input farms operated by foreign and local capitalists with the aim of expanding their operation as far as their profit margin could permit. The additional investment that the government made on farm machinery and on the supply of plant protection and fertility enhancing inputs as well as on the installation of new irrigation infrastructures certainly put these farms at a distinct advantage than the age old peasant managed subsistence farms dominating the landscape. No doubt the government was hopeful in converting them into a model farms were the service and producer cooperatives would one day emulate. To this effect not only did the government maintain these farms as state owned and operated, but dramatically expanded them to areas hitherto known as suitable for grazing.

By the year 1981, the size of the state farms expanded to 231,000 hectares from mere 70,000 hectares during the outbreak of the revolution. The soviet style state farms had all the necessary resources at their disposal and employed close to 250,000 laborers at one time. This of course is on top of the hundreds of thousands of peasants who regularly participated as free labor during the peak of a cropping season. For example in the 1981/82 crop season close to 40,000 peasants were rounded up from Godjam, Begemdir and Tigrai to participate in hand weeding and sickle harvesting and manual threshing operations of sesame crop in the Humera Region. Some who took part in the forced labor campaign were old, infirm, pregnant and few blind individuals. In this regard the adversities were too much for many to survive the first cropping season. In a not different scenario when labor shortage became critical in 1972, 3,000 unemployed workers were forcibly rounded up from the streets of Addis Ababa and air lifted to work in the private farms of Humera. Here again 70% of them died between the first harvest and the new planting season.[38] Forced labor was traditionally an entrenched practice in Ethiopia that even the socialist government used it flagrantly.

In addition to the free labor provided, the State farms absorbed 75% of the entire investment in agriculture while they contributed about 5% of the

37 Review of Agricultural Cooperative System in Ethiopia, Annex 2.1 and 2.2, (Addis Ababa: MoA, 1987)

38 Thomas Killion, Workers, Capital and the State in the Ethiopian region, 1919 – 1974, Vol. I. (Stanford University, 1985), 539.

country's food requirement.[39] By the time the *Derg* fell from power, the expense incurred on state farms was about 300 Million USD.[40] In real term this amounted to ¾ of the total loan expended by the Agricultural and Industrial Development Bank (AIDB) to the agriculture sector. At the height of the socialist fervor, that is, between 1982 – 1987 where the state farms were once viewed as a panacea to Ethiopia's' agricultural ills, the farms relied heavily on government subsides amounting to half of the state expenditure allotted to the agriculture sector.[41]

The Land Reform and Food Security

Following the revolutionary measures taken in early 1975, cereal production slumped to all time low primarily because of the unprecedented level of violence that soon erupted. As a result substantial amount of the previously cultivated fields left unattended. In particular, this was noticeable during the period that the government was in life and death struggle with the Northern rebels and the invading Somali army in the East. During this intense period where violence was the defining factor, production area was reduced on the average by half. Moreover, cereal production shrunk from 5.2 million MT before the big famine of 1973, to mere 3.9 million MT in 1978. The lowest was in 1975 when the land nationalization decree was issued and the armed conflagration intensified. Beginning 1979/80, however, not only was the cultivated area expanded but also the total cereal production begun to improve reaching a dramatic level of 6.4 million MT.[42] It is important to note here that by this time the invading Somali forces were repulsed, thanks to the support provided by the Cuban and South Yemen governments, the turmoil inside the country subsided and the war in the north reached stalemate, hence for the modest increase in cereal production. Moreover, to reassure the Ethiopian people that the era of insecurity, internal feuding and power struggle has ended and the period of reconstruction and food production dawned; in late 1978 the *Derg* launched the National Revolutionary Development Campaign and Central Planning and urged the people to support it. It also was sending clear signal that normalcy is back to the land where arbitrary detention and 'revolutionary measures' against those deemed counter revolutionary elements was frequent in the past.

---

39 Desalegn Rahmato, Agrarian Change and Agrarian Crisis: State and Peasantry in Post-revolutionary Ethiopia. Africa: Journal of the International African Institute. Vol. 63 (1) (1993), 42.

40 James C. McCann, People of the Plow: An Agricultural History of Ethiopia, *1800-1900,* (Madison: The University of Wisconsin Press, 1995), 251.

41 World Bank Country Study and Guide. (1991).

42 FAOSTAT Data base (2004).

In spite of the unfavorable weather that triggered the big famine of 1985 cereal production remained steady throughout the 1980s. This is partly due to the increase of cultivated lands in areas hitherto considered unproductive or used as traditional grazing grounds and not due to the introduction of new innovations to boast production. In 1980 alone, 13% more land was added for cultivation and production of crops. In real terms it is an increase of 68% from the base year 1975. In a nutshell during the entire period of *Derg* rule, cereal production fluctuated between 5 and 6 million MT, save for the 1982/83 cropping season, which showed a phenomenal production increase never attained before. As record shows this was the year were rainfall was reported to have been in abundance and distribution above optimum. To the bewilderment of the many in the government, the generosity of this particular rainy season has been technically explained to the occurrence of Group Two type of El Nino events that made relatively positive impact on the main rainy season, the *kiremt*.[43]

The famine of 1985, the one considered vicious by any famine standard in Ethiopia in the past century, had left close to 8 million peasants in the verge of death.[44] Tragically, one million of those died in a span of few months. The precipitous decline of agriculture production after the merciless drought hit the land in early 1985, triggered the *Derg* to tinker on settlement projects to fulfill the most elusive but cherished objective; achieving food security. After the events that draw the generous support of the international food aid agencies, celebrities and school age children worldwide, cereal production begun to recover steadily. In the years following the cruel famine, cereal output went up by 60% from the previous year while areas cultivated with cereal crops increased by 12%. This trend continued right into the period where the EPRDF assumed power 1991. Again, it is important to note here that; the increase in total cereal production can be attributed to the expansion of cultivated lands throughout the 1980s. Despite the increase of cereal production over the years, the yield per unit land remained the same and in some particular years falling below average. In real term, the cereal yield per hectare has remained the same since the early 1960s. Safely put during the same period the yield per hectare for Teff was 0.6 MT/ha, while maize and sorghum fluctuated between 1.5 and 1.6 MT/ha respectively. For wheat and barley the yield was 1.1-1.2 MT/ha for perhaps one century. In other words output per unit land remained pathetically stagnant. Whatever benefit is gained the decade, it was definitely canceled by lack of technical breakthrough and unchecked population growth.[45]

43 M. Ward and Yeshanew, World Sea Surface Temperatures and *Kiremt* Rains in Ethiopia. (Addis Ababa, NMSA, 1990).

44 Relief and Rehabilitation Commission, Ethiopian Government, Early warning & Planning Services, (Addis Ababa: April 1985), 6.

45 US Census Bureau, international data base http://www.census.gov/index.html. According to the census population increased by 70% between 1975 and 1990.

Therefore, this seemingly phenomenal increase in total cereal output in the 1980s could be attributed to nothing more than an extensive cultivation practice pursued and to some degree the limited use of fertilizer to boast production which begun to pick up slowly in early 1980. In a nutshell, the area covered by grain crops during the entire 15-year socialist rule in Ethiopia was in the range between 4.2 and 5.0 million hectares, that is, an increase of 800,000 hectares from the base year, 1975. Again this clearly suggests that it is not fertilizer as such that was responsible for increasing yield per unit land but the extension of cultivated lands. Generally, fertilizer consumption in Ethiopia increased from per capita of 0.8kg/ha in 1970 to 7kg/ha in 1975 to 12kg/ha in 1987.

It is ironic to note that the perceived increased in the level of cereal production did not occur because of the participation of the small holder but of the massive application of fertilizer to state farms and PC fields. The majority of peasant did not get preferential treatment by the government agencies to effectively use fertilizer as production boasting component. According to a paper presented at the National Research Review Symposium at Nazareth in 1988, the greater bulk of the fertilizer imported during the two quarters of the 1980s was largely consumed by the State Farms and PCs. However, some SCs with means to purchase farm inputs have also capitalized on the use of the highly subsidized urea and super phosphate fertilizers to boast production. In particular those specializing in commodities that fetch high price in the local market such as Teff (*Eragrostic abyssinica*) and pepper (*capiscum*) grown in the central and south central highlands benefited modestly.

Total cereal output went up from a little less than 4 million in the early years of the revolution to a little over 6 million MT by the time the revolution lost steam in late 1980. Although such a deduction is based on the available data and therefore, might not accurately reflect facts on the ground, nevertheless it is a good indication that the land reform act of mid-1970s failed to address the chronic food shortages plaguing Ethiopia for decades. Thus, one can argue that the effort put by the government to improve production and elevate productivity to a level where the nation shuns food aid from abroad did not happen. The notion that Ethiopia has been or closer to being self-sufficient in foodstuff during this period can be easily dismissed by studying at the figures of food aid shipment to the country during the period.

A cursory look at the total food aid inventory of the World Food program shows that between 1974 and 1979 Ethiopia received food assistance worth 353 Million dollars. The aid continued with same tempo throughout the decade of the 1980s with cereals and fortified food as the main components. During this period the country received close to 4 million MT from `WFP alone.[46] This

---

46 World Food Program (WFP) 1993.

is on top of the 6.7 million MT donated between the years 1978 - 1991 by others.[47] All throughout the decade of 1980s, the country has continued to be one of the top food aid recipients in the world out competing other food deficit regions in sub-Saharan Africa.[48] During the transition period, that is, between the ignominious exiting of the *Derg* from power 1991 to way beyond the new era of the ruling EPRDF in Addis Ababa food aid continued to flow unhindered.

Implementation Challenges

Given the country's low level of development at the time of the outbreak of the February revolution, it was quick to observe numerous factors that were destined to slow down if not derail the successful implementation of the land reform program. Certainly, these were not extraneous factors but inherently Ethiopian caused by its long slumber oblivious of its surrounding. They range from lack of progress in technology to low level of literacy. In short Ethiopia, at the time of the revolutionary upsurge was a classical feudal state of African brand that in essence has close resemblance to the Ethiopia that existed in the 19th century.

The countryside where 90% of the population resides was frozen in time. If by divine providence or some kind of miraculous act, the generation that passed a century or two was to return for a visit, they would have found their way without the aid of compasses or GPS. Practically little has changed since their departure from this earth for them to celebrate. Grant the permanency of Addis Ababa as a capital of the nation was assured and the consolidation of few provincial towns become reality. Roads and other transportation facilities were at bare minimum to traverse the most difficult terrain created by the natural rearrangement of the East African land mass some 5 to 20 million years ago. It contains rugged mountains soaring to 4,300 m.a.s.l and areas situated at 150 m.b.s.l. Together with the arid plains east of the chain of mountains and the relatively wet lowlands west of it constitutes what one calls the modern geography of Ethiopia. Other than that, the peasantry continued to use salt to sweeten his coffee drink, apply herbs and holy dirt (pinch of soil from renowned 1000 plus years old monastery) to cure a variety of ailments ranging from wounds to malaria and syphilis, spray holy water to ward off evil vibe, chew shrub roots as a good luck charm or apply to entice a prospective suitor. In a nutshell the peasantry remained locked in vicious cycle of poverty and ignorance until the outbreak of the February Revolution.

---

47 Transitional Government of Ethiopia (TGE) National Program for disaster Prevention, Preparedness and Mitigation, (Addis Ababa: 1993). Some figures (1987 to 1997 are obtained from WFP and UNDP source).

48 World Food Program, 'Food Aid review' (1992).

This as a background, therefore, is not hard to pin point the would be major impeding factors working against the smooth implementation of the reform announced soon after the ungraceful exit of the Emperor in September 1974. Among those considered major obstacles to translating the messages of the revolution to good ends were fragmentation of land, insecurity of tenure, and the use of unimproved farm inputs and tools. To this, one adds the low literacy rate and ignorance so much prevalent among the peasant sector that made the task of reform quiet impossible to tackle. Failure to accessing resources in more fair and equitable way has dumped the will of the ordinary person to work hard leading to fatalistic thoughts of no fundamental change is possible. Tragically it was also one of the most inaccessible countries in Africa, perhaps no better than the Belgian Congo that have a rudimentary road infrastructure to allow farmers interact or exchange commodities as they please while. Between the Italian forceful ejection from Ethiopia in 1941 and the defeat of the *Derg* by EPRDF in 1991 a mere 20,000 Km roads mostly paved rather than lined with asphalt was built in the country. Moreover, the size and the composition of the agents of change such as the bourgeoisie, the working class, the intelligentsia were not adequate enough to synthesis these historic events either in unison or as a specific power group. The army like the working class was composed of mostly illiterate that have barely shed its most oppressive rural background to navigate the difficult terrain of social upheaval that will soon engulf the country.

Unable to grasp the dynamics of the mass movement at the time and completely overlooking the objective conditions existed in the country, the *Derg*, mostly made up of noncommissioned officers, declared a swift reform without the slightest regard or rationalization to their decision. Moreover, there was no mechanism in place to assess the impact let alone put a contingent plan in place to remedy should the impulsively designed plan halt to a ground. In real sense, the land reform decree that purportedly made the Ethiopian peasants owners of their destiny had in fact turned them into dependent subjects controlled by well-organized centralized body served by numerous overseeing agencies. It should be understood here that the land reform merely acknowledged the usufructuary right of the tiller not the ownership of the land, indirectly perpetuating somewhat similar system of serfdom long abhorred by the peasants. Throughout time, serfdom was the major impediment to initiating a long-term soil and water improvement programs. It did not allow the serf to decide for his own good. That explains why the barrages of land reform directives coming from the top fell into deaf ears. The trick to entice the beneficiaries of the land reform to engage in greater food production, participate in land conservation practices and adopt wide range of afforestation programs did not succeed either.

The rural-urban dichotomy which although was less conspicuous before the Italian invasion of 1935, had begun to show a modest differentiation after the return of the Emperor from his forced exile. Obviously, this exacerbated the situation further where landlords extract maximum surplus to satisfy their newly acquired habit of consumption. As has been going on for ages, any disobedience could invite a retribution that the peasantry is advised to avoid at any cost. The rigidly enforced feudal laws, driven by inflexible management of the economy together with peasant passivity made food insecurity more probable. Eventually, these led into deeper food crises in the 1980s triggering a comical hide and seek game between the peasant and the government agents chasing to extract surpluses in the form of taxes and shady contributions in the name of lofty nationalistic slogans such as 'duty call for the motherland'.

It should be made clear that the land reform campaign was launched with few opposing it. The measure was universally acclaimed except for few daring feudal lords who stood their ground and fought guerrilla war to reclaim what was theirs for a century.[49] However, the majority including the urban population was on the side of the government unreservedly in as long as the objective was to distribute land equitably to those who work on it. Contrary to the enthusiasm and the hope shared by many among them the serf, the rural poor, and the downtrodden, their vanguards the students were skeptical of the military group's ability to carry the revolution to its 'consummation'. One is that, there was few technically competent strata of people to guide the rural development efforts and fewer technocrats to formulate policies, devise strategies and follow up action programs. To the students and especially to the EPRP, the monumental problems that Ethiopia faced cannot be overcome without the formation of provisional people's government that will include all.[50] This annoyed the *Derg* and decided to squash the EPRP through armed terror adding more fuel to the fires burning all around the country. Soon, the applicability of the land reform without taking into account soil variability, the human density per land size in given locality and the capital required to execute some of the ambitious projects to everybody's satisfaction was put under question mark.

Lack of unanimity among those elements set out to effect change further jeopardized the implementation of the land reform and the development work associated with it. Not only did contradiction between the army and the civilian contenders for power slowed down the momentum for change but also the bloody in-fighting that soon ensued between the various forces claiming to belong to one or the other left leaning political groupings. No doubt this

49 The Enqueselassie brothers all of Shoan nobility fought the *Derg* heroically and lost. So was a Texan look like noble the youngest son of Ras Birru, Merid met his fate in Ok Corral type shoot out.

50 Prison confession of Berhanemeskel Reda a prominent Ethiopian student leader in the 1960s and one of the founding and politburo member of EPRP. http://www.assimba.info/uploads/files/dosse/Berhanemeskel_%281%29.pdf (14)

affected the implementation program seriously. Sadly enough, most of those perished in the inter-factional fighting and many of those exiled technocrats came from various professions including agriculture. In the course of executing their duties be it in the countryside, provincial towns and even at the headquarter level many were wrongly targeted for execution and incarceration. Persons with views different than one or the other group were branded as spineless 'middle roader' with no role to play in the revolutionary process.

The development program launched by the *Derg* required a huge capital infusion to succeed. The investment capital expected from the Soviet block was limited while foreign capital from the West was unthinkable given the *Derg*'s close relation with the socialist world.[51] There was no or if at all little internally generated capital to speak of. Moreover, businessmen, traders and entrepreneurs were viewed as anathema to socialist Ethiopia to allow them make a difference.

In addition the largely mechanized state run farms with 230,000 ha under their disposal required a substantial amount of manpower to run smoothly. Furthermore, to boast production, agricultural inputs had to be available and at an affordable price. Neither government resources nor peasant purchasing power made that possible. As a last resort, the government opted to focus on areas that amounted to less than 8% of the total cultivable land that is, the state farms and PCs to solve the chronic food shortages in the country. This bizarre policy hurt the agriculture sector by postponing the time and the resolve needed to combat food insecurity. Nevertheless, the land reform decree brought about various socio-economic inspired associations hitherto unknown in Ethiopia. Among those established to accelerate national development was the PA with declared membership of about 6 million peasants organized under the umbrella of 20,400 PAs, 11,000 SCs and 3000 PCs everywhere in the country including in the most troublesome northern region.

Another aspect of the impediment related to the reform program was the spontaneous, unplanned and cruel decision to bring together scattered hamlets and dwellings into a compact villages and settlements in span of few months. In the mid-1980s, the *Derg* begun to undermine the very reform it claimed to have espoused fervently by pushing an untimely and ill-prepared resettlement program in both the marginal and potentially productive areas of the country. Together with the forced villagization program launched soon after and the human resources spent to win the unwinnable wars spreading throughout the land made the reform a caricature of its stated objectives. Although there is no independent study carried out to shade light on the human cost during the fratricidal war of 1960 –1990, estimate run into hundreds of thousands. It is

51 The shift of alliance from West and in particular the US to the Soviet Union targeted Ethiopia. It became convenient for the Super Powers to change place while the region bleed to death.

however safe to speculate close to a million Ethiopian soldiers including Eritrean, Tigrean and Somali civilians with their armed guerillas perished in the war. Those left behind in the villages and hamlets of Ethiopia proper were also indirectly affected by the war as they were socially obliged and sometimes coerced to cover farm chores of those absent but fighting wars in the North, East and Southeast.[52]

Tragically the movement of people from one place to the other in fulfillment of the government's wishes has always coincided with the active planting season and in rare case during the time of harvest. The villegization program was especially cruel to farmers in Hararge region where farmers almost universally were accustomed to planting a stimulant shrub, the *Chat edulis* in their backyard dwellings, as cash generating enterprise. They were forced to abandoned that and live in clustered villages far away from their past residence. By the end of 1985, another half a million people were uprooted from their birth place in Tigrai, Wollo and Northern Shoa to be resettled in Wollega, the Blue Nile basin, Keffa and Illubabor governorates. By mid-1988 over 12 million Ethiopians or 33% of the farming community took part in the government's villageization program.[53]

The extreme dislike that the Western countries harbored toward the *Derg* and their refusal to cooperate with it in any shape or form has made the implementation of the reform difficult. In the face of the formidable opposition by the western countries in particular the USA and more significantly as a result of the internal upheavals spearheaded by the Eritrean and Tigrai nationalist movements; the reform program suffered a serious setback before it took off the ground. Dependency on the USSR and the Democratic Republic of Germany (GDR) had only plunged the country into abyss. It should be noted that, the support that the government received from the socialist countries did not go beyond few agricultural hardware and inconsequential and unappreciated technical expertise.[54]

It is safe to argue that both the institutional and technological bases as well as internal capital needed in the transformation process were either lacking or rudimentary at best to bring the desired change. Here an attempt is made to identify the major factors that might have rendered the land reform move by

---

52 Aregawi Berhe, A Political History of Tigray people's Liberation Front 1975 – 1991, (Tsehai Publishers and Distributors,2010), 251. All the same, the TPLF promised members of the Militia and other cadres that the people in their respective localities would be able to farm their land and take care of their families while they are in duty.

53 Teffera Haileselassie, The Ethiopian Revolution 1974-1991. From monarchial autocracy to military oligarchy. (New York: Kegan Paul International, 1997) 270.

54 Save for Italy's lukewarm support of the regime in some aspect of the economic development program, the *Dergh* remained isolated much of its volatile tenure in power.

the *Derg* ineffective and perhaps regressive in as far transforming the Ethiopian agriculture into becoming self-sustaining and dynamic.

Institutional Inadequacy

To declare that "all rural lands to be the collective property of the Ethiopian people" is one thing and to have the necessary institutional base (managerial and entrepreneurial skills, and the will power) and the material capacity (technology and capital) to realize the objective is another. The famous *zemecha* carried out under DTCC in early 1975 by about 70,000 university and senior high school students in rural Ethiopia, was significant in so far as acquainting students to the reality of rural Ethiopia and to the condition that the peasant was forced to live for centuries. Indeed it is an eye opener for a city dweller whose lifestyle is different in large measures from that of a peasant who toils from dawn to dusk or watch peasant woman fetch unsanitary water from a stagnant source and from a considerable distance. In such circumstance it could be unfathomable to an ordinary person witness a beat-up peasant stood upright and carries daily chore with little complaint. Facing such a reality has certainly left the student somehow dumbfounded, angry and with a sense of urgency to transform the country.

Moreover, the students might have also attempted to introduce peasants to many fairy tales about Marxism, socialism, Imperialism and stories of peoples of far off lands such as North Korea and China whose citizens have performed miracles on the land.[55] One might add that living among the populace, albeit for a limited period of time and sharing experience with them must have had a profound effect on the majority of students of urban background. No doubt, living among the peasants gave the students an opportunity to validate their Marxist doctrine of class-based societies. This also coincided with the *Derg*'s obsession to bringing closer the two contrasting social settings, that is, the countryside with towns through revolutionary indoctrination. In this case the students were found as useful tools to perform the task of bridging the gap. Without doubt, such sudden but voluntary interaction of the peasantry with the student *zemecha* must have had a practical implication to emboldened few peasants defend themselves against feudal backlashes. Given the insularity of the Ethiopian peasants to many things

---

55 Aregawi Berhe, A Political History of Tigray People's Liberation Front, 1975 – 1991, (Tsehai publishers and Distributors, 2010) 158. The veteran of the guerilla war against the *Derg* writes of the early stage of TPLF where discussions were conducted in abstract idiom and parameters of the Leftist student movement rather than being focused on reality supported by a solid grass-roots appraisal of the needs and aims and expectations of the local (largely peasant and non-literate) population.

outside their immediate environment, this encounter is indeed a positive development.

Whatever the degree of the interaction that the student had during their brief stint in the countryside, the role that the *zemecha* purported to have played in providing leadership for peasants to establish their own associations is very much exaggerated. The various socio-economic association including peasant self help, musical troupe, cottage industry of weavers and bricklayers etc. begun to spring in the countryside and in the provincial towns way after the *zemecha* participants exited the scene. This is not to deny the contribution made by the *zemecha* students before and after the land reform but to underscore the fact that as an agent of institution builders they were minuscule in size and socially incapable for such a task. Moreover, the division that soon emerged among themselves made their work ineffective and nullified what small credit they owe for reorganizing the countryside. However, if one speaks of legacy that is still remained entrenched in Ethiopia (Eritrea as well) forty years after their debut is the local administrative structure the PAs and the urban Neighborhood Associations, the *kebelles*. Radical students and the intellectual left championed vigorously for this enduring institution to exist.[56]

There are good reasons why institutional base in Ethiopia was weak. Modern education in Ethiopia started rather late. School bearing the name of the crown prince Teferi Mekonen was established in Addis Ababa in 1925, two decades after the first school, Menelik II opened its gates for students with high pedigree. In the same period, there were few mission schools located in the east and in Addis Ababa. In addition the Italians build three schools in Eritrea catering for indigenous boys. Despite the limited scope that these schools had in terms of preparing students to pursue higher education of learning or providing technical knowledge of immediate application, graduates of the schools in Eritrea played a pivot role in helping shape post-Zeweditu Ethiopia[57].

Generally, modern schools were viewed with suspicion by the traditional society to discourage students enroll in sizeable numbers. In fact when the Teferi Mekonen School opened its doors for the first time half of its registered students failed to attend of fear that the new school will keep student astray of

56 Governments who superseded the *Derg* used the *kebelles* as a vehicle to keeping records, transmitting directives from bottom up and vice versa, compiling data for urban planning and development. In the past *kebelles* were also used to suppress the collective will of the people to benefit the ruling group. Nevertheless this is novel institution not common in Africa but a standard feature in many of the socialist countries.

57 Graduate of these colonial schools in Eritrea served Ethiopia before and after the war in many capacities some reaching ministerial post.

their traditional values.[58] To the majority of the feudal elements modern school were regard as sinister ploy for European expansion and eventual occupation of the country. For the young and the progressive such as the crown prince, however, education was the means of protecting the country from falling victim to European aggrandizement and one secure way of lifting up the country from its backwardness. In 1935, that is, a year before the Italian aggression, the student enrollment in 20 public schools stood a mere 8000.[59] In any case, the introduction of modern schools in some timid ways did not come before the 1940s. The period between the brief Italian intervention in 1935 and the unsuccessful coup deta't of late 1960, almost all provincial capitals had high schools for students of humble background. In the eve of the revolutionary upsurge (1974), a country of 29 - 30 million people was reported to have 1,300 primary and secondary schools in its public school system. At the outbreak of the revolution the number of students attending primary school reached 1,226,124 from about half a million in 1967/68 scholastic year.[60]

The absence of qualified and skilled manpower was apparent from the outset. Take for example the number of the students graduated annually from the schools of agriculture. During the imperial regime, the college of agriculture at Alemaya was less equipped to prepare more than 60 students per year. To make matters worst some of the graduates end up in jobs remotely connected with their field of training because of the exigencies to fill vacant managerial posts created in the process of institutional formation.[61] In the eve of the revolution a record 26 students graduated from plant science alone. The number is by far less for graduates in agriculture economics and engineering. Almost half of the graduates from the two junior agriculture colleges of Jimma and Ambo go straight to Alemaya for a degree program without acquiring some field experience in the area of their training. Similarly, the agricultural research institutions had fewer personnel at their disposal and were even less prepared to tackle subjects pertaining institutional changes useful for the reform to

58 Zewde Reta, Yeqedamawi Haileselassie Mengist. (New Delhi: Laxmi Publications, 2012), 57 (Amharic).

59 Damtew Teferra and Philip G. Altbach, eds., *African Higher Education: An International Reference Handbook* (Indiana University Press 2003), 316-325.

60 Education in Ethiopia *Strengthening the Foundation for Sustainable Progress* Table 2.1. Enrollments by Level in Government and Nongovernment Institutions, Ethiopia, 1967–68 to 2001–02. 2005. P26). With only one university at the time college enrollment was limited to selective few.

61 Richard Dowden, Africa: altered states, ordinary miracles, (Public Affairs New York, 2009), 526. Unfortunately, this is not a phenomenon peculiar to Ethiopia as many African 'engineers were (also) running government ministries, historians turning to be politicians, scientist becoming diplomats' in the 1970s and in 1980s. Richard Dowden. 2009. Africa: altered states, ordinary miracles. Public Affairs New York. P. 526.

succeed. In the period preceding the land reform the total number of students pursuing college degrees was put at 6,731. Insignificant numbers indeed, when comparing students pursuing higher education in Egypt at the time.[62] An additional one third of this figure was reported to have been studying abroad. In a country with only one university and few technical colleges just before 1974, high school enrolment stood at less than 74,000.[63]

Unable to expand institutions of higher learning and faced with the dilemma of accommodating high school leavers in the one and only University of Haile Selassie I, the government designed a system to train students for vocational jobs other than managerial. The Sector Review as came to be known brought wrath against the government by students and teachers because it encapsulated an education strategy that places emphasis on basic and non-formal education including apprenticeship, community practices and continuing education. The Sector Review was all what it took to spark an uprising leading to the 1974 February Revolution which the Imperial regime never lived long to regret it. In more precise term, the educational system was never designed to tackle problems of national development.[64] Based on these facts, therefore, it was doubtful whether students coming out of this system were sufficiently equipped to meet the social challenge let alone understand the peasant psychic they set to change.

During the revolutionary period of the past two decades and before the ascension to power by EPRDF, there had been some effort made by the *Derg* to increase the number of agriculture graduates from less than 60 in 1974 to 600 in 1986. The number of students attending primary and secondary school also jumped from 1,042,900 in 1975 to 3,926,700 in 1989. However, the attempt was not complemented with the complete overhaul of the curriculum or the incorporation of subject matters to reflect the country's need. Thus, the system left the student graduates ill prepared to tackle the problems confronting the country.[65]

The seventy thousands secondary school and university students sent in the early period of the revolution to herald the change sweeping Ethiopia were like a drop in a bucket not only from the stand point of numbers, but also from a professional preparedness as well. Being the product of a traditional society, which had little modern political institution crafted in its past history, the students lacked political astuteness to act as a catalyst for orderly change. In typically idealistic fashion to change the society overnight some if not all the

62 Alan Richards, Higher Education in Egypt. Policy Research Working Papers. WPS 862. (World Bank, 1992), 27. In 1974 the Public Universities in Egypt enrolled some 301,325 students of which 46% where pursuing science education.

63 Ethiopia Observer, 1967 EC (1974 AD), 56.

64 Tekeste Negash, 1990. The Crisis of Ethiopian Education: Some Implications for Nation-Building.. (Uppsala: Department of Education, 1990), 5-7.

65 Ibid., P. 11.

measures taken by the students alienated the low and middle level landowners. Many reacted instinctively against the measures and put question mark on the land reform laws just promulgated. For example, the untimely agitation by students to expropriate indiscriminately all properties belonging to the landlords in some areas of the south led to the destruction of farm tools and the slaughtering of farm animals by owners. Such pre-emptive action by landlords to avoid the forceful seizure of their property hurt production seriously.

The attempt to forcefully collectivize farmlands was not received well by peasants either. In addition, their premature call for the immediate installation of a provisional civilian government in place of the military provoked the army to unleash terror throughout the urban areas of the country. The dictatorial tendency and arrogance behavior displayed by some student campaigners toward the toiling masses was in complete contradiction to what the student movement stood for during the years of campus struggle.[66] The role of the student would have been more useful had their activities were strictly confined to agitation while at the same time campaigning for the implementation of the land reform, no matter how unpalatable to their liking. But as events begun to unfold the students took it upon themselves to stir peasants engage in the call of changing the government and to violently oppose regional officials representing the provisional government. As the result few peasant activist lost their lives fighting for and against the government. In retrospect, it is fairly safe to state that the formidable task of land distribution would have never taken place so swiftly without the initiative and conscious participation of the peasants themselves. Indeed it would have required thousands of rural agents travelling through the impassable and rough rural tracks to complete the task. Unfortunately, the state did not have the human capability to do just that. Worst of all lacking a clear implementation procedure so as to follow government directives in effect put the students and the peasants in tenuous position forcing them to resort to a trial and error method.

From the discussion above it is safe to conclude that an institutional response to social and economic demands cannot be conceived in isolation to the cultural condition, literacy level and technical achievements of the society in question. Then and only then does one expect the emergence of well-developed managerial and entrepreneurial capability of a nation? By the same token, the strength of the professional or cultural associations, the power of labor unions and the maturity of political parties of a country are but the reflection of its economic growth stage or the overall developmental level that it has achieved. In this case, Ethiopia was at the bottom of human development

66 Alula Abate, Peasant Association and Collective Agriculture in Ethiopia: Promise and Performance. Seminar paper presented at the (Addis Ababa University Geography Department March 1981), 3.

index to realistically expect many of the elements necessary for social and economic change to be near the grasp. Unfortunately, Ethiopia's opportunity to travel along the path of an all-round development was tragically cut short by the brief but brutal intervention of fascist Italy in 1935 which had deliberately pulverized the indigenous thought processes, vision and the will to act independent of Europe by those elements most appropriately named as the Japanizers by Addis Hiwot.[67] Many of the educated, the skilled and the critics of the feudal system were deliberated hunted and murdered by fascist Italy after it has temporarily occupied the country.[68] An eyewitness to the deliberate hunting of educated Ethiopians by fascist officials relates 'Abyssinians are backward, do you think that they will progress better if the learned and qualified among them are exterminated?'[69]

Obviously, the question of designing institutions relevant to the land reform and beyond was left to the Non Commissioned Officers (NCOs) and low ranking military officers and their few intellectuals and technocrat advisers who emerged as important players in post 1974 Ethiopia. The role of the latter group was pivotal in charting the course that the revolution took and therefore, partially responsible for what followed next. In this case, the socialist path of development was selected over indigenous common sense, historical experience and collective wisdom. Tragically, the institutional model adopted was European, and in particular that of the Eastern brand. Considering the level of frustration of the Ethiopian students and their disdain toward anything connected with the country's feudal past, the development path chosen was less surprising to say the least. It is no secret that the Ethiopian student body at home and abroad had for long being infatuated and at times enamored with the socialist ideologies of the far off lands.

Equally it is not mysterious for the army to endorse the ideologies espoused by the student and intellectuals because it lacked one of its own. But the NCOs and their officers were canny enough to play one Marxist group over the other in order to retain power permanently. Determined to dictate and adopt brand of ideology rooted in other people's experience is tantamount to ignorance of the country's history and culture. It also implies that one is lacking the basic understanding of the dynamics of social change at work in Ethiopia. Logically, therefore all the proclamations issued with regard to production relations during the revolutionary period failed to address the long-term interest of the peasantry. Expectedly, many of the agriculture related issues generated

---

67 Addis Hiwot, Ethiopia: from autocracy to revolution. (London: Occasional Publication No1. Review of African Economy 1975).

68 Bahru Zewde, The Ethiopian Intelligentsia and the Italo-Ethiopian War, 1935-1941. (The International Journal of African Historical Studies, Vol. 26, No. 2, 1993) 294.

69 Laszlo Saska, *Fascist Italian, Brutality in Ethiopia 1935 – 1937*. (Trenton: The Red Sea Press, 2015), 113.

little or no sensible debate among independent learned citizens of the country or failed to attract the attention of the urban based intellectually oriented political groups that mushroomed at the outbreak of the revolution. The initial period of the revolution was marked with realignment of forces that soon turned to bloody resembling the Bolshevik revolution of the 1920s.

One of the achievements of the Ethiopian revolution was the spontaneous formation of PAs in areas where the landlord/tenant relationship predominated. Although literature dealing with the functions and duties of the peasant initiated associations before being legalized by Proclamation No. 31 is scanty, the indication was that each association went ahead in defining its geographic boundaries and put set of rules to administer members and the land under its jurisdiction. Dictated by the circumstances unfolding and unperturbed by threats from the feudal lords, the peasants, albeit briefly, replaced the power structure which existed in some *wereda* and villages before the formal pronouncement to establish the PAs came into effect. Suffice is to say that, the brief autonomy that the associations enjoyed was lost for good with the subsequent proclamations issued after December 1975 dealing on matters of PAs and their Consolidation (No. 71/75), Rural Land Use Fee and Agricultural Activities Income Tax (No.77/76), AEPA Establishment (No.130/77) and finally the formation of Agricultural Cooperative Society (No. 138/78). Without a doubt, these decrees imposed from above have had a stifling effect on the producer to act independently.

As a result of the unprecedented decrees issued by the government and the armed conflagration that engulfed the country, agricultural production slumped before it started picking up again in early 1980s. Certainly, the incorporation of the peasant majority into the political process on their own volition would have released enormous energies and greatly simplified the task of economic reconstruction. Instead the PAs were further bureaucratized if not marginalized and in the name of safeguarding the motherland their members subjected to all sort of abuses by the power structure. Having said this, it would be useful to deal in detail with some of the negative by-products of the institutional deformity -inherent in the Ethiopian state- that had become fetters to production instead of becoming beneficial tool to agricultural transformation.

The PAs and their Weakness

The PA in a given *kebelle* was well structured, centrally run and vertically integrated corporate body. With certain limitations, the PA was empowered to redistribute land, administer justice and engage in economic development. Any person residing within the legally constituted *kebelle* boundary was entitled to be a member of the association, hence the right to work on the land. A peasant who chooses not to be a member of the association forfeits the right to work on *kebele* land. The AEPA headquartered in the capital spoke on behalf of the

20,367 unit of associations with an approximate membership of six million households on matters that directly concern the peasants with the central government. In turn the government used the AEPA channels to transmit directives to all units of PAs throughout the provinces, districts and the *weredas*.

From the surface the PAs appears to be the perfect vehicle for development that the anti-feudal and pro socialist elements hankered for long time. This indeed might have been the case had the PAs were assisted to flourish democratically and encouraged to grow on their own. Unfortunately, this was not found to be the case. Desalegn in his well-researched book described the situation,

> ....where many PAs are devoid of initiative, where the resources of their communities are improperly used, and where problems that could easily be overcome by collective effort remained unsolved. PA officials believe that their organization's main task was to implement government directives... PAs often expect the initiative to come from outside and their officials perceive their role to be one of transmitting decisions made elsewhere to their membership[70].

The sterility of the PAs also stems from their basic lack of expertise and experience to run the new rural institutions based on model of East European experience very different from the home grown associations. This was particularly true with the PC who aspired to look and acts like the Chinese commune or the Soviet collective farms

As the peasant serfs before them were required to surrender their surplus produce in the form of rent, that is, on top of the unaccountable services that they gave to the landlord if they wish to remain on the land, so were the PA members after the land reform. Under the Damocles sword of the PA authorities, the peasant were obliged to make payment (tax, dues, special contributions) and perform certain task supposedly for the common good if their right to work on the land is to be upheld. Although Peasant's contribution in the form of cash or in kind was relatively small, the sum total effect of the many extractive programs that the peasants are required to participate made it quiet onerous. Some of the extractive dues devised by the government to siphon cash from the society at large include the contribution to rehabilitate the war affected areas in the Ogaden, or in response to the call of the Motherland, to combat the famine of 1985 and in support of the Literacy Campaign etc.

In addition peasant had to pay annual membership dues to the PA and support the Revolutionary youth and women organizations in their *kebeles*. The peasants shouldered rural public work undertakings including the beautification

70 Desalegn Rahmato.1984. Agrarian Reform in Ethiopia. (Uppsula: Scandinavian Institute of African Studies, 1985) 77.

of many of the small administrative towns. It was also peasant's revolutionary duty to alternatively work on lands owned by members of the militia who for reason only to satisfy the whim of the authorities were shipped off to the war fronts.[71] In all fairness subjecting the peasants to endless activities that have little or no bearing to production and obliging them to support causes that did not directly promote the welfare of their community had to a greater extent contributed to the decline of agriculture production and productivity in Ethiopia.

Disruption of the everyday life of the peasant farmer in partial fulfillment of the government's request to contribute toward the defense of the country was painful indeed. But more so was the insecurity of landholdings fostered by the reform itself. Once the land distribution was in effect it became apparent to local officials that subsequent redistribution was necessary to either accommodate newcomers or to address equity on landholding size. In the face of a looming drought or adverse weather conditions this practice has caused havoc on production. The process of redistribution and reallocation of land continued, as land remained fixed while the number of people seeking land increased. This practice is aptly described as "a distributive leveling reform based on fixed but increasingly diminishing rural asset".[72] The practice of redistribution of land was not common throughout Ethiopia but in some districts of Wello peasants were persuaded to exchange their plots of land many times. Earlier studies on land redistribution indicates that peasants whose plot has changed hands ones amounted to 240, twice 271, and those whose land was rotated like spinning wheel of more than three times reached 112[73]. After the ouster of the *Derg*, in mid-1990s redistribution of land in Gojam has sparked unrest among farmers that some districts send delegates to Addis Ababa to protest the action taken by the Zonal administration. In some areas of Tigrai, Amhara, Oromia and Southern Region an average of 52% of the land was redistributed to accommodate others and ostensibly to arrive at an equitable share.[74]

In a nutshell, the desperation and helplessness of the Ethiopian peasant was best described by an elderly man of 85 from Wello who was on record to express his dismay to the land distribution and redistribution, land allocation

71 Aregawi Berhe, A Political History of Tigray people's Liberation Front 1975 – 1991, (Tsehai Publishers, 2010), 251. Similarly, the TPLF also promised members of its Militia and other cadres that the people in their respective localities would be able to farm their land and take care of their families while they are on duty.

72 Desalegn Rahmato, Agrarian Reform in Ethiopia. (Uppsala: Scandinavian Institute of African Studies, 1984)

73 Yaraswork Admassie, Twenty Years to Nowhere: Property Rights, Land Management and Conservation in Ethiopia (Trenton: Red Sea press, 2000), 195.

74 Klaus Deininger, Daniel Ayalew Ali, Holden Stein, and Zevenbergen Jaap. Rural land certification in Ethiopia: Process, initial impact, and implications for other African countries. (World Bank Policy Research Working Paper 4218, 2007) 22.

and relocation and the highhandedness manner by which directives are transmitted from top to bottom. "Our main problem is the taking away of somebody's land and giving it to someone else. The *Derg* did not suit us and neither do these new ones (EPRDF). The poor accepts when he is told to fetch water or to split firewood, but does not like when one messes around with his wife or his land. The peasant is impoverished…".[75] Although isolated case, this is reminiscent of the *Derg*'s action where lands were redistributed numerous times and where peasants from the drought stricken North and Central highlands were accommodated in what was termed at the time *as sigsega* literary translated into 'move over and make space for somebody'.

Absence of Production Support Systems

If cereal production is to grow at a steady pace it must as the matter of priority be supported by an elaborate research and extension system. This has been the practice since the dawn of modern agriculture. The cultivated plants and their progenies in use today and the various methods of cultivation practiced in many cultures of the world plus the remarkable water conveyance systems and the techniques employed to draw water in defiance of gravity was the result of man's conscious effort and at times of sheer serendipity to better his lot. Thus, as demographic change occurs and population grows so was the quest of meeting the new challenge. Certainly, the urge for developing new things in order to cope up with the new challenges has been the paramount concern for each passing generation. But, the tempo differs from one region or country to the other. Generally, it is here where divergence between nations of the Northern and Southern hemisphere is clearly noticeable. In the turn of this century with the advances made in science, thanks to the industrial revolution, a new dimension was added to agriculture production i.e. the dissemination of those research results to the actual users in an organized fashion, the extension. Today, the vitality of nations and the level of their agriculture development are judged on the bases of how effective their research and extension systems are interlocked.

The truth of the matter is that Ethiopia was not even closer to resembling the nineteen-century Europe when the new millennium dawned. In pre-revolutionary Ethiopia there were seven research centers mandated to conduct crops and animal research. The number of these poorly equipped research centers was expanded to fourteen soon after the *Derg* assumed power. In addition many scattered testing sites formerly used as mere observation plots were upgraded to the level of sub-centers to cover as much distinct areas as possible. More significantly those research centers were restructured to accommodate various crops of regional importance. Following the restructuring of the research institution in 1986, the centers were designed in

75 Yaraswork Admassie, Twenty Years to Nowhere: Property Rights, Land Management and Conservation in Ethiopia, (Trenton: Red Sea Press, 2000) 209.

such a way as to represent specific agro-ecologic condition so as to focus on crops and livestock very common to the area in which it is located. The agriculture colleges were also mandated to conduct a collaborative research with the research Centers across the country. Although the research institutions were fashioned in the Euro-American image, most of them lack basic facilities to conducted experiments and therefore produce reliable information. Despite the modest assistance provided by the bilateral and multilateral agencies, the research centers suffered from serious lack of human and material resources.

Quality notwithstanding, the research findings generated out of these centers were not always helpful to small farmers. That is because they rarely addressed the production problems that the peasants face. Most of the researches efforts in Ethiopia were expended on generating innovations that require considerable resources to apply at peasant production level. Thus, by default, the main beneficiaries of the innovations became the few and economically insignificant mechanized farms run by the state. In 1981 alone the state farms which constituted 4 percent of the total cultivated land received 76 percent of fertilizer and 95 percent of the improved seeds generated at the research centers, leaving only a quarter of the fertilizer import and almost nothing of the improved seeds to the peasant sub-sector which by government account was about 94 percent.[76] Up to the end of the 1980s, the State continued with its biased policy of favoring the large scale farms instead of the peasant farms with the greater potential to fulfilling the objectives of food self-sufficiency.[77]

The research efforts would have been more relevant to the peasant had the thrust of the investigation being focused on soil and water conservation practices, adaption of minimum tillage as part of the farming system using locally improved farm implement, selection and cross breeding of draught animals. Crop breeding strategies should have aimed less in evolving cultivars that perform well in favorable environment instead of those with special attributes of drought/pest/disease tolerance, early to mature or moderately to low fertilizer responsive ones.

In the past, research scientists had one major weakness unable to overcome. By virtue of their training and the professional bonding that they had developed through time with major regional and international centers, they seemed to have lost sight of the specific need of the peasant in Ethiopia. The

76 (1) Ajit Kumar Ghose, Transforming rural agriculture: agrarian change in Ethiopia since 1974, (Journal of Development Studies 22 No 1, 1985) 135 (2) Ottfried C. Kirsch, Fred V. Goricke, Johannes F. G. Worz. 1989. Argicultural Revolution and Peasant Emancipation in Ethiopia, (Saarbruken, Germany: Verlag Breitenbach, 1989), 141.

77 Girma Kebede, The State and Development in Ethiopia. (Humanity publishers, 1992), 8-9.

areas of research were usually defined not in term of their applicability but acceptability in local and regional conferences. In almost all cases, experiments are conceived, proposed and carried out without peasant's involvements. Moreover, results were rarely verified in farmer's field for their appropriateness. Instead middle level technical cadres test the innovative packages in a fenced verification plots. Little or no attempt was made to draw farmer's attention to these wondrous plots to witness the agricultural "miracle" in what is commonly practiced in India and south Asia as "field day". The worst part of this exercise was that promising innovations were not immediately made available to farmers on a trial base. Instead, by mimicking the highly developed agriculture systems of the world, improved seeds or other farm technologies developed at the centers are made to undergo rigorous testing before release. In Ethiopia a single innovation or a package required at least 7 to 10 years before approved for wide scale use. After so many years and plentiful of resources spent to generate the packages, it was less than certain if they can be of any use to the peasant under his management.

Whatever the size of the extension department might have been, its services was more tilted toward serving the State and collective farms the same as was the case in pre 1974 Ethiopia where agricultural services including loans and credits were accessible to the landlords and commercial farms alone. Unlike the TFYP of 1967 formulated to invigorate the peasant sector, the post 1974 agricultural policy put heavy emphasis on the state sponsored cooperatives and state farm ventures.[78] As has been pointed out earlier these ventures were small and economically unviable when compared with the vast peasant sector, which had been starved of all the production support services. Over and above the government policy of favoritism, their compact size and aggregate structure made the state farms and cooperatives attractive to researchers and extension personnel to deal with. Through their Research and Advisory Department, the Ministry of State Farm Development (MSFD) also pursued an aggressive search and adopt policy for inputs, farm techniques and training. At times the link between some of the research centers and the state farms go deeper than simply carrying out joint research programs to solve specific problems or exchange technical information. But in what is termed as a triangular conspiracy it also involved private input dealers ready to sell recommended inputs to the state farms who were known to be fond of using the so called "magic inputs". The eagerness to use pesticides, fungicides and fertilizers at unreasonable rate and at all cost to succeed also stems from the fear of being scolded or even

78 In the years before the government launched the mixed economy policy in 1989, the Ministry of State farm Development absorbed 50% of government investment in agriculture, 75% of loans extended by the Agricultural and Industrial Development Bank (AAIDB), and shared great number of college and technical school graduates as a supporting staff in its various enterprises.

subjected to revolutionary discipline by the authorities for being economic saboteurs.

In Ethiopia the link between research and extension has always been very weak. The reason is that the mechanisms that were in place at the time were not strong enough to bridge the wide gap that exists between the two. For one, the extension was viewed as a separate entity not a complementary to research and vice versa. Most significantly they both operated separately and were under different institutional guidance. This basic weakness made both institutions hardly responsive to the peasants need. More significantly, the extension agents in Ethiopia lacked incentive to perform their task efficiently. They were unmotivated and pessimistic to the extent some became indistinguishably similar in their lifestyle and thinking with the peasant they set out to change. By nature of their training they were assigned to far away district lacking basic modern facilities sometimes electricity. Their number was few but required to cover large area with no logistic support behind.

In Hararghe region, during the early days of the revolution, one extension agent was expected to cover three to four PA areas sometimes on foot and in occasion by motorcycle. The term of reference of a typical agent was too vast to make an impact on the farming community. It ranged from advising on the best techniques of growing crops or maintaining poultry or managing improved beehive to advice on health issues to socialist production and home economics. For many agents with or without peasant upbringing such hardship was too painful to bear. Once assigned to certain duties in a given *kebele* they are rarely called back for consultation, feedback or to receive further training. In a real world, though, with properly oriented research system, the extension agents hold the key to production improvement. To remedy the flaw in the system and make it relevant to farmers the government, in 1986 introduced the Research and Extension Liaison Committee at both regional and national level. However, the committee was unable to bridge the gap that it was designed to achieve for various reasons among which lack of human resource was one.

Past government's monumental failure to build a sound agricultural research and extension institutions is reflected in nowhere than in the poor performance of the agriculture sector itself. One possible reason for this is that the reckless response of the *Derg* to the perceived or actual threat (dissent) by the educated class that drove many out of the country. As a result seasoned research scientists was either working in countries that guarantee their safety or turned into useful asset of the various liberation fronts fighting the government. Those remaining few were no match to the massive problem the country faced and worst of all were less inclined to serve at remote centers, which have no amenities. In a way this attest the fact that education as it had existed in Ethiopia might have been an irrelevant component to the economic and social development needs of the peasantry.

Agricultural Marketing and Pricing

In an effort to regulate production and maintain fair prices of food products in the urban areas, the *Derg* commissioned a grain purchasing agency with a mandate to purchase, store and sell food grains to consumers whenever necessary. Implicitly, this will also benefit peasants by purchasing their produce in time of abundance or bumper harvest. That of course never happened. Such measure, no doubt, was intended to protect the producer from unscrupulous grain dealers who might otherwise take advantage of the elementary law of economics, supply and demand. In reality, the Agriculture Marketing Corporation (AMC) as the new agency came to be known arbitrarily imposed grain quotas on each farm household, and squeezed everyone for their produce at gate price values not commensurate with the cost of production. Meanwhile, the peasant was permitted to sell his produce at the open market only after satisfying the quota imposed on him or her by the agency.

Naturally, the peasant lacking market outlet for his produce sells whatever is left to the grain dealer who in turn made profit by selling to the consumer at exorbitant price. The difference between the farm gate price and the open market was roughly 100 percent and sometimes more. For example in Gojam the farm gate price for brown Teff, a highly desirable grain by urban dwellers was 37 dollars per quintal but when sold in the open market it fetched 94 dollars.[79] Under the new mandate the AMC was also required to purchase 50 percent of the wholesaler's grain purchases at approximately 5 percent more than the farm gate price. Any grain dealer found operating before fulfilling the quota imposed on him by the AMC runs the risk of license revocation. Thus, to avoid the trap set by AMC in the grain business, as indeed in many other businesses dealings in Ethiopia at that time, it was normal for many to enter into shady deals as a means of survival. The most tragic of all is that the AMC never lived up to its expectation and in fact raised havoc on the traditional market channels that the peasant was accustomed to deal comfortably.[80]

The AMC did not only impose quota on the purchase of grains normally cultivated by peasant associations in a given area, but also demanded crops never grown or long ceased to be important in the area. As an illustration in a Tef or Wheat growing region, the peasant associations were forced to make available a certain amount of oilseeds never grown or due to reasons only known to the peasants ceased to grow. Such was the dilemma faced by farmers in the Sendafe area outside Addis Ababa that many were forced to buy pulse from local markets to fulfill their quota. This bizarre way of enforcing

79 James C. McCann, People of the Plow: An Agricultural History of Ethiopia, 1800-1900,(Madison: The University of Wisconsin Press, 1995) 251.

80 Desalegn Rahmato, 'Cooperatives, State Farms and Small holder production' In S. Pausewang, Fantu Cheru, S. Bune and Eshetu Chole(eds) Ethiopia: Rural Development Options. (London and New jersey: Zed Books 1990), 6.

government regulation has caused more confusion than the AMC can handle that the practice was discontinued in the early 1980s.

There was little doubt on the mind of the local experts including those working for the agency who argued time and time again that it was this unwarranted interferences by the State that lend to production mishap. But the government continued to insist that following such policy would guarantee market stabilization and by extension ensure supply to the consumer. Moreover, in as long as the State's priority remained on bolstering the war efforts and maintaining government and party functionaries on cheap purchased grains at the expense of the peasant, the role of AMC was regarded as necessary if not crucial. [81] On the other hand the initiative to form associations that protect their basic interest long dashed, peasants were left with no option but to sell their produce at whatever price the state agency and the grain dealer offered. But there was growing conviction that they were not offered commensurate price for their produce. At last the peasants begun to play a hid and seek game with the AMC by partially trashing their crops as a showcase and stashing what is left in the haystack to be sold in the black market. Such was the condition of the state of agriculture in Ethiopia during the two decades rule of the *Derg*.

Ethiopian Technological Base

In Ethiopia, prior to the revolution there were no well-developed institutions to help in the transformation of society through the creation and/or adaption of new technologies. The absence of this certainly hampered technological growth necessary for societal transformation. The history of many nations notably China with similar features as Ethiopia decades ago attests that building institutions to aid in the advancement of technological innovations is essential. China as backward agrarian as it might have looked at the time has all the basic ingredients for takeoff. In the agricultural transformation of China one deduces that technological progress took place after the institutions long established had matured, not instead of it. It is because of those institutions that workable land reform was achieved in china.[82]

Contrary to what seemed to have occurred in China and other nations with similar circumstance of feudalism and feudal landholding system, the institutions that developed following the land reform act in Ethiopia did not

81 Richard Dowden, Africa: altered states, ordinary miracles, (Public Affairs New York, 2009), 271. Similarly in Zambia the agricultural marketing boards bought agricultural commodities from farmers way below the world price so as to make it available in cities and towns to preempt looming rebellion, unrest and worker strikes in the country.

82 John Wong, Chinese land Reform in Retrospect. Land tenure Center, (Madison: University of Wisconsin, 1974), 4.

pave way for any homegrown technological novelty to appear in the scene. The reason was that many of the institutions that came into being and the rules, regulations and directives that accompanied them did not reflect the national priority in agriculture. For example, emphasis that should have been put on peasant farming was instead placed on large-scale farms including the collective farms. To run such kind of farms even at semi-mechanized level entails employing large machines and equipment. Unquestionably, the road to agriculture transformation begins with importing/and or manufacturing appropriate tools and equipment necessary to uplift the lot of several million peasants. It is precisely because of this misguided development strategy that focus on state enterprises and cooperative farms that the government begun importing modest amount of agriculture hardware mainly farm machinery and construction materials. The net result of this faulty strategy was reflected in food production decline. It also had an indirect but depressing effect on research and development efforts to manufacture appropriate farm tools and implement. This largely explains why the Ethiopian landscape remained for a long time bare of any signs of "modern" agricultural tools, implement or inputs essential to increasing production and productivity at peasant level.

The TFYP of 1967 was formulated on the recognition that the country's state of agriculture was backward and therefore, need to be transformed into dynamic and sustainable one. The reason for the technological stagnation has been universally blamed on the moribund feudal system. By the mid of the 20th century it become abundantly clear that the agriculture backwardness of the country was laid bare for all to see. It is no accident, therefore, that the 1960 coup plotters begun to transmit radio messages to the public underscoring the '*continued ignorance and low standards of living*' of the people.

Almost as if responding to the demand by the failed coup plotters, the Imperial Government in collaboration with international agencies launched few pilot projects in the central and southern regions of the country with the specific aim of generating minimum and comprehensive package program targeting peasants surrounding the project areas. For the first time in the country's history research results quiet applicable to small scale as well as large-scale farmers emerged at pilot projects of CADU and Wellamo Agricultural Development Unit (WADU). In particular at CADU farm tools and implements were designed to suit low level farming conditions. The Unit's attempt to develop new crop varieties and the necessary inputs to raise crop yields at peasant level gave further impetus to production. The speculation is that, if only CADU and WADU were left to flourish and their activities duplicate all over the country, the fate of the Ethiopian agriculture would have been slightly different.

However, the feudal institutions of pre 1974 Ethiopia were not prepared to deal with such a challenge where the peasant serf is in control of the production process. No sooner did research results start trickling down and thus, making a difference in the life of the peasant serfs that the big landowners begun pushing them out of the land.[83] After evicting many of the serfs, in order to maximize production, the landlords relied heavily on imported inputs and heavy machinery instead of those tools and implements developed at the Unit. Sadly enough, the experiment, which started less than decade ago to ostensibly empower the peasant by means of improving his production tools was instead systematically dwarfed by those giant machines and equipment, imported from abroad. In a short time the landscape in and around the project areas were superficially transformed into modestly mechanized Canadian prairie farms look like. The sad part is that many of the serfs who for generation toiled to support their families were gradually uprooted from the land. The rough estimate put by experts at the time was that one tractor power replaced 20-peasant household easily. It is not hard to imagine, therefore, that as the result of the introduction of the new innovation the landlords who owned the land in the first place evicted many serfs out of their lands. Beginning the late 1960s up to the mid-1970s where the revolution erupted, businessmen begun to acquire lands originally cultivated by serfs. It is ironic that when land reform law came into effect in 1975 the developers and the landlords lost their land to the revolution.

In the mid-1980s, consistent with its agriculture policy of establishing bigger and partially mechanized farms to transform the Ethiopian agriculture, the government begun to give more emphasis on the PCs. The transformation of what were once subsistence farms into state of art with agro-industrial enterprise as the ultimate objective was projected to be met through the distribution of locally assembled tractors and farm equipment of the Soviet brand. The outcome of this ill-conceived venture was disastrous to say the least.

Among many agricultural subsistence farming the world over, the Ethiopian agriculture has some distinct features of its own, That is, the greater part of the agriculture system is based on animal drawn iron plough culture. This unique position of the Ethiopian farmer in the world of subsistence farming was not exploited to the fullest extent by either inventing or modifying implement such as deep ploughs, manure sprayers, land levelers, knapsack sprayers, harvesters, binders, threshers etc. etc. Furthermore since the use of oxen as a source of traction power is as old as the country itself, it could not have been very hard to introduce or convince farmers use other draft animals including donkeys, camel, horse, mule or even cow to undertake farm operations. It is important to recognize that the country is blessed with close to 53 million head of cattle and the same amount of ruminant and draft animals[84]. It was not before the government realized the futility of its past

83 Micheal Stahl, Ethiopia: political contradiction in agriculture development. (Uppsula: Publications of the Political Association No 67, 1975), 79.

84 CSA, Agricultural Statistics Abstract (2012).

actions namely, turning the Ethiopian agriculture into modern socialist enterprises by means of mechanization involving the state and a minute sections of the peasant sectors, that it turned its attention to the small scale farming and the appropriate technology needed to develop it.

IV

TOWARD FOOD SELF-SUFFICIENCCY: THE PREDICAMENT

No nation can live with pride unless it has the capacity to feed itself
—Prime Minister Indira Gandhi 1970s

Ethiopia on the Eve

For nearly two decades, the military regime in Ethiopia had battled variety of forces wanting to overthrow it. Few more than others were formidable enough to pin the entire army to the ground rendering most government structures in the rural area dysfunctional. The Eritrean Popular Liberation Front (EPLF) and the Tigray Peoples Liberation Front (TPLF) both operating in the north challenged the *Derg* seriously to pose as an alternative to it. In addition there were other armed insurgents in the Ogaden, Oromia, and even in the Blue Nile basin fueled by the government's inability to deal with the internal contradictions rooted in Ethiopia's past. This further led to subtle and sometimes open interference by foreign forces specifically the Republic of Somalia and the Arab world. Unquestionably, one of the most burning issue at the time was the nationality question where each historically constituted ethnic group felt that the unitary system in Ethiopia did not meet and in fact worked against its aspirations. Despite the nationality issues taking forefront in the struggle against the *Derg*, the problem of underdevelopment manifested in chronic food shortages remained the core issue needing serious attention.

Ominously, by mid-1980s even the so-called surplus producing areas of the country such as North Amhara and some of Oromia region begun to experience serious food shortages

Furthermore, the *enset (Ensete ventricosum)* culture long believed to be superior to the cereal farming practices of central and northern Ethiopia was found to be no longer true as tens of thousands of *enset** dependent peoples from Wollaita, Hadiya, and Kambatta were forced to leave their land of birth to resettle in relatively fertile lands of western and Blue Nile basin of Ethiopia. In all fairness, the misery that gripped the *enset* dependent population of southern Ethiopia in mid-1980s can be attributed to the scarcity of land more than the crop itself.[1] To illustrate the point the man to land ratio in Kambatta in 1984 was 475 persons per square kilometer. The same is true for the adjoining districts of Hadya and Woliata where landlessness in particular among the young was rife prompting many to migrate to larger towns in search of job.[2] In the past, the cultivation of *enset* gave Southern peasants an edge against famine better than the cereal dependent northern compatriots. In a subsistent mode of production, the cultivation of tubers, corms and underground root crops guaranteed peasant from adversity. Earlier studies show that 120 to 240 plants of this large sized false banana stem, occupying a mere 360 to 720m^2 of land can feed an entire family of 6 for a year. The plant is not only moderately drought tolerant but also produces an easily storable 40kg of high consistency paste per stem. In terms of providing food security without taking much of the cultivable space, *enset* is by far superior to cereal.[3]

Obviously, drought and famine continued with vengeance in the traditionally food deficit areas and the war torn northern region. With these massive problems heavily weighing on the government, it was a matter of time for the *Derg* to give way to yet another political force destined to shoulder the twin problems of the twentieth century Ethiopia, food insecurity and national stability. It is ironic, however, that the new rulers who through the barrel of gun took the mantle of power from the *Derg* were not spared from the verdict of history as drought and famine continued to pound on their door step throughout the decades of the 1990 up to the end of 2010s. Little did they suspect that despite their revolutionary rhetoric and finger pointing at the *Derg* and its predecessor the imperial regime, they themselves were unwilling victims of Ethiopia's historical legacy of, underdevelopment, civil war and famine that had raged for centuries?

As pointed out earlier the root of this complex problem is the opaque and

---

1 Diseases common to Enset, such as the bacterial wilt and Fusarium wilt was widespread beginning the 1980s to aggravated the situation at the time as well.

2 Ethiopian Population Census, (1984), 161.

3 E. Westphal, Agricultural Systems in Ethiopia Agriculture Research Reports. (Wageningen: Center for Agricultural Publishing and Documentation, 1975), 126.

archaic feudal system that suffocated the national sprit and damped enthusiasm for progress for a very long time. In this regard the Derg, the EPLF and the EPRDF inherited a nation slightly better than what it was in the turn of century but by far insecure in having to deal with the nagging issues, that is, the inability to provide 'daily bread' for its citizens while maintaining its cohesiveness as an ancient nation state. In part the separation of Eritrea from Ethiopia is nothing more than a symptomatic of the century's old problem of regionalism, food insecurity and a gamut of factors causing national instability.

Grant, underdevelopment, food insecurity and misrule by successive regimes have been scourge in Ethiopia but what groundbreaking measures would the EPRDF take to address them. Would they take extraordinary steps to exorcise the country from its past demons that kept it under perpetual poverty for centuries? How would the new rulers in Ethiopia redress past injustices stemming from mal administration and an inequitable distribution of resources? In that case it is important to look deeper into the efforts put by post feudal governments. Suffice is to point out that each in his own way tried to address the burning issue regarding governance, food security, internal cohesion and inter-regional relations. People are the product of their historical circumstance and any measures they take or pose to undertake is subconsciously influenced by the actions of past generation that preceded them. Certainly they are influenced by many factors with similar denominator to claim any originality.[4]

From the above argument one concludes that the *Derg's* accession to power did not stop or diminish opposition to central rule nor did EPRDF's intention of divesting power from the center to the periphery avoided the threat of separation two decades later.[5] Moreover, the failure of the EPLF to resolve the lowland-highland dichotomy in a just way or its inability to narrow the Christian-Moslem divide further proves that indeed the history of past generations weighs heavily on the present. Understanding such a fact will certainly minimize ones wild expectation from taking over and at the same time injecting some sense of realism on ones actions so as to add more but useful

---

4 Karl Marx, Eighteenth Brumaire of Louis Bonaparte (1852). 'Men make their own history, but they do not make it as they please; ......but under circumstances existing already, given and transmitted from the past. The tradition of all dead generations weighs like a nightmare on the brains of the living'.

5 Oromo Liberation Front Mission statement declares that The fundamental objective of the Oromo liberation movement is to exercise the Oromo peoples' inalienable right to national self-determination to terminate a century of oppression and exploitation Likewise the Ogaden National Liberation Front believes The revolution of the People of Ogaden is based upon their absolute rejection to the unauthorized disposition of their territory in early 20th century by the British Government; and subsequently on their constant appeals to obtain the right of self-determination.

building blocks on the existing edifice. For this to materialize, however, the young in particular the educated should be anchored to traditional values of his/her origin and show a sense of humility, dedication and incorruptibility. In this case the departure point should be not in mimicking developmental paths based on alien experience but on home growing ideologies easily identifiable by the people.[6] The urban terror unleashed after the revolution has all the markings of revolutionary experience from abroad both recent and past.

Ethiopian feudalism has remained firmly entrenched to negatively affect the lives of people for the most part of the twentieth century. Hypnotized by its place in history and disguised and sustained by self-deceit people in the helm refused to admit that change is overdue. They had vowed that the feudal system is eternal and if at all requires minor modification to function. On the other hand, the writing on the wall was clear that change was inevitable. As opinions and views were never expressed in the open, it was a matter of time before unforeseen forces are unleashed to resolve those outstanding issues so much important to the country. Obviously, there could not have been more rational act than to allowing citizens access to land, address regional and ethnic issues to promote national stability, provide unlimited social services to build a robust society and declare in a true sense of the word that 'religion is private while a country is common to all'.

As time advances the chorus for change was getting louder. It was clear, however, that, neither the proponents of change nor the opponents were prepared to budge an inch from their stated goals to make transition as smooth possible. Finally the choice was narrowed down to two, that is, either to give face lifts to feudalism in the interest of the entrenched feudal group (the choice that had gradually developed after the abortive coup d'état of 1960) or use extreme means to overthrow it. In the interim there appeared a creeping social unrest spreading from the mountains of Sahel in Eritrea to the arid Ogaden region and many in between. The sporadic but deadly revolt by the Bale Oromo and the peasant revolt in Gojam were of some concern to the Imperial regime during this period. In a sense the February Revolution of 1974 and the scattered peasant uprising that preceded it together with the urban led-armed rebellion in Eritrea and Tigrai is the product of the new generation facing the old squarely in the eye. Unfortunately, the intermittent pang of the 1960s gave way to an excruciating pain in the new millennium for the peoples of the region to bear. The result of the 'wild expectation' allude above is the emergence of a new state such as Eritrea and the clan fractured Somalia were both have yet to meet the minimum standard of statehood required.

6 It useful to refer to the ideological literatures of the vanguards for change mainly students and intellectuals that begun to emerge after the failed Coup d'état of December 1960.

Post-Feudal Economic Reform

The seizing of power by group of military men popularly known as the *Derg* in 1974 had upset the status quo previously guarded jealously by the *ancieñ regime*. The exploitation of the peasantry as a whole and the degradation of the minority nationalities and the Moslems in general were for once and forever made illegal and forbidden by law. Before the advent of the February Revolution, Muslims lacked the basic religious rights that the Christians in particular the followers of the ancient Orthodox Church took for grant.[7] Together with the question of nationalities that Ethiopia has underplayed for ages, the feudal regime whose power base was rooted in amassing huge tract of land and the extraction of free labor from the peasantry finally succumbed to the forces gathered against it. Following the unprecedented popular upsurge that toppled Emperor Haile Selassie's government, the *Derg* declared the National Democratic Revolution (NDR) of the classic socialist brand with centrally planned economy as a core but also allowing national entrepreneurs to operate side by side with the public enterprises.

Despite earlier promise by the *Derg* to give an ample room for the private enterprise to play a role in the economic development of the nation, many of the budding entrepreneurs were shut out from participating in buying and selling commodities, their license restricted and at times out rightly revoked. Judging by the various economic measures taken during the first two years in power, the *Derg* made it clear that no economic system that run parallel let alone overtake the one prescribed by the socialist ideologues in the government was permissible. A short time after the NDR program was proclaimed, the government begun taunting merchants and traders of hording essential goods. It blamed many as counterrevolutionary for expressing their views on universal issues regarding land and governance, and threatened to administer revolutionary justice against those openly challenging the new order.

In theory the NDR permits the anti-imperialist national bourgeoisie to play if not prominent but secondary role in nation building. In practice, however, this proved to be difficult to realize. As a result a wedge between the government on one side and the student/business class on the other soon developed to culminate into bloodshed. To drive the point home, the government executed few businessmen accused of economic sabotage and imprisoned people in jail for illegally carrying foreign currency. It became clear for the skeptics that the *Derg* has no intention of entertaining a laisser-faire economy in place of a centrally planned one. This action was followed by the

---

7 The Imperial Constitution of 1955, Article 126. 'The Ethiopian Orthodox Church, founded in the fourth century, on the doctrines of St. Mark, is the Established Church of the Empire and is, as such, supported by the State. The Emperor shall always profess the Ethiopian Orthodox Faith. The name of the Emperor shall be mentioned in all religious services'.

random imprisoning of big and small merchants throughout the country sowing fear on the business class. In a sense this ill-fated measure continued to define the government's true nature up to the end of its life. No wonder why many people cast serious doubt on the mixed economy (capitalist/socialist) launched by the government in early 1989, two years before its ouster.

During the seventeen years of *Derg* rule, the economy was highly regulated with many restrictive policies in place. In the agriculture sector, the policy ranged from making peasants follow orders on the types of agricultural commodity they are supposed to produce and at what price they are expected to dispose. In addition peasants were asked to cultivate unfamiliar crops on their fields or buy from the market to meet the quota demand placed upon them. The policy also put more emphasis on the tiny farms run by the state and cooperatives, which amounted to less than 4% of the total cultivated lands in the country. These farms in turn received the lion share of the meager resource available but at the same time failed to mitigate famine constantly knocking on the doorstep of the peasant folks. Under the *Derg* rule, the Ethiopian economic continued to suffer due to the excessive meddling of the state bureaucracy in the affair of the producer. In the late 1980s, average GDP growth remained very low at less than 2%.[8] Food insecurity and sporadic famine continued to hover over the country almost unceasingly throughout the 1980s, threatening the fabric that holds the society together. In short, the spark that caused the Ethiopian revolution to erupt in 1974 was never fully addressed to save the socialist government from its ignominious defeat by yet another force, the EPRDF.

The EPRDF government, an amalgam of diverse nationalist parties, upon seizing power in mid-1991 abolished by a stroke of a pen many of the vestiges of the old regime.[9] To win western confidence and calm the fear of many Ethiopians already feed up of the 17-year rule under socialism, the EPRDF slowly but surely negated its long held socialist doctrines, as inappropriate and perhaps inappropriate to Ethiopian condition. Thus, pursuing an open door policy except for some vital sectors of the economy remained under the control of the government. The coalition force at the helm boldly challenged the unitary system of government that took shape in late 19th century Ethiopia. Instead they proposed decentralized system of government composed of political units based on ethnic homogeneity. Furthermore, they begun to

8 Eshetu Bekele & Mammo Muchie, DIIPER Research Series Working Paper No.11 (Denmark: Aalborg University, 2009).

9 An environment quickly turned hostile by remnants of the old regime and demagogues of urban brand as well as from genuinely apprehensive citizens who viewed the Front as ethno-nationalist with sinister motive bent of destroying of multi-national Ethiopia. These groups together with the student body in the capital tried to inject ethnic sentiment and foster power struggle in what is already a vacuum space that the EPRDF came to fill.

drastically curtail the operation of the State Farms, and dismissed the notorious AMC, which has been unbearably onerous on many subsistence farmers. It also reviewed price mechanisms in place and abolish internal trade restrictions, strictly enforced by the previous regime. Judging by the series of measures that the EPRDF undertook one couldn't miss the direction that the new government was heading. Soon, it made its intention known by declaring that the strategy of Ethiopia's development would be Agriculture Development Led Industrialization (ADLI). The ADLI, which came into effect in April 1994, was assumed by many to mean that privatization of land and the expansion of commercial agriculture is in the plate. The reason for such misconception was that the EPRDF presented itself as an antithesis to the *Derg* at least in appearance if not in substance. Rightly so, the *Derg* was blamed of failing to solve the chronic food shortages that the country had experienced for ages but also for plunging the country into economic chaos. The new rulers accused the military regime as, unresponsive, demagogic and tyrannical at best.

The first economic reform came into effect at the end of 1992 exactly two years before the adoption of a new constitution as the supreme law of the land. The new economic policy targeted the countryside as vehicle for future development of the country. Many of the pronouncements made on the economic development strategy had defined the course but within the frame work of the new economic reform. This includes the rationalization of tariff structure, stabilization of domestic market and the elimination of export tax. Within few years the GDP grew by 6.4%.[10] Indeed the emphasis placed on the countryside with its 85% population and producing over 50% of the country's Gross National Product (GNP) cannot be considered as out of line. Thus, the countryside symbolized by the peasantry who work on the land become the cornerstone of the country's new economic policy.

The constitution embodied one important clause believed to be essential for the new development strategy to work. In it, issues related to land tenure system and the mode of land utilization was laid out clearly. Article 40 of the newly adapted constitution defines land as belonging to the "government and the people" of Ethiopia.[11] Quiet a leap from Article 2 of the 1932 imperial constitution, which stipulates that 'the land, the people and the law belong to

10 Reddy Prabhakar Tada, "Economic Reform and Structural Changes in Ethiopia Since 1992: an inquiry" (International Conference on African Development. Archives Paper 18, 2001), 4. http://scholarworks.wmich.edu/africancenter_icad_archive/18

11 Ethiopian Constitution of 1995. Article 40/3 "The right to ownership of rural and urban land, as well as of all natural resources, is vested in the State and the people of Ethiopia. Land is a common property of the Nations, Nationalities and Peoples of Ethiopia and shall not be subject to sale or to other means of exchange."

the emperor'. However, Article 40 did not signal a dramatic shift of policy from the previous government, which also placed land under "the collective ownership of the people". In this regard, the 1975 decree on land reform and the constitution that soon followed appeared to be similar if not identical to that of 1995 constitution adapted under the EPRDF supervision. In reality, however, land under *Derg's* rule was far from being the collective property of the people but an alien entity that benefited less the section of the society that the reform was formulated to help. When the reform was put into practice, peasants once again found their lot if not worst but a little bit better than the serfs before them. This time though, the peasantry had to deal with better-organized state than the one before it.

This does not mean that the 1995 constitution is radically different from the rest when it comes to major land issues. They all forbade the peasant farmer to make full use of his property up to including mortgaging and selling land in the open market. What makes the new policy on land slightly different from the one drafted by the *Derg* in early 1987 was that it gave a breathing space to the peasant who for generations lived under oppressive environment. The new constitution allowed him to operate freely on the land allotted to him without the heavy hand of the state on how to manage his/her affair. According to the new land utilization directives put out by the government, it became legal to enter into business arrangement with outsiders of modest income to jointly work on the land for mutual benefit. It is also permissible to employ labor whenever necessary. Thus, short of selling or transferring the title completely, peasants in many areas of the country were allowed free hand within the prescribed law on the land.

The facts remains, however, the peasant lacking a secure tenure over the land may not be enticed or obligated to engage in conservation practices and productivity enhancing investment on his holding. Politics around this issue has locked the government in a debate with the opposition forces, mostly of the urban background who insisted that there would be no breakthrough in Ethiopian agriculture unless land is traded like any commodity. The government continues to advocate against land sale/mortgage. In this regard, the core of EPRDF's land policy is no different than Article 5 of the Rural Land Reform Proclamation No. 31 of 1975, which prohibits the sale, exchange and mortgage of land. Unlike, section 5 of the proclamation, which severally limits the use of, hired labor, the new policy put no restriction on whom and how much labor one can hire to do a job.[12]

The EPRDF led government is unequivocal when it comes to the question of land and its beneficial use by peasants. It argued that land is the

---

12 Laws governing land were set earlier by the Land Reform Decree of 1975. The *Derg* sponsored 1987 Constitution served for only three years before it was abrogated by the EPRDF led government.

only asset available for the majority of Ethiopians to a secure livelihood, no matter how precarious. It made it clear that tampering with the land where its access to ordinary citizens is curtailed by insensitive laws is more harmful than beneficial. True enough four fifth of the population is of rural background and will remain so for a long time to come. The reason cited is that the prospect of the country becoming industrial or a service-rendering nation in such short time so as to absorb displaced or landless peasants is unrealistic. If for some fatalistic reason, regulations on land are lax or allowed for the market forces to play freely, the peasant will face an imminent eviction-causing catastrophe of biblical proportion.

The fledgling Ethiopian economy may not be in the position to considerably devote resource to diversify the economy so as to be able absorb the displaced fleeing the countryside; the choice taking thus far is prudent. The proponents of land 'not for sale' argue that the 1974 revolution although may had its own dynamics to take the route that it took, nevertheless, it was the eviction of peasants in the Arsi and Shashemene area by landlords that gave impetus if not trigger the eruption of the popular revolution.[13] It was this involuntary displacement of the peasantry by egregious landlords including commercial farmers who pushed for mechanization instead of the usual share cropping arrangements supported by traditional production methods employed for centuries that aggravated the agrarian problem in the country. Indeed who in his right mind will let history repeat itself by allowing the tragic situation turn into complete farce? It is therefore, understandable for the government to argue that, devising formulas that would give peasants guaranteed use of the land for life. It is hoped that this will rectify the problems associated with the current land tenure system. Thus, short of land title, a certificate of operation will suffice to alleviate any apprehension that the peasant might have with regard to use of the land more securely.[14] The land certification process that begun in 2003 assuaged the fear of many who saw sinister plan in the commodification of land as satisfactory solution to making land productive and against land fragmentation. While many saw the measure as a gimmick that it may not solve the fundamental problem of land tenure insecurity.[15]

Not satisfied with the certification program started in early 2003, the advocates of 'land for sale' insisted that under the current law, land

13 Micheal Stahl, Research Contradiction in Agricultural Development: A study of three minimum package projects in Southern Ethiopia. Report # 4. (1973), 5.

14 Land certification begun in Tigray in late-1990 to mitigate land dispute and fragmentation. The long-term implication of land certification is that it could eliminate tenure insecurity so deeply embedded in peasant psyche and encourage investment on the certified land.

15 Berhanu Abegaz, "Escaping Ethiopia's Poverty Trap: The Case for a second Agrarian Reform". (Journal of Modern African Studies, Vol. 42, No. 3, 2004), 313-342.

fragmentation is inevitable to a degree that working on miniscule size fields for so many millions would be futile to say the least.[16] Although certification of land may serve as a deterrent tool for further division of land, ultimately tradition may weigh heavily on the peasants (PA) to make room for those coming of age, unless the state ruled against that. It was common in the past that the PAs were duty bound to accommodate new members reaching the age of manhood in their localities. In today's Ethiopia, the downside is that population taken as an independent variable may play against this egalitarian approach of allowing members access to land without affecting the livelihood of all members of the PA. This is because of the diminished size of potentially cultivable land to go around. In as long as land remained fixed and the population grows at its present rate of 2.8%, it is probable, that reshuffling of land may take place anytime making the reform meaningless. The process meant not only smaller farms but also the fragmentation of it. As aptly observed by Desalegn, the decreed on land reform was less than satisfactory to the peasant and can be described at most as 'distributive leveling-reform based on fixed but increasingly diminishing rural asset.'[17]

Without going into details, the land policy in Eritrea as affirmed by the law drafted in mid-1990s, gave peasants in the traditional mixed farming area of the plateau usufruct right over the land. Article 23 sub-article 2, in the non-ratified and ultimately shelved constitution states that 'All land, water and natural resources below and above the surface of the territory of Eritrea belongs to the State. Usufruct rights of citizens shall be determined by law'.[18] In the case of Eritrea, an additional factor that seem to complicate issues related to land is the government veracious appetite to usurp lands in the traditional grazing areas where agro-pastoralist eke a living by rearing livestock and by pursuing an opportunistic mode of life including crop cultivation when it suit them. Relying heavily on the inexhaustible free labor at its disposal, that is the National Service recruits, the government of Eritrea continually expands into the mid and low altitude areas known to receive scanty rain and with fragile soil structure prone to various agents of erosion. Given, the precarious position of the agro-pastoral population to making a decent living out of the age old practices of cattle rearing supplemented with the cultivation of stress resistant crop species, land alienation in whatever guise would remain a very contentious issue in Eritrea.[19]

---

16 Desalegn Rahmato,Ethiopia: Agricultural Policy Review In: Digest of Ethiopia's National policies, Strategies and Programs Ed. Taye Assefa(2008).

17 Desalegn Rahmato, Agrarian Reform in Ethiopia. (Uppsala: Scandinavian Institute of African Studies, 1984), 63.

18 The unimplemented Eritrean Constitution 1997.

19 *Yebio Woldemariam*. LAND IN ERITREA: a contentious commodity. Taking on current political issues, (London: Proceedings of CDRiE'S Conference, 2010), 101. http://ssrn.com/abstract=1664873

Taking Stock on Agriculture

With the formulation of relevant set of policies and the exertion of political will to implement every bit of them make the task of challenging the obstacles associated with food production easier? However, some are intractable requiring good governance to tackle. Here we shall discuss the main obstacles to the economic growth of Ethiopia in particular as it is related to the agriculture sector. In order to appreciate the predicament in agriculture production and suggest ways of circumventing them it may be necessary to visit past history and probe at policies and actions taken by governments in the 20th century.

Fragile Statehood

Among the many inconspicuous but serious problems that stagnated agriculture development in Ethiopia was the fragile statehood inherited from the past. Without doubt the long sought but perpetually eluding sense of national security and stability that the country had hopelessly hankered to achieve contributed to the low level of agriculture activities in the country. This nagging concern has come into sharp focus in modern times more than in anytime of its long history partly because of the power play by the international and in particular regional powers vowed to weaken the centripetal force to hold strong. Obviously these had hindered or slow down efforts in the 20th century to centralize the socio-political life in the country.

Thus, serious challenge to its territorial integrity came from the Somalis in the east and Eritreans in the north. Later the Oromos joined the chorus to embrace the popular demand dearly held by numerous left leaning political grouping, that is, 'self-determination up to secession'. Regardless of who was in the helm and at what historical epoch, Ethiopian rulers were less fortunate to focus solely on matters related to social and economic well-being of the people while striving to maintain the center as crucial link to meet internal and external challenges. At times the push-pull factor exacerbated by regional and sectarian feelings is so strong to threaten the very fabric that was holding the structure of the semi-unitary state. This phenomenon, therefore, appeared in every turn of Ethiopia's' history decelerating the forward march sought by various emperors among them Emperors Tewedros, Yohaness, Menelik and more recent Haile Selassie.

The modern Ethiopian state is believed to have evolved from a sort of diffused outlaying regions into a country with a clearly defined boundary in the turn of the 20th century. Since the lurking and continuous peeping of European powers and their junior partner Egypt in late 19th century in the shores of the Indian Ocean and the Red sea, there was not a day for an Ethiopian ruler that national stability and security was not of a major concern. Since Italia's interest in the Horn of Africa dates back before the scramble of the African continent in 1984/85 into different spheres of European influence, colonial agents disguised as priests, explorers, diplomats, social observers meticulously studied the country's societal formation and the contradictions within as well as the style of governance to cause much concern to the feudal state eternally haunted by incessant inter-regional rivalry. Thus, following the short lived military conquest of Ethiopia by fascist Italy in 1935 many of the contradiction that was particular to the country begun to surface in earnest. Well aware of the existing contradictions, the Italian colonial authorities begun to exploit them to their advantage. First they parceled the country into five regions favoring the minorities over the majority ethnic groups. They also found it convenient to favor Muslims more than the Christians and the Catholics more than the Protestants. In some cases, the serfs were promised free land of their own in as long as they renounce their allegiance to Haile Selassie and support Italian pacification effort.[20] They provided arms to groups deemed as enemies of the 'Amharas'. Some encouraged by the advancing Italian expeditionary forces, for example the Raya Azebos of southern Tigrai took the bait and harassed the retreating Ethiopian army from the northern front in the early days of the war. The Oromos in Jimma under the powerful chief Aba Jobir not only cooperated with the Italians to rout out patriot activities in their district but also put reward money for the head of any person caught fighting against the Italians[21].

Earlier, the Turks who had occupied the entire Red Sea rim for over three centuries attempted to exploit the Muslim/Christian divide and eventually put an effective blockade by restricting movement of people and goods from the interior to the coast. In effect the occupation of Ethiopia's natural outlet, the port in Massawa, by the Turks sealed the country's fate to interact freely outside its borders. One can also argue passionately that Ethiopia was on the edge of facing serious calamities throughout the 20th century ranging from locally generated and proxy supported rebellions to pockets of famine and pestilence taking its toll among the population. All throughout the second half of the 19th century and up to the audacious aggression of Fascist Italy 1935, the feudal state was unsure of the European intentions toward the country. Their insatiable appetite of territorial aggrandizement at the expense of historic Abyssinia did scare the wit out of the rulers of the day.[22] The two major wars against a European power and the numerous skirmishes that occur in between

20 Ironically, two decades earlier Lij Iyasu who was in life and death struggle with the Shoan aristocracy employed the same ploy to win the heart and mind of the peasantry.

21 Tadesse Mecha, Tikur Anbessa--The Black Lion—(Asmara: Corriera Eritrea Printing Press, 1943) 77. (Amharic)

22 The Tripartite agreement between the three major European powers in 1906 and the open intention of the British to hold the country as protectrate after its forces helped liberate Ethiopia from the fascist in 1940 is proof enough for Ethiopian rulers to be worrisome of foreigners in any shape or form.

in the late 19th century and the middle part of the 1930s bear testimony of the country's legitimate concern of existentialism. Not to mention the blockade strictly enforced by the Turks on the Red sea coast for more than three centuries and the failed attempt by The Egyptian Khedive Ismail, a European vassal to conquer Northern Ethiopia.

During the early period of the twentieth century, the state was ferociously squashing regional rebellions and engaged in the business of cajoling recalcitrant feudal lords to submit to the central authority. After the death of the farsighted and most able Emperor Manlike II, the state fell into disarray where various contending Feudal Houses mainly the Shoan and Wello vie for power through their favorite candidates for the vacant post. The ascension to power by the Emperor's daughter, Zewditu did nothing to appease the contending forces. However, the appearance of a young foreign influenced Ras Tafari in the political scene shortened the struggle between the old Abyssinian stock fearful of European machination and Islamic domination and the new young Ethiopians led by Teferi.

The old guards with sizable private army at their disposal were not particularly fond of the Ethio-Europe relationship that Teferi wanted to nurture. They saw no good in Ethiopia having an enduring relationship with Europe. They believed unlimited interaction with the outside world might ultimately work against their personal vested interest and the country at large.[23] On the other hand, Teferi cautiously courted the young without becoming overly critical to the entrenched feudal system. However, his attempt to wrestle power from the traditionalist was never successful. At any rate, the period between Menelik's death in 1911 and the Italian invasion of Ethiopia in 1935, was not particularly marked with groundbreaking reform of national significance, but events in that period may had imparted a sobering effect on the nation left behind by any standard measures of development. Nevertheless, some of the steps taken by the court toward opening the country to foreign influence and hence for modernization did encourage few of the foreign educated elites, known as 'Japanizers' to advocate for more change. Be that as it may, the movement for change championed by Ras Teferi, later Emperor Haile Selassie was not exactly a replica of the Meiji dynasty in Japan where truly national government was established in place of the divided regional political power. In Japan the concentration of power under the visionary Meiji Dynasty

23 1/ Zewde Reta, Yeqedamawi Haile Selassie Mengist. (New Delhi Laxmi Publications and Boston USA: 2012), 52 (Amharic). Crown Prince, Ras Teferi who paid an official tour to Europe in the 1920s must have felt serious despondency when two third of the feudal lords who accompanied him were unmoved by what they have seen in Europe or express their unwillingness to copy anything useful from it. Thus, his effort to impart impression of Europe's technical advancement on his retinue failed miserably. 2/ Harold Marcus, Haile Selassie I. The Formative Years 1892-1936, (Trenton: Red Sea Press, 1995), 41.

symbolized the end of feudalism/Shogunism, the demise of artificial controls over personal choice and livelihood, the end of national seclusion and the end of economic stagnation.

Contrary to the widely held views, that Haile Selassie played a vanguard role and therefore was supportive of the radical thoughts espoused by many especially those travelled outside the country, the truth of the matter is that he broke his solidarity with the so-called modernizers and returned to his base of feudal constituency.[24] Sadly enough the modest modernization process begun by Emperor Haile Selassie came into abrupt halt by the Italian intervention of 1935 to resume in snail pace after the liberation of the country in 1941. Exacerbated by the intermittent provincial wars waged by gamut of nationalist forces in the second half of the past century, the agriculture sector among other sectors of the economy suffered greatly. Nearly all the armed confrontation that took place in the past century affected the countryside more than the urban centers. Moreover, the meek posture displayed by Emperor Haile Selassie toward the landed gentry and his complete disregard of the peasant's wellbeing exposed the country to further hardship.[25]

Following the abortive coup d'état of 1960, schism begun to develop among the ruling circle.[26] The coup being perceived as anti-feudal, many among the people including the students blamed the country's backwardness to the feudal system with the Emperor at the hub. In a sense the 1960 coup brought to the surface the trùe face of Ethiopia unmasking the 3000 year old independence myth. After the event, the feudal government shaken from its foundation began to take significant measures in regard to agriculture development. In 1966 it provided land for some 6,000 landless with fewer than 3 hectares each at their disposal. The MCD also sponsored the development of farms here and there for the unemployed or landless later to be supplanted by private developers at the Shashemane plain, the Awash Valley and the Tekazze

24 Bahru Zewdu, Pioneers of Change in Ethiopia. (Addis Ababa University Printing Press, 2002), 55.

25 Zewde Reta, Yeqedamawi Haileselassie Mengist. (New Delhi Laxmi Publications and Boston USA: 2012), 567-580, 627. (Amharic). Recently published book by a well-connected ex diplomat to the Shoan court and the government in general reveals that the feudal lords who remained central part of the imperial regime were jealous of their privileged position to scrutinize the work of the young and enlightened members of the government. It is amazing to learn that the self-made of humble ground Mekonen Habtewold has to be reshuffled from one directorship post to the other 5 times in less than 10 years. His advocacy for fair play, introduction of proper taxation and promulgation of laws to protect the serf brought him a wrath from the feudal lords to rotate like a spindle from one post to the other.

26 The challenge to the imperial regime by two prominent Shoan personalities Dejatmatch Takele WoldeHawariat and Dejazmach Alula in the 1960s is viewed by the student body as the beginning to an end.

Basin. Few years after the attempted coup, land was granted to civil servants by royal decree. In addition all *maderia* land operated by members of the Territorial Army was converted to free hold in 1967. However, the major bottleneck to agriculture development, that is, landlord vs. serf relationship left untouched until the mid-1970s.

The land tenure system, long recognized by progressive elements including the students as the primary cause of Ethiopia's economic ills, was challenged in earnest. In 1966 the government instructed MLRA to commission study on how best land is shared equitably to all. At long last, the government begun to debate, no matter how mute, regarding the very system that sustained the ruling class for centuries and impoverished the population at the same time. Obviously, the first draft was rejected by landowning upper house as well as by the majority of the lower house immediately after it was tabled. It is no surprise that both houses were filled with landed gentry and hereditary rulers to openly welcome a reform that will ultimately undermine their interest.[27] Suffice is to say that six years after the submission of the first draft law governing land to the parliament, the issue was left hanging on the air until it was finally addressed by the Ethiopian Revolution in 1974.

The sweeping land reform decree of March in 1975 also brought to the surface issues or matters some with unpredictable consequences. Soon after the declaration of the land reform decree, fierce wars ensued between the government and the radicals on one hand and the feudal lords and landed gentry on the other. The revolution also heightened the expectations of nationalist, separatist as well as regionalists to wage war between and among themselves and with the government of the day. The weakness of the Ethiopian state became glaringly clear when urban violence begun to threaten the semblance of law and order in cities and towns throughout the country. The 'white terror' and its response the 'red terror' was a mixture of cruelty and savagery where thousands of students, skilled and educated people perished, further weakening the state apparatus to Pre-coup period of 1960. Ironically, the revolutionary terms used to justify actions taken by the forces unleashed after the dethroning of the Emperor was a carbon copy from earlier revolutionary episodes in the Soviet Union and perhaps China.

The *Derg,* which carried the mantle of power for seventeen years, left the political scene unceremoniously to yet another political force whose resolve to maintain what is left of Ethiopia through ingenious means is legendary to say the least. Nevertheless, the EPRDF itself had to wage a variety of low-level wars in the east, the south and with its northern neighbor Eritrea for twenty

---

27 Seyoum Haregot, The Bureaucratic Empire: serving Emperor Haile Selassie. (Trenton: Red Sea Press, 2013) 70.

years. Tragically, the event of the past twenty years has an aura of past history and may be repeating itself. Although many of the wars conducted with regionalists and Islamic oriented groups did not match that of the full-fledged war with Eritrea, nevertheless the mini wars contributed to the bumpy road that the coalition government was forced to take. However, the border war with Eritrea not only diverted much needed resources from the agriculture sector but also injected yet a familiar sense of insecurity in a fractured region.

The new government also inherited much of the longstanding intersectional strife within its borders and outside it to forgo the role of mediation and arbitration to restore peace. Areas particularly known for occasional troubles are the Rift valley, the Borona and the Ogaden plains as well in parts of the Gambella administrative region. Recently, the clan strife between the Nuer and the Anyuak in the south left the government vulnerable in as far as keeping peace within the various nationalities is concerned. Similar incidents between the Boran Oromo and Garri ethnic groups not to mention the frequent armed clash between the Afar and the Issa draw the attention of the government. Unlike the *Derg*, however, the EPRDF led government almost always used persuasion mixed with force to deal with the age-old contradictions that exist within the different nationalities and between it and the society at large. In all fairness, the degree of military response to quell uprisings or restore peace between warring national groups is reasonably tempered when compared with the regimes preceding it.

Today the EPRDF's opponents are the Diaspora Ethiopians most often derived from the Amhara and Oromo national groups, who view the regime as representative of a section but a minority among the Ethiopian population. The government is blamed for the audacious measures it undertook such as granting the right of nationalities to form local governments anchored on their historical make up that includes language, physical boundaries, and modes of production. The administrative reform, thus far instituted did little to assuage the apprehension of many urban Ethiopians and those in Diaspora whose presence in the national politics cannot be dismissed as lunatic fringe. They were part of the formidable forces that called upon the government of Ethiopia to encourage and facilitate the consolidation of peace and security, respect for human rights, democracy, and economic freedom in the country.[28] The 2005 national election reflects the anger of urbanite Ethiopians and their formidable supporters in the Diaspora toward the EPRDF.

Meanwhile, the federal arrangement and the laws crafted thereafter seem to encourage the minorities to remain within Ethiopia. For restive region such as Ogaden and the resourceful nation of Oromo, the option to secede as put in the article 39 of the constitution give more assurance to stay in the fold. The fear is that this could play in the hand of ethnic demagogues ready to exploit

---

28 US HR 2003, Ethiopian Democracy and Accountability Act of 2007.

any resentment against the regime in place. Sub-nationalism is serious threat to national development in societies where literacy rate is low and resources limiting.[29] Thus, this double sword provision in the constitution could revive sectionalism and even encourage breakaway movements by few of the 80 different nationalities that the country is made of. This will only go to show that the EPRDF like the *Derg* before it is the inheritor of the old state, which had battled tirelessly to quell, crush, and obliterate any national dissension and separation.

It is not debatable to submit, therefore, that one of the primary drawbacks to Ethiopia's inability to stand equal among the developing but somehow adequately fed nations of the world is its fragility as an organized state. Thus, it may be inaccurate to put blame entirely on the *Derg* and the EPRDF alone for Ethiopia's agricultural ills or under development. Suffice is to say that the economic malice is deeply rooted in the country's past. Both governments succeeding the feudal regime are simply the product of the most "ignorant and underdeveloped nation of the world" to borrow Asbe Hailu's excellent commentary about a century ago.[30] The feudal system with the emperors at the hub was preponderantly responsible for the pathetic situation that Ethiopia finds itself today.

Land, the Primary Asset

Land is primary asset and a source of livelihood for more than 80 percent of the population. It is no surprise, therefore, that the EPRDF lead government and the *Derg* before it choose to give closer attention to land issues. Thus, putting land as a centerpiece of their policy both governments emphasized economic reform based on agriculture development, albeit each taking separate path. While the *Derg* pursued socialist model of development the later gradually opened the country to capital intervention to the extent of allocating 3.2 million hectares for lease. Unlike the *Derg*, it gave slack to the peasant to operate on his usufruct land without undue interference from the state, thus, giving the impression of ownership.

It took few years before the EPRDF begun focusing on land, though. With the proclamation of ADLI it begun to direct its attention on the trimming

29 In contradistinction to the violent course taken by many nationalist groups in Ethiopia to achieve their objectives, the Scottish movement for separation or even the aspiration of the Basque, the Catalans, the Flemish sub-nationalist groups assume nonviolent means or choose the ballot box to secure any deferred dream. In the final analysis this European tribes are educated, adequately fed and have stakes in the resources that the nation provides to get involved in physically destructive projects.

30 Asbe Hailu 'sile mengst gudat', (Addis Ababa: Brhanina SELAM Printing press July 22, 29 1927) (Amharic).

or disposing state owned farms, expressing its intent to improve production at peasant level, and deal diligently with the huge livestock sector. In retrospect, the EPRDF's approach to governing the country was best described as extremely cautious. Having a narrow base of political support when seized power, it felt insecure and therefore allowed the state apparatus to function in more or less the same way as before save for drastic modification in the area of national security. [31] For example, the most controversial, the *kebele,* a community organization with deep root in both the countryside and the urban areas was left intact. Secondly, unlike the EPLF of Eritrea which hunted and killed opponents outside its territory and dealt harshly with those associated with the *Derg* internally, the EPRDF by comparison was reconciliatory with the bulk of the past functionaries of the regime except those involved in corruption and terror act. In short the new rulers who came to power with a barrel of gun did not have a political capital to dispense let alone squander, thus, found it expedient in their part to go slow. To ensure that no potential adversary arise from the ash of the fallen regime, it rounded up all high government and party functionaries for safe keep in various prisons and detention camps.[32] More significantly it dismantled the notorious security apparatus that kept the people under constant fear of detention and sometimes death since the declaration of Ethiopian socialism in 1975. Judging from its political posture at the time of its induction to power, therefore, one can deduce that the EPRDF did not have ready-made recipe on civil administration let alone a policy aimed at transforming the country radically. It was slow learning process by threading over a fine line.

It is for this reason that the new government refrained from taking drastic actions to reform the agriculture sector. The rationale provided for such inaction was that until the constituent assembly is elected and deliberate on the land issue; it found it unwise to take any action at all. The overriding concern of the government in early 1990s was to politically stabilize the country more than working to eliminate the specter of famine. Understandably, such exigency was viewed by many as less opportunism and no callousness in the part of the

31 Prime Minster Meles Zenawi in his intimate address to Eritrean Intellectuals in 2011 reveled that unlike the EPLF that had the support of its people when entering Asmera, the EPRDF lacked political capital to be overly cautious when dealing with many issues that may antagonize the people.

32 It is tragic and comical as well to see those detained members of the old regime which no doubt many had committed unspeakable crimes have to wait for a verdict while the trial dragged for almost two decades with their right for bail denied. On the other hand it is unparalleled in the annals of African history where new rulers did not follow the path of arbitrary detention and execution as witnessed by the *Derg* regime in 1974 and the coup plotters of December 1960. In both episode hundreds of former government officials, landlords and suspected opponents were cold bloody massacred without due process of law.

government but an act of prudency in the face of uncertainty and extreme hostility shown by sections of the urban population. And when it attempted to take piecemeal administrative measures on land there arouse serious confusion in some districts where peasants were emboldened to claim lands taken away by the previous government to accommodate land shortages experienced by neighboring PA. It is important to note that the EPRDF far from dismantling the PA structures established by the previous government, it strengthened them further to serve as part of administrative appendages to the center.

The government had an ambivalent position toward the relatively well-organized state farms in the country. While many downsized and few completely abandoned to be reclaimed by the original owners, still the government was slow to take actions on them. In some case and perhaps for political expediency the government took swift measures without waiting any legislative decision from the constituent assembly. To illustrate a point, it is worth mentioning the case of the semi-mechanized and partly gravity irrigated farms located in the Middle and Lower Awash valley. This traditionally Afar grazing grounds were taken by large developers over half century ago, hence for their partial return to the rightful owners. In other words, without much fanfare it gave back portions of the most productive lands to the clans inhabiting the area while for the time being tens of thousands of prime lands were left fallow and unattended. By conservative estimate only a third of the forty thousand hectares of irrigated land in the Middle and Lower Awash valley remained under government supervision planted with cotton. Once these highly productive lands were returned to the pastoralist who decades earlier were unjustly deprived of their grazing lands, clan elders begun entering into various kinds of working relationships with private developers many with verbal agreement only.

After the informal but partial handover of the irrigated lands to the pastoral population, clan leaders unable to operate and manage the farms practically resorted to extorting money from potential investors through shady partnership deals. Many of the investors lacking an institutional backing to legally demand an end to the interference from the clan leadership became helpless and the farmlands were rendered almost useless for many years. In this regard, productivity loss was estimated at 25,000 MT of raw cotton and an additional 40,000 MT of wheat per one growing season, that is, should double cropping operation was contemplated on only less than ten thousand hectares in the Middle Awash valley.[33] As a result many developers lost millions of Birr lacking government protection for their investment. The absence of clear-cut

33 Based on experimental data obtained from Melka Werer Research Center, the author was able to arrive at reasonable estimation of the amount of wheat that could be reaped during the off-season, that is, after the harvest of the primary crop, the cotton.

directives coming from the government on the *modes operandi* as to how the partnership between the national investors and the Afar beneficiaries ought to work seriously hurt the prospective investors ready to share half of their income with the nominal 'landowning' clan.

It is ironic that the return of the land to the rightful owners was not officially made public. The handover was not accompanied by rules and regulations for investors to seek recourse from justice. Suffice is to say that the opportunity time lost to maximizing benefit from working on the most productive lands in Ethiopia was staggering. Gradually, the situation improved by early 2000s and few among the many early entrants reaped the fruit of their investment. Lately, some of the irrigated lands are parceled among clan members and are operated as small farms. It must be borne in mind that this was the original intention of the government when handing back the tens of thousands of irrigated lands in early 1990s to the pastoral population. Nearly two decades after it consolidates state power, the government sold 140 million dollar worth of state farmlands to private developers in 2012.[34] Many more farmlands were sold after the push to make land available to expatriate developers was made officially an acceptable investment strategy[35].

The handing over of the semi-mechanized state farms to the local population or communities was not limited to the Afar region alone. Because, Similar measures were also taken in other regional states such as Oromia, Amhara and the Southern Region. Although the handing over of portions of the state farms to the local population resulted in production declined nevertheless, the government's lack of enthusiasm or interest to manage the left over state farms did not help either, hence, for its decision to dispose them to the highest bidder.

The return of lands operated by the MSFD to their original owners, the pastoralists and agro-pastoralists, are everything but freehold. They cannot be sold or exchange but used for eternity as usufruct. The new constitution adapted 1995 does not encourage and in fact forbid land to be sold and mortgaged as any commodity does. As noted earlier, the rationale behind the governments' firm stand against the privatization of land was that peasants were feared into rushing to sell or mortgage the land and may end up being hapless creatures that no one envies. Conversely, if privatization of land is permitted willy-nilly the chance that it will be in the hands of few will be certain,

---

34 Reuter News Agency. (March 29, 2012).

35 At Zewai in the Rift Valley farms formerly managed by the State was leased to foreign investors to produce wine and flowers. In Gambella another company with foreign connection is posed to develop 250,000 hectare of land for the cultivation of maize, teff, oilseeds and sugarcane.

making the majority of the peasantry landless.[36] The consequence of privatization, according to EPRDF would be dire where huge peasant population could be forced to leave the land and flock to cities and towns in search of job. Understandably, industry and the service sector being modest each contributing about 14 and 45 to the GNP respectively, the EPRDF argument appears plausible.

There is a concern on the part of the government that the land holdings system currently in place may be inefficient and slowing the tempo for rapid food self-sufficiency. Nevertheless, such concern has not led the government into revising old policies. In its resolve to making the country food secure, the government continues to argue that the peasant should be the central focus of any effort aimed at achieving food security and must be provided with the necessary inputs to utilize his labor and land efficiently.[37] This implies that if production-enhancing elements mainly inputs and technical information is readily available the peasant will ultimately succeed in changing the agricultural state of the country. Thus, consistent with its policy of 'don't mess up with the land', the government remained adamant not to structurally alter the utilization method of land by its unwillingness to visit the provision in the constitution dealing with land. Instead it continued to focus on production-boasting strategies such as inputs, land certification and micromanagement on land ownership short of mortgaging.

The land policy currently in place has some features that permit peasants do what is best for them. For example he can rent his land, employ farmhand, and utilize water under public domain. However, for many critics of the policy, the change so far made appears to be cosmetic, but in reality the measures taken so far are encouraging.[38] Notwithstanding, the increase of total cereal production registered in 2012 the yield per hectare also went up by half to a quarter of ton minimum during the past decade.[39] It is important to underline here that the government without completely doing away with Article 40 of the constitution that prohibit the sale or mortgaging of land was able to free the peasant from having to worry on matters regarding where and how to sell the produce, the exercise of his usufructuary right on land and the lifting of many of the onerous duties imposed by the state.

The protagonist of land for sale/mortgaging, on the other hand claim that

---

36 Gebru Mersha and Mwangi wa Githinji. 2005. Untying the Gordian Knot: The Question of Land Reform in Ethiopia, (ISS/UNDP Land, Poverty and Public Action Policy Paper No. 9. 2005), 7.

37 Ethiopia Observer. Media and Communication Center. Speech by Prime Minster Meles Zenawi in a Workshop at the New Coalition for Food Security in Ethiopia. 10/03 (2003)

38 Stefan Dercon, Ruth Hill Vargas and Andrew Zeitin; Revisiting Agriculture Led growth in Ethiopia. (DFID-UK., 2009), 8.

39 CSA, (2012)

until and unless land is made a free and convertible commodity it is unlikely that the livelihood of the people and in particular that of the peasantry will change for the better.[40] They argued that it is through aggressive individual investment and unhindered expansion of the sector that real improvement will come. In the process not only will food become available and in adequate quantity but also the sector would be in a position to absorb the displaced persons in the processing industry. It is further lamented that the refusal of the state to consider the sale, mortgaging and/ or long term renting of land is simply a justification and in line with the age old practices of monopolizing land and shrewdly using it as a political leverage to deal with the peasantry. Indeed, in the absence of an empirical data that show that land sale actually work against the interest of the peasant, the assertion by the government is subjective and at worst conjectural. To the contrary some peasants believe that land should be "moved to where it is employed most productively".[41] Conversely there is clear indication from peasants in the most densely populated regions of Amhara, Oromia and Tigray that the current arrangement of land utilization is acceptable.[42]

Stretching the argument further, economists suggested that the way out of the impasse is to devise a superior alternative method of recognizing plurality of ownership including private, corporate and collective ownership rights.[43] Part of this argument is based on granting land as freehold outside the traditional seed-plow culture into the 'tribal' semi-nomadic regions of the country. It is assumed that these measures will help stem land fragmentation, allow peasants invest in their farms and enjoy access to collateralized credit from financial institutions and provide the means for the courts to enforce judgment on defaulters. It may appear that the land reform laws as they exist today are inadequate and need further reforming. Persons outside the government are also calling for the privatization of land if turnabout is expected in the agriculture sector.[44] Recently some of the major food contributors to

---

40 GebreHiwot Beyakedagn one of the most articulate of the Japanizer argued on similar line.

41 Dessalegn Rahmato, 'Land tenure and Land Policy After the derg' ED. Harold Marcus. New trends in Ethiopian Studies Vol. III: 12th International Conference of Ethiopian studies, (Trenton: Red Sea Press, 1994).

42 Brhanu Nega, Berhanu Adenew and Samuel Gebreselassie, Land Tenure and Agricultural Development in Ethiopia. Ethiopian Economic Association/ Ethiopian Economic Policy research, Workshop. Addis Ababa: EAA November 13, 2002).

43 Berhanu Abegaz, Essays on Ethiopian Economic Development. (Avery Publishing, 1994) 320.

44 The head of a newly formed political opposition party Dr. Admassu pledged wide-ranging land reform as the key manifesto strategy for the 2005 Ethiopian elections (Addis Ababa: IRIN, 14 Jul 2003)

Ethiopia have joined the debate in advocating for reforming state ownership of the land to boast farming and end food insecurity.[45]

On the surface the argument against the current land policy and for complete privatization of land appears innocent. After all the land reform decree of 1975 and the pronouncement thereafter to regulate its use, have not created conducive environment to stimulate production and improve productivity. This was attested by the government's periodic call for food aid since the land reform of 1975. Although the frequency and the scope of emergency food aid to Ethiopia have decreased substantially, it has not totally vindicated current government officials who from time to time prematurely declare that the country is on the verge of attaining food self-sufficiency.[46]

However dismal the outcome of past reform might have been, it is not totally justifiable and in fact unwise to advance an idea of freehold outside the plateau where the agricultural practices is predominantly pastoral and agro pastoral. Putting aside of the environmental consequence that it may have on the areas traditionally reserved as grazing grounds, the notion of expanding of ones operation into hitherto uncultivated areas implies increased production through extensive rather than intensive cultivation. In other words, to argue in favor of expansion rather than intensification without paying due attention to the social and environmental ramification on the lowland regions of the country is ill advised.

The perception among many highland people is that the vast plains in the lowlands are sparsely populated and therefore open for grab. But to advocate for a free holding rights so as to ease overcrowding in the highlands is morally wrong and politically dangerous. As seen for decades in the Awash Valley and in the Eritrean lowlands today, resentment has been simmering over lands taken by settlers, investors and in the case of by the government's adventurous actions and practices.[47] Understandably, the people living in the so-called tribal, semi nomadic to nomadic areas are numerically small and

---

45 OXFAM joined in the chorus for land reform. (Reuters: November 14, 2003).

46 1/ (Michelle Riley, EMORY REPORT March 3, Volume 49, No. 23, 1997) "We estimate that we will have a surplus of between one-half million and 1 million tons of maize, roughly the equivalent of the grain shortfall we had in 1984 and 1985," Meles Zenawi's words. 2/ BBC Talking Point, Interview with PM Meles Zenawi (January 2005). "Since agricultural growth has been in double digit figures for the past two years we might achieve that (food self-sufficiency). Hopefully by 2007 may not need food aid" 3/ Press conference of Meles Zenawi at AA August 11 VoA Correspondent Peter Highnline Meles Zenawi remarked that in 5 years' time (within the five year economic development plan that begin in October 2010) Ethiopia will be sufficient in food supply and will export.

47 Ali Salim, National Unity Requires More Than Mouthing Slogans. (Awate.com September 3, 2009) http://awate.com/national-unity-requires-more-than-mouthing-slogans

demographically well dispersed. But they also occupy environmentally fragile areas to support an additional human and animal migration from the densely populated areas of the plateau. Incidentally, 35% of Ethiopia's animal resource and almost 90% exportable livestock come from the low altitude areas targeted for expansion.

Moreover, what makes the proposition preposterous is that almost all the areas located below 1000 m.a.s.l receive inadequate rainfall (200 - 350 mm annually) to be of any use to peasant farming as practiced in the high plateau. Agro-climatic conditions on those areas classified as marginal are such that the amount of rain received is small and erratic to be useful only for the regeneration of natural pasture. Even then a well-planned community supported range management regime should be in place to argue in favor of that. In exceptional case, that is when rain intensity is above average and the distribution rational, it may be possible to grow drought tolerant and early maturing sorghum and pearl millet. It should be noted here that Article 40, sub. Article 5 of the constitution granted the pastoralist the right to free land for grazing without the threat of dispossession. In actual fact, if the greater part of the lowlands were to be of any use to agriculture, they would have to be supplemented with adequate moisture and as frequently as possible. The reason is that the evapotranspiration rate in the low laying areas is higher than the cool and moderately moist plateau to require supplemental irrigation water to raise cereal or forage crops.

Because of infrequent rain coupled with the uncontrollable pressure exerted by herd size, the status of the rangelands is at an alarmingly poor state. The riverine plains, which in the past served as an excellent grazing ground reserve for livestock, are taken over by large-scale irrigation schemes to be useful to the pastoralists during the emergency period. In almost all schemes there is a constant battle for survival between the relatively wealthy and powerful developers/government on one hand and the pastoralists on the other who needed to reclaim back the land lost to them.[48] Unlike in Egypt or Israel where water is diverted to irrigate field crops hundreds of kilometers away from the source, most of the irrigation schemes in the Ethiopian region are confined along the river banks creating an effective barrier for the river to overflow its banks, denude the plains and upon receding leave luxuriant plant life for livestock to graze during the dry season. The farms are protected by strong levees to prevent rivers overflow their banks and do unwarranted damage. Given this fact it is not surprising to continuously witness, that at every drought incidence that the region has to experience, it is the population

48 Certainly, the race for land grab currently underway will aggravate the food security situation further. At the moment foreign conglomerates among them Indian and Saudi financiers are busy acquiring lands to grow food and industrial crops in the South and western part of the country including the most fertile valley of them all, the Awash.

inhabiting the low altitude areas in general and the pastoralist in particular that are gravely affected.[49]

The technology to minimize the impact of drought and other adverse climatic conditions in Ethiopia are nonexistent or at their infancy stage. In this regard the situation of the highland dweller is also as precarious as that of the lowlander. It must be noted here that most of Ethiopia's mixed farming with crop cultivation as a central activity is practiced in the high plateau. Together with the mid altitude areas, it is here that all the crop based agricultural activities are carried out. The cultivated lands in Ethiopia constitute roughly 12% of the total land mass. It is here that the 80 percent of the subsistence farmers eke a living.

Other important factor to consider is that given the social makeup of the peasant and the values and emotions associated with it, the idea of wholesale lands sale and/or mortgaging could be a hard proposition to sell.[50] This is despite the casual but honest remark by peasants who sometimes leave an impression on the proponents of privatization that they believe that land should be "moved to where it is employed most productively".[51] It is prudent therefore, to consider that tampering with land could have a far unintended social consequence on the largest section of the society, the peasantry and the agro-pastoral. In the traditional north despite the insecurity that may have existed throughout the time arising from occasional regional conflicts and inter feudal rivalry, the peasant has always been assured of a parcel of land upon which his livelihood greatly depended. In those areas, by and large, every claimant duly recognized by village members to be *wediabat* or a descendent of *akniabat* has shared the traditional communal lands from the beginning of times.

To agitate or even advocate for absolute privatization of land where approximately seventy million people eke a living from is suicidal. The irony of it is that land sale may not well be the panacea to Ethiopia's agricultural ills. Nevertheless, if privatization of land appears so appealing to any government in the future it is certainly a break from the centuries old practices where the state or to be precise where the nobility reign supreme over lands and the matters pertaining to land allocation and most certainly its utilization. As have been pointed out earlier, in the past state power emanated from monopolizing

---

49 Yebio Woldemariam. A Report on the Socio-economic Conditions, Production Status and Institutional Strength of the Afar. (Addis Ababa: UNDP 1993).

50 Berhanu Nega, B. Adenew and S. Gebreselassie. 2003. Current land policy issues in Ethiopia. Land Reform, 2003/3, Special Edition. (Addis Ababa: FAO Repositoy, 2003).

51 Dessalegn Rahmato, 'Land tenure and Land Policy After the Derg' ED. Harold Marcus. New trends in Ethiopian Studies Vol. III: 12th International Conference of Ethiopian studies, (Trenton: Red Sea Press, 1994).

and dispensing land to whomever it wishes. Thus, in a political term this suggests that the strength and cohesiveness of the state in ancient Ethiopia was measured by the degree to which the rulers were able to control land firmly and indisputably. This phenomenon become more pronounced and rigidly enforced in the centralized system of government established during the last century where peasants serfs and in rare case even the landowners were forced to toe line in fear of losing the land. This is in contradistinction to the situation in the Savannah of West Africa where population up to the dawn of 20th century was relatively small and land for cultivation was abundant to discourage centralism and in turn depriving rulers' power over the administration of land. Nevertheless, it is doubtful that putting land on the market will solve the land tenure system in both the highlands and lowlands without engendering far more difficult and complex social problems. Nor will it boast production to the extent of making the country food secure. Given the low level of inputs used today and the paucity of comprehensive agriculture research packages to augment the effort excreted by the government it may well be a while for the nation to be food self-sufficient.

Finally it boils down to how the policy makers and the stakeholders meet the challenge and devise means by which land is valued not in the classical terms of being bought and sold as commodity but in terms of its unique feature pertinent to Ethiopia's stage of development. It is imperative, therefore, to give the experiment underway-ample time to work in order to achieve the goal of food security. In other words it may be necessary for the state to acquiesce the power and privilege invested on it as the nominal overseer while guaranteeing the peasant unlimited access to land without depriving the local entrepreneur the opportunity to make the best out of the land is sufficiently enough to consider as break from the past.

After the February Revolution, the serfs in the south and west were also given the usufructuary right on land for life. The right to use land for life without the provision 'land for sale 'attached to it might at one time appeared flawed. But to prescribe the 'land for sale' in such traditional setting is not good choice either. It may provoke opposition by corporate or PA members reluctant to accept outside elements in their midst as new landowners. Many could resent and harbor fear to co-exist with people coming from far off place with nothing but land aggrandizement motives in their mind. As pointed out earlier, since the fall of the *Derg*, spontaneous land leasing activities has been in progress until it was essentially sanctioned by the government to cover many of the sedentary areas of the country. Such an arrangement may turn to be useful in increasing productivity especially if the renter is assured a prolonged stay on the leased land to recover his/her investment. However, to permit land leasing or renting to continue without obligating both partners in this arrangement to devote part of their resources in land amelioration, conservation and rehabilitation is technically unwise. For production to continue at a sustainable level it is imperative that the primary source of livelihood, that is, land and water need to be conserved and safeguarded from abuse.

As indicated earlier, the government approved and helped expand the practice of land certification pioneered in Tigrai to other regions of the country namely Amhara, Oromia and the Southern Region. As of now the certificate on "Right of Land" possession was issued to millions of peasants across the country. Suffice is to say that within 5 years after the process begun, certificate was issued to some 20 million parcels of land guaranteeing men and women alike free access to land. Equally significant is that married couples are issued a joint certificate ensuring equal access to land. Such recognition is significantly important to women because it empowers them to run their affairs as they see it fit without depending on their partners for decisions that affect their lives. Indeed this is a dramatic shift from the traditional norm where men and men alone were the ultimate arbiters when it comes to the question of land. This novel idea hitherto unknown in Ethiopia could ultimately benefit the new generation of peasants whose reference point is not the serf system of yesteryear but the one that accorded them the right to work on a certified land and to rent it for a considerable time if they desire so. According to World Bank survey of 2005 the majority in the sample perceived that 'certification [of land] increased incentives to investment in trees (88%), soil and water conservation structures (86%), and sustainable management (66%).[52]

The issuance of usufructuary certificate to peasants is entrusted to the MoA, with regional governments playing an active role in the execution of the program. The land certification program is designed to provide peasants security to work on the land for as long as they live and hopefully an access to credit services in the future. On top of increasing production and productivity, land certification is also hoped to reduce litigations arising from land dispute, facilitate land use planning and management as well as encourage investment on the land. However, it is more relevant to mention here that the key to self-sustaining agriculture activities in Ethiopia lay on the extent of how effective measures are taken to rehabilitate and pamper the land long neglect by bad governance. It is somewhat an accepted fact that soil lost due to erosion amounts to 42 tons per hectare per year or about 182 Million M3 per year by 2005 estimate.[53] Empirical data so far show that land certification appears to

52 Klaus Deininger, Daniel Ayalew Ali, Holden Stein, and Zevenbergen Jaap. Rural land certification in Ethiopia: Process, initial impact, and implications for other African countries. (World Bank Policy Research Working Paper 4218, 2007), 16.

53 EfD Newsletter, Green Accounting puts price on Ethiopian soil erosion and deforestation. (2010)

have a positive effect in Tigrai where investment in tree plantation, soil conservation and other land productivity enhancing factors increased by 45%.[54]

Size of Landholdings

Undoubtedly, land size is as important to low input peasant as the timely spray of pesticides, herbicides or fertilizers are to intensive cultivator. In other words while the former hopes to reap more harvest from extended land, the later strives to maximize output from a fixed land through the use of production enhancing inputs such as high yielding cultivars, fertilizer, row planters and irrigation facilities. It is not secret, therefore, that poor peasants tend to compensate crop yield through extensive cultivation, that is, labor permitting. This also lends to one plausible explanation why poor people in particular peasants in the developing nations are less persuaded by government agencies to limit the number of children they should have. Notwithstanding, religious beliefs and cultural taboos, in the short run limited parenthood advocacy, plays against the interest of poor peasants who require large pool of labor to perform farm chores sometimes impossible to handle with few in the household.

However, extensive cultivation has a physical limitation in the Ethiopian highlands. In the first place, there are little or no more of extra lands available in the highlands for extension. Hence, for the urgent need of periodic redistribution of land to accommodate new members of the society. This in turn makes land fragmentation the norm rather than the exception. The 1997 land distribution regulation adopted in Tigrai was therefore, designed to address the problem of fragmentation by declaring moratorium on further parceling of the land. Similar measures taken in other regions further proved that parceling of land neither improved productivity nor increased production. In this regard, the Amhara region did not distribute land since 1998 while no such thing happened in the Oromia region since 1994.[55] It should be made clear that, land in the Ethiopian highlands is no more an infinite commodity to continue parceling and expect miraculous results. One of the primary objectives of land certification is to stop further parceling the land into unproductive lots.

As far back as early in 1970 the landholding unit per family in the *rist* areas as well as in the predominantly share cropping regions were known to be less than a hectare.[56] Tragically, 50% of the peasants fell under this category of land holding. The number of peasants operating on fields with size ranging 1-2

54 Holden T. Stein, Klaus Deininger, and Hosaena Ghebru, Impacts of low-cost land certification on investment and productivity (American Journal of Agricultural Economics, vol 91 #2 (2009).

55 Fitsum Hagos, J.Pender & Nega Ghebreselassie, Land degradation in the Highlands of Tigrai and strategies for sustainable land management, Working Paper 25 (ILRI, 2002).

56 CSO Report, Survey (1966-1968).

hectare was by far lower than those owning less than a hectare. It is clear, therefore, that the handful of landowners including the royal families and their closest allies the nobility, oversaw fragmented and small size of millions of parcels of land across the country. In the south and the west these smaller plot units were distributed among the serfs who labored independently but with the obligation of sharing their fruit of labor with landowners to the tune of 1/3 to ½ of the produce. In the eve of the revolutionary upsurge, the commercial farms owned and managed by capitalists were miniscule in size.[57] Despite their size, the effect that these modern farms had on the livelihood of the people living nearby was enormous. The reason was that the commercial farms were conveniently located in and around the traditional grazing areas and in particular in the strategic livestock feed reserve grounds along riverbanks so useful to the pastoralists during the lean times. Their negative effect was significantly felt in the Awash and the Rift Valley areas of Assaita-Afambo, Gewane-Amibara, Wonji-Metahara, Blattie and Arba Minch. Decades earlier, that is, beginning 1960s up to the period where the revolution erupted in mid-1970s, modest sized commercial farms in the highland and in particular at Bako area in Wellega, the Cheffe in Wollo, the Chillao and Shashemene plains in Arssi and Sidamo respectively begun to grow at the expense of serfs who were pushed out by well-organized developers and their partners the land owning feudal class.[58]

With the nationalization of land the situation did not improve either. Proclamation No. 31 of the Land reform Act of 1975, acknowledged the usufructuary rights of an individual with each receiving not more than 10 hectares of land. But as studies after studies show this figure was far from being realized. Practically there was no single member of PA who owned more than a hectare. Indeed one can surmise that this was an arbitrary figure put by the enthusiastic executioners of the land reform decree without giving serious thought of what was on the ground. It also confirms that in spite of the concentration of land in the hand of few landowners, still rural lands were never administered in a grand scale very common in many Latin American countries or the white owned farms in Zimbabwe before the implementation of Fast-Track Land Reform Program. One can definitively conclude that the relationship between the landowner and the multitude of serfs working on the land was based on one to one relationship based on primitive undertakings and concerned only in siphoning the surplus produced without allowing the partnership to flourish and invest in production boasting inputs or capital layouts. The Ethiopian landowning class and the nobility were no match to their counterparts in other parts of the world of similar standing where they

57 CSO Report, (1975), 4.

58 Micheal Stahl, Research and Contradiction in Agricultural development: A study of three minimum packages projects in Southern Ethiopia. Report #4 (1973), 5.

made full use of the land under their possession.

Statistics compiled in late 1970s from random sampling of 10 districts nationwide later revealed that 26% of the land holding surveyed was between 0.5 ha to 1 ha. Individual households with a possession of 1ha to 2ha amounted to 33%.[59] Earlier study further indicated that after the readjustment made to compensate some unfortunate PAs operating in less than 800 hectares as permitted by article 8 of the Land Reform proclamation No. 31 was complete, the national average for those operating on 0.5 hectare and below was 32%, while those owning 0.5 to 1 hectare were accounted to be 26%.[60] Because of the dissimilarity in land size, soil fertility and population size that lived between and within regions, many PAs in the southern region ended up with land far below the 800 hectares than the law permits. As a result some of their members received less than 0.2 hectares on the average.

An attempt by the Relief and Rehabilitation Commission (RRC) to categorize land holding sizes in each region revealed that except for Bale, Arssi, Wellega, and Gojam administrative regions where land size was above 1 ha, the rest of the provinces had an average of 0.5 hectare per person. Sidamo of course has the smallest of 0.21 hectare per person.[61] There was similar report in the mid-1960 indicating that 30% of the peasants own less than 1 hectare with three-quarter of the Sidamo and Gomugofa peasants working in less than 0.5 hectare of land.[62] The 2001/2002 statistics on land holding further reveals that peasants with land holding ranging 0.1 to 0.5 hectare amounted to 30% while those operating 0.5 to 1.0 are no more than 26%. Surprisingly, 24% of Ethiopian peasants own land size ranging between 1.0 to 2.0 hectares.[63] The case of Sidama, Hadia and Kambatta is dire than anywhere else in Ethiopia where peasants in these districts are believed to own parcels of land ranging in size between 2,100 to 5,900 M^2.[64] The above discussion points at the fact that land fragmentation were serious problem for Zonal Administrators to take drastic measures to mitigate it.

In general terms, however, since the early 1980s land holding per farmer seemed to hold steady or declined slightly. This is despite the overall population increase in Ethiopia from 44 million in the mid-1980s to what is now about 90 million. This can be attributed to one or the combination of two things; the expansion of farms in hitherto virgin but less productive lands for example the

59 MoA/FAO, Ethiopian Highland Reclamation Study, (1983).

60 Alula Abate and Tesfay Teklu, Land reform, Structural Changes and Rural Development in Ethiopia: case study of two widely different regions. Institute of Development Research. (Addis Ababa University, 1978).

61 RRC, Food Supply Prospect: Crop and Livestock Dependent Food Supply Systems. Early Warning and Planning Services (Addis Ababa: 1986).

62 CSO Report, Survey on selected provinces between 1966 and 1968 (1968).

63 CSA, Ethiopian agricultural sample enumeration 2001/2002. (Addis Ababa: 2003).

64 CSO, (1984).

escarpments and/or the ban on redistribution of land to accommodate new members in the PA is holding firm. However, one needs to be cautious not to confuse the almost steady land holding size to massive influx of the rural population to urban areas because the rural to urban migration has not changed significantly. While the population census of July 1999 showed that the rural population was about 52.6 million, the one projected for 2014 is estimated at 71.3 million. In this case the increase is about 35%. At the same time the urban population census for both periods was found to be 9.1 million and 16.7 million, an increase of 15% in a span of one decade.[65]

Vegetation and the physical status of land

According to the 1960 estimates by the MoA, cropland made 9.5%, that is, 11.4 million hectares out of the total landmass of 120 million while meadows and natural pasture constituted 50% or 59.8 million hectares. In the early 1960 forest and woodland was believed to have occupied 7.6%, which was about 9 million hectares.[66] This figure is in total agreement with those obtained from a study conducted by Food and Agriculture Organization (FAO) experts in late 1950, which put the forest and woodlands at 7.2 million hectares.[67] Recently the Ethiopian Biodiversity and Tropical Forests came up with the figure so alarming that unless drastic measures are taken to substitute wood as a bulk of building materials and source of energy, the survival of the soon to be 100 million inhabitants is in serious jeopardy. Six decades after the first forest inventory was made by the FAO, the area occupied by forest amounted to 3,337,988 hectares down by more than a half of its original size.[68] The current rate of deforestation in Ethiopia is put at 1,630 Km^2 per year.[69] Based on this fact, therefore, one estimates that forest cover since the 1990 were reduced from 4.5 million ha to 1.9 million ha in 2006.[70]

Obviously such reckless onslaught on the forested land inevitably accelerates the rate of soil erosion and land degradation to the extent that the country is now more threatened by the poor state of land than the absence of rain that even many of the well-meaning observers fail to notice. Vegetation cover is the key to slowing down or completely preventing agents of erosion from going amuck. It also facilitates the replenishing of moisture to the soil by

65 CSA, Report (1999) and CSA, Population Projection of Ethiopia (2013).

66 Agriculture in Ethiopia, Yemastawequia Minster Addis Ababa (1964), 3. (Amharic)

67 Hufnagle, H.P., Agric in Ethiopia (FAO: 1961).

68 Ethiopia Biodiversity and Tropical Forest, 118/119 , USAID assessment (2008)

69 Reusing Matahias, Change Detection of Natural High Forests in Ethiopia by Remote sensing and GIS techniques, Vol. 33 part B7, (Amsterdam: International archives of photogrammetry, 2000), 1253.

70 Tsegaye Tadesse, An overview of the forest ecosystems of Ethiopia: Functions, trends and future directions. In. Proceedings of the first green Forum Conference, Ed. Seyoum Mengistu and Camille De Stoop, March 2007. (Addis Ababa: 2007).

playing buffer between the descending rain droplets and the surface of the land. Vegetation is also a natural windbreak preventing from serious wind erosion taking place.

Soil productivity has been in serious decline for at least a century. Twenty years ago, about 3.7% of the total cultivated land was regarded as highly degraded and therefore useless to produce crops of any sort while a little over 50% are on the verge of being degraded to be of any use unless rehabilitated as soon as possible.[71] Years later, the Ethiopian Forestry Action Plan put the seriously degraded land to 28%, with 23% more as moderately to barely useful for agricultural purposes.[72] Lately, 14 million hectares of actual and potentially cultivable land in the country are suspected to have less than 50 cm depth of top soil hardly deep enough to support the vertical progression of deep rooted crops much less withstand wind and water induced erosion.[73]

Depending on rain intensity, land gradient and relative vegetative cover, soil loss in the highland due to various agents of erosion is estimated at 32 and 130 Ton/ha/year for the average highland soils with some sort of vegetative cover and cropped lands respectively.[74] Generally, soil loss is put at 42 t/ha/year on cultivated lands and a maximum of 300-400 t/ha/year in highly erodible and intensively cereal cultivated fields. By extension nutrient loss as a result of soil erosion is enormous. Nitrogen, an essential element for crop development is spent at the rate of 30 to 36 Kg per hectare per year while phosphorous loss is estimated to be between half Kg to 20 Kg per hectare per year.[75] This notwithstanding, the studies including the one conducted by the College of Agriculture in early 1960 had established that Ethiopian soils are deficient in phosphorous and therefore, need to be replenished with either organic or inorganic fertilizer.[76]

The beginning for such grand scale devastation is not hard to trace. For one it is not preposterous to suggest that for a long time the country was not up to a challenge to come up with an alternative source of energy or invent a substitute to building materials to spare the wanton destruction of the wood

---

71 FAO, Ethiopian Highland Reclamation Studies Final Report, (1986).

72 Ethiopian Forestry Action Plan, (Addis Ababa, 1993).

73 Chilot Yirga Tizale, The Dynamics of Soil Degradation and Incentives for Optimal Management in the Central Highlands of Ethiopia. Ph.D Dissertation, (University of Pretoria), 2007.

74 Berry, Leonard. 2003. Land Degradation in Ethiopia: Its Extent and Impact, (World Bank, 2003), 3.

75 1/ UNDP, World Development Report, (Oxford University Press, 2002).
2/ FAO Integrated soil management For Sustainable Agriculture and Food Security in Southern and Eastern Africa, (Rome: FAO/Harare Proceedings of the Expert Consultation, 1999).

76 H.F. Murphy, Fertility and Other Data on Some Ethiopian Soils. Experimental Station Bull. No. 11. (Dire Dawa, Ethiopia, 1963), 511.

resources. Moreover, the *ancieñ regime* did not give priority to afforestation work or draw a plan to harvest forest products rationally. The State was for centuries, warlike, with high degree of propensity for violence and habitually dispensing justice and/or safeguarding territorial integrity from a roving mode. No doubt, such nomadic characteristics fancied by the ruling class in a land were agricultural practices had an imprint for over a millennium had a devastating effect on the landscape.

In the past, the feudal system hops from one region to the other so as to make its presence felt through violence and demolishing everything and anything on its way. This was its *modus operandi* for many centuries. The absences of an administrative center radiating from well-defined towns or settlements in the 18th and 19th century worked against the evolvement of modern state. As attested in Europe after the industrial revolution, indeed cities and towns become the necessary vehicle of development.

Tragically, Ethiopia never had a permanent sit of government until the early 20th century when Addis Ababa by sheer luck was selected to be one. Thanks to Empress Taitu who stumbled at Addis Ababa, a spot endowed with soothing natural hot springs believed to be a cure for arthritis that the emperor suffered at the time to finally decide to reside in a permanent spot and dispense justice. This in turn attracted the nascent merchant class to trade their hardware in large settlement without having the need to move from one place to the other. Thereafter, the peasants begun to converge on the emerging towns following the establishment of Addis Ababa, the imperial city, to exchange their primary commodities for consumable goods including some imported ones. The brief occupation of the country by fascist Italy must have added an impetus to the formation of semi-urbanization in the country where peasants felt the need to exchange their produce to manufactured goods. Large urban settlement is therefore, late phenomenon in Ethiopia since the decline of Axum and Adulis in the eighth century. The attempt to make Gondar the permanent capital of Abyssinia in the 17th century was unfortunately, thwarted by the ensuing civil war which lasted close to one century.

Without doubt the roving nature of the imperial court at earlier times had a devastating effect on the environment. There is no evidence to suggest otherwise. It is said that whenever the Emperor and his retune descends on a region they mercilessly consume what is in front of them exposing the landscape to real danger of long term environmental degradation.[77] Pankhurst describes that the last of the roving Emperors, Yohaness the IV at the height of his power frequently marched from Mekele to Debre Tabor; Gonder to

77 Merid Welde Aregai, Gonder and Adwa the tale of Two Cities, (8th International Conference of the Ethiopian Studies Vol.2, 1989), 41.

Adwa and Gojam to Lasta, at the head of 200,000 standing army.[78] Such an army of men and most often accompanied by women and children has definitely an adverse effect on the landscape including vegetation, soil and water. This is a phenomenon repeated throughout the centuries. Alvarez, a member of the Portuguese embassy to Ethiopia in the early 16th century witnessed an enormous size of people accompanying the court along the road from one encampment to another and described it as 'scarcely credible' where entire population of men, women and children, on foot and on horseback, carrying tents and provisions moving at ease'.[79] A scene such as this was common throughout the highlands when the 20th century dawned. Thus, it is not hard to imagine what menacing horde such as this will do to a land that only a year ago was the victim of raid by regional feudal chief or was hit by drought, affected by locust invasion or perhaps recuperating from the effect of localized infectious disease.

Relatively, the South was not spared from the same effect that the North had to experience throughout the medieval and up to the modern times. Beginning the 1800s, the Abyssinian rulers in particular the Shoan ruling house turned the South into a backyard where resources both human and material can be expropriated at will. By the end of the 19th century the south was turned into hot bed of contention not only among the Ethiopian ruling class with the Shoans warding off their rivals but also Europeans who cast their eyes on the abundant resources including forest products, coffee, gold, and exotic animals.[80] Today many parts in the southern region are too densely populated to even support subsistence life.[81] The soil completely degraded and forest habitat dwindled to no more than quarter of what it was 50 years ago. The blanket destruction of the forestland in the Gambella region in 1984/85 to make room for victims of drought from the north is too fresh for Ethiopians

78 Richard Pankrhust, Economic History of Ethiopia 1800-1935 (Addis Ababa: Haile Sellassie I University Press 1968).

79 Francisco Alvares, The Prester John of the Indies; A true relation of the lands of the Prester John, being the narrative of the Portuguese embassy to Ethiopia 1520. The Translation of Lord Stanley of Alderleyu (1881), (London: Published for the Hakluyt Society, University Press, 1961), 320.

80 Al Gore, Earth in the Balance: Ecology and the Human Spirit, (Plume publisher, 1993). He, writes the victimization of Ethiopia by two forms of dysfunctionality in this century. In the 1930s, the Italian Fascists invasion and after World War II western civilization in its way-searching for resources-did so as well. "The millions who have starved to death are, in a real sense, victims of our dysfunctional civilization's expansionist tendencies."

81 In the districts of Sidama, Hadia and Kambatta the concentration of people per square kilometer was in the range of 168, 178 and 475 respectively decades ago. Today population density per square kilometer in Hadia is about 336. http://ingeniousplc.com/index.php?option=com_content&view=article&id=106&Itemid=81

of the late generation to forget.[82] It is this mishandling of the natural resources which begun early and continued to this day that compound the food security in the country. To ameliorate the degraded landscape, today the Ethiopian Environmental Authority has come up with the reforestation plan to establish and manage natural forests and parks, deciduous and woodlands to the tune of 26 million hectares within the coming 5 years beginning 2011.[83]

Production Support Systems

Inputs necessary to empower the Ethiopian peasant produce adequate food for him and ultimately for the market was never in the agenda of past governments. Until recently, the low input applied and the primitive tools used to produce crops had put Ethiopia at the lowest rung of the ladder even by African standard. FAO data also suggest that as recently as mid 2000 cereal yields in Ethiopia were below most countries in the region such as Uganda, Malawi and Kenya. Recently though, Ethiopia is closing the gap and in fact its total production output is less by 20% than India where in 1993-2003 was half of that.[84] The average yield per hectare for wheat now fluctuates between 1.8 to 2 tons per hectare an astounding increase of more than half ton per hectare from a decade ago. This can be attributed to the wide scale use of fertilizer and the rapid increase of improved seeds utilization by peasants. In this case the technical inputs including the advisory aspect of the support system cannot be underestimated.

Tools and Machinery

Despite Ethiopia's early introduction to the ox-plow culture, peasants continued to operate under extreme pressure to perform normal agro-technical operation such as plowing, harrowing or cultivation using animal power. The majority of peasants have been accustomed to using draught animals for plowing, hauling or performing farm chores. In prerevolutionary Ethiopia oxen

---

82 1/ Ethiopia is accused of setting aside some 3,619,509 hectare of land for lease to foreign companies by the Oakland Institute, a Land Grab monitoring group in the world. At present an Indian conglomerate, the Karuturi Global has begun working on portion of the 100,000 hectare of land leased in 2006. To facilitate mechanical cultivation the land was made bare of its natural vegetation. 2/ In April 2013, at press conference in Brussels, Ethiopian Prime Minster Hailemariam disputed the allegation that Ethiopia had made available millions of hectares of land to foreign corporations. He made it clear that not more than 400,000 hectares were contracted to foreign developers.

83 EPA, Environmental Management Program, an Accelerated Sustainable Development to Eradicate Poverty 2011- 2015, (Addis Ababa: 2010), 17.

84 Stefan Dercon, Ruth Hill Vargas and Andrew Zeitin; Revisiting Agriculture Led growth in Ethiopia. (DFID-UK., 2009), 8.

ownership by peasants was as follows; 30% owned no ox whereas those with one or pair of oxen amounted to 50 percent. Today the situation has not improved significantly as peasants without oxen continued to grow in number. The 1980 estimate put peasants owning pair of oxen at 29% and an equal number had no means to plow the land except through share system or on temporary hire bases.[85] A study conducted in mid-2000 in the central highlands arrived at similar conclusion where the proportion of oxen less peasants was one-third of the entire population sampled.[86] This is an indicative that situation has not improved drastically to call for a celebration.

Recent estimate put the livestock population as 111 million with cattle alone making close 53.3 million. Obviously this puts Ethiopia in number two positions in Africa and perhaps the seventh on world scale.[87] It is ironic, though a country reputed to have an excess of livestock per head than any country in Africa except the Sudan has to suffer from lack of an important means of production in subsistence farming, draught animals. Unlike in many parts of Africa where *Trypanosomiasis* prevails in grand scale, in the Ethiopian highlands and in large parts of the lowland this deadly cattle disease is virtually absent, hence for cattle to flourish spectacularly in this part of Northeast Africa. Generally speaking the Ethiopian region in particular the plateau has a pleasant climate to favor livestock rearing. Indeed it takes one good rainy season after a prolonged drought to witness balance restored among the seemingly destitute livestock population in the country.

Ethiopia's agriculture has been as archaic as one can guess. Traveling across the countryside, one rarely encounters an improved and modern means of agricultural production such as tools and equipment in action or sitting idle. Occasionally and in few pockets of farmlands where ones operated under state farm administration, it was not uncommon to see tractors, combines and even chemical sprayers operated by engine power.[88] However, the equipment that the peasant employs to work the land still remains crude. The iron plow, which incidentally is within the means of every peasant to own, a wooden yolk and other wooden accessories to hold everything together are the primary tools of the peasant today. Except for the sickle, ax and hoe made by local blacksmith

85 Provisional Government of Socialist Ethiopia, Distribution of land and farm inputs among the farmers of Ethiopia, with special reference to the distribution of traction power, Bulletin 4, (Addis Ababa: MoA, 1980).

86 Tewedaj Mogues, Must drought and poverty go hand-in-hand? (Madison Wisconsin: College of Agriculture and Life Science University, March, 2004).

87 CSA, (2010/11).

88 Floriculture and horticulture production begun to gather steam in early 2000 in the central and southern part of the country with the majority of the farms operating in greenhouse and glasshouses settings. At the same time outside this sub-sector the use of huge machineries, farm equipment, and tractors are in the increase at spectacular rate.

the rest of farm tools and accessories are of non-metallic nature. Thus, it is unthinkable to expect well-ventilated granary silos in rural towns or greenhouses producing off-season vegetable, ornamental or medicinal crops in large scale. The use of knapsack sprayers or *shadoof* to draw water for supplemental irrigation is nonexistent. Indeed it is distressing to witness countryside devoid of hand-operated harvester/binders or animal driven row seeders. Furthermore, no human or animal drawn irrigation water system long adapted by their neighbors in the distant time, the ancient Egyptians and the Nubians, nor donkey or mule dragged wheel cart are in the scene to attract the attention of casual observer.

The implication of not developing or at least adapting simpler technologies had cost the country an opportunity time hard to redeem. In turn this has caused loss in productivity of an imaginable proportion and also triggered deterioration of the physical resources. Because of improper or lack of soil conservation practices supported by proven engineering techniques, Ethiopia is known to have lost 130 and 182 million metric tons of top soil in 1995 and 2005 respectively. In monetary terms this amounts to 639 million Ethiopian Birr (EB) in 1995 and 766 million in 2005.[89]

Acutely aware of the problem that it has inherited before its ascension to power, the *Derg* in late 1970s imported hundreds of tractors from the GDR and the USSR to be distributed among the state farms and PCs. The East Germany tractor the Fortschritt ZT brand, were the first to appear in the scene with the hope that it will make a difference in the cultivation of large-scale farms found at the Awash, Bebeka, Sinana, and Anger Gutin. To complement the importation of agriculture equipment from abroad a tractor assembly plant was established with the cooperation of the Soviet Union in 1984 at Adama, formerly Nazareth about 100 km south of the capital. The aim was also to manufacture farm implement to replace the traditional tools used by the Ethiopian peasant for centuries. Despite the good intentions of the *Derg* and its ally the Soviet Union, production of these novel machines continued in haphazard fashion satisfying neither the state farms which favored western manufactured tractors and implements nor the impoverished cooperative farms, the pet project of the socialist government. Yearly production target for the tractors at the Nazareth Tractor Assembly Plant (NATP) was generously set at 1000 but the assembly plant produced less than two tractors a week to make a dent in the country's overall machinery inventory. With the subsequent decline of soviet power and with the intensification of the civil strife in the country, the assembly line almost came to stand still. Unfortunately, the opportunity of buying or leasing these assembled tractors was severely restricted for the larger section of the peasant community to take advantage. It

---

89 EfD Newsletter, Green Accounting puts price on Ethiopian soil erosion and deforestation. Environment for Development Initiative (2010).

should be borne in mind that peasant sector is responsible for producing 90 percent of food crops in the country

The machines, the equipment, and many useful tools of agriculture from the Eastern Block countries were in for surprise as soon as they hit the ground. From the start technicians begun complaining about the ineffectiveness of the tractors manufactured in socialist countries further lending ammunition to those who timidly advocated for the importation of western goods in particular the British manufactured Massey Fergusson (MF) or the American John Deer (JD). The very few tractors of this type introduced prior to the revolution were of western made with the impeccable reputation when put on the ground.

Furthermore, the circumstance under which the tractors made their debut in Ethiopia did not settle well with operators who were so much used to western toys. In this regard, the government dictated how many of the tractors and to which farm should be distributed without prior consultation with those concerned. Their aversion to the East European tractors was further intensified by the pitiful sight where tens of hundreds of them found idle in all government supported farms because of minor technical mishap. The major problem of the ZT tractors as was the case with NATP was their hydraulic system, which renders the machines instantly, obsolete the moment they are put in the field. Tested on various soil types and operated under different climatic conditions the ZT as well NATP was no match to MF, which had an excellent performance record across locations. This is despite the favorable track record that the Belarus tractors, the parent brand of NATP had in the 1960s in few areas of Ethiopia in particular the Setit Humera with its heavy vertisolic soils. All being equal, the traction power of these tractors was admirable by any standard. As pointed out earlier, the primary complaint lodged against the Soviet and East German made tractors was that they had faulty hydraulic system and over heat too quickly. In reality, it was more perception or outright bias against the machines, equipment, and tools imported from the Eastern Bloc countries than purely mechanical problem that turned off bureaucrats and technicians alike. Skilled people were less eager to modify parts or improvise to make the machines run ones stalled as they normally do with those tractors imported from the west.

At the time when the tractor assembly line went into operation in 1984, the government turned its attention to address the bottleneck affecting agriculture production and productivity in the peasant sector. It begun to mobilize resources for the design and development of appropriate tools and equipment to benefit the peasantry at large. To accomplish this in the early 1985, the socialist government through the Institute of Agricultural Research negotiated with the United Nations Development Program (UNDP) to setup research and implement development workshop at Melkasa, few kilometers south of Adama, where the Belarus assembly plant, the NATP was located. The mission of the Agricultural Implement Research Center (AIRIC) was to solve the age-old problems that the peasants faced in the production of agricultural commodities. In that case, research work was made to focus on three activities namely, the development of farm power and machinery, Improve soil and water conservation practices and the third was the design of proto types for developing post-harvest technology and food engineering. The goal was to produce mechanically driven equipment (e.g. motor mounted stationery wheat thresher) while the greater part of the plan was to introduce animal powered sprayers, spreaders' dusters, planters, threshers and water conveying equipment. Few years into existence, AIRIC developed many prototypes waiting for mass production. Among the implements worth mentioning were improved oxen drawn plows, thrashers and corn shellers, tie-ridger, inter row cultivator and row seeder[90]. The massive reproduction of these tools and implements was realized more than two decades after AIRC was established. In 2005/2006 alone the government distributed 100,000 moldboard plows to poor peasants.[91]

When the EPRDF took power in mid-1990 it abandoned many of the agricultural policies pursued by the *Derg*, but was keen to continue supporting some projects deemed useful such as those tool and implement manufacturing enterprises. While many of those condemned insolvent enterprises, such as the Kaliti Tools and Metallurgical enterprises were privatized. However, during this time readymade tractors were imported in unprecedented scale. As private farms begun to spread and many of the state owned farms went to the highest bidder, it was apparent that the EPRDF government gave some attention to the existing tractor assembly plant and looked beyond. This is one of the reasons why, in mid-2000 the government begun to explore a joint venture with an Indian firm, the SONALIKA, to establish a tractor assembly plant in the south. No doubt AIRIC, as public institution did get more attention by the government but was not empowered enough to substitute the importation of tools from abroad. Despite the existence of AIRIC for over three decades and the importation of tools and implement in a scale not seen before, the impact of these novelties on the countryside is neither impressive nor hopeful. Simply put, the level of agricultural development in Ethiopia is still low to give credence to the notion that there are some quantitative let alone qualitative change occurring in the agriculture sector. In point of fact, it will need more than manufacturing tools and implement, importing and assembling of tractors to expect the threshold of food self-sufficiency is reached. It will require more

---

90 The author was one of the early beneficiaries of the AIRC invention. With the help of the engineers at the Institute, I was able to secure mechanically operated cottonseed cleaning machine for my business. The mechanical machine had the capacity to produce two quintals of seed per hour (1/5 of ton).

91 Personal communication with officials of the Ministry of Agriculture and with one entrepreneur Abraham Haile Selassie.

of grass root involvement in the manufacturing of tools in a cottage scale, an abundant and cheap energy, concerted land rehabilitation exercise and the curbing of the burgeoning population to a manageable level to expect qualitative change in the Ethiopian agriculture.

Fertilizer Utilization

The use of fertilizer as production enhancing agent came rather late to Ethiopia. Before the revolution, fertilizer application was not a common practice even in the semi-mechanized large scale but rainfed farms of Humera or the irrigated cotton farms in the lower Awash Valley. In this regard the commercial farms were not in any way better than the peasant farms. Generally speaking, the per capita consumption of fertilizer in Ethiopia was one of the lowest in the world. Even by African standard Ethiopia ranks below Kenya.[92] This being the case, it did not escape peasant imagination that, if properly applied fertilizer could double or triple yield.

Fertilizer as such was first introduced to cultivators in Ethiopia in 1966/67 in the Shashemene area of southern Ethiopia and later in the north in a program dubbed as Freedom from Hunger Campaign. The project was supported by FAO and the World Bank and had a limited coverage. Although totally biased in favor of State Farms and PCs, fertilizer consumption increased during the *Derg* era in a substantive way. The consumption of fertilizer steadily increased from 27,000 tons in 1976 to 46,000 in 1982. In this regard only 7% of the farmers were known to have used fertilizer as production enhancing element.[93] By the time that the *Derg* was ousted from power in 1990, fertilizer imports jumped to 102,000 tons.

In truth the use of fertilizer by peasants improved by leaps and bounds with the taking of power by the EPRDF. Between 1993 and 2003 the new government distributed a total of 1.5 million metric tons of fertilizer to farmers. In 2007/2008 cropping season a little over 6 million hectares of cereal, oil and horticultural crops were fertilized using 0.4 million ton of nitrogenous based phosphate fertilizers and urea. Be that as it may, the number of peasants using fertilizer increased from 14% in 2001/2002 to 34% in 2007/2008.[94]. It is an impressive feat by the new government to simply overlook or not acknowledge.

Although the importation of fertilizer increased modestly when the socialist government was in power during the 1980s, the main beneficiaries at that time were the co-operatives and the state farms, which command less that 4% of the total cultivable lands. During the *ancien` regime* the per capita

92 FAO STAT, (2007).

93 FAO, Fertilizer Year Book, (Rome, 1988).

94 1/ CSA, Statistical report on Socio-economic characteristics of the Population in Agriculture household. Landuse, area and production of crops (Addis Ababa, 2003) 2/ CSA, (2008).

consumption of fertilizer for the rural population never hit the mark of 1 Kg per hectare. Over the years, fertilizer application to cereal fields increased modestly from 0.204kg /hectare in 1961 to 1.9 Kg in the 1980s to 16 Kg/ha in 2000.[95] In the 2004/2005 cropping season an estimated 3 million hectares or an equivalent to a third of the land cultivated with cereal crops were fertilized by nitrogenous and phosphate fertilizers.[96] However, this amount to mere 28kg/ha of fertilizer applied to the fields occupied by cereals in both the *maher* and the *belg* seasons. It is however, reassuring to notice that by 2013 the total amount of fertilizer distributed as well as consumed is about 700,000 MT of which cereals received 640,000 MT.[97] This is an increase by about 100% from 2006 or six fold since the early 1990. Despite the impressive increase of fertilizer use in Ethiopia, the fact remains that consumption per capita when compared with many developing countries such as Kenya or India is low to propel the country's cereal production level to any degree of food security let alone export.[98]

There has also been a concern that the blanket use of fertilizer without the proper assessment of the physical condition of the soil is unproductive and may not have impact on yield. To address that concern the MoA with the support of one transformation agency in the country begun drawing *Wereda* based soil fertility map which would guide in the efficient use of fertilizer and may also encourage the introduction and use of other forms of soil enriching agents other than Urea and Diammonium Phosphate (DAP). This is an important first step in the transformation of the agricultural sector, which for a little over half a century relied on Nitrogen and Phosphorus based on blanket recommendation rather than supported by appropriate soil tests and plant requirements.[99]

Improved Seed consumption

Although there is no reliable data showing the exact amount of improved seeds used by peasants prior to 1990, estimates put that less than 2% of the entire farming community in Ethiopia employed selectively developed improved seeds for cultivation. This is separate from the industrial crop, cotton, which was accorded the highest priority in terms of inputs including the supply of 90%, improved seeds. Earlier estimate on the use of improved seeds by private

---

95 Country profile, Agriculture and Food-Ethiopia. Earth Trends (2003).

96 CSA, (2005).

97 CSA, (2013/2014).

98 In 2009 the fertilizer consumption (kilograms per hectare of arable land) in India, Kenya and Ethiopia stands at 166, 32 and 8 kilograms respectively, (World Bank Report 2010).

99 Personal communication with Dr. Tadelle Gebreselassie, soil fertility expert and Regional Project manager of EWUAP of the Nile Basin Initiative (NBI).

holding shows that cereals covered a mere 0.75% of the cultivated lands and another 0.64% for the permanent crops.[100] However, the use of improved seeds by peasants increased modestly since the late 1990s. The 2004/2005 statistical report show that the areas covered by improved seed ranged between 3.2 and 3.8 % of the total cultivated hectares of 10.8 and 9.6 million hectares respectively.[101] Although no dramatic change is recorded, nevertheless the use of improved seed has been steadily in the increase with 465,000 hectares covered during the 2008 season alone. This is equivalent to 4% of the total cultivated area of the country. All in all the total amount of improved seed used during this period was 21,951MT. Almost 90% of the seed distributed was cereal with maize, wheat and Teff taking the lion share. During the 2010/2011 cropping season for both the *meher* and the *belg* season an estimate ¾ of million hectares were covered with 37,368MT of improved seeds.[102] Latest information show that since 2011 areas covered by improved seeds toped one million hectare[103]. It is very impressive when compared with the previous two decades of the *Derg* rule and certainly the *ancien regime*. Still Ethiopian agriculture is in lag phase and need to change pace to meet the enormous challenge. After all, the total cereal seed requirement to cover the entire farmlands in Ethiopia is in the range of 400,000 to 500,000 Metric tons.[104] At the moment less than 10% of the cultivated lands are planted with seeds other than landraces.

As we shall see later the problem of improved seed utilization may not be the lack of it but the inability of the state agencies to multiply, package and deliver it to users. Except for few hybrid maize and selective horticultural seeds, all are handled by the National Seed Industry Agency, predecessor of the Ethiopian Seed Corporation, a parastatal organization established in 1978. For example from the early 1960 up to the dawn of the twenty first century, 44 varieties of bread wheat has been approved for release of which the most significant in terms of popularity among farmers were *Laketch, Dereselign, Enkoy, Pavon 76, Chenab 70, Blue Jay, Dashen, HAR 407, HAR 416 Mitikie and Shinna.* Although many were developed as location specific and therefore not recommended to cultivate in areas deemed unsuitable for their development. However, few proved to be stable across locations. Hence, the major drawback to their widespread use lays on the lack of mechanism in place to multiply and distribute them way beyond their theoretically recognized adaptive locations.

100 Mulat Demeke, Ali Said and T.S. Jayne, Promoting Fertilizer Use in Ethiopia: the implications of improving grain market performance (Addis Ababa: Working paper 5. Ministry of Economic Development and Cooperation, 1997).

101 CSA, (2005).

102 CSA, (2010/2011).

103 CSA, (2013/14).

104 Techane Adugna, Birth of The Ethiopian Seed Trade Association, Seed Info, (Addis Ababa: Ministry of Agriculture, Ethiopia. Newsletter of the WANA Seed Network - June 2007).

Thus, the newly released varieties remained confined in few localities to be of any use to farmers at large. Many of those promising materials therefore, outlived their usefulness while farmers continued to employ the indigenous or landrace season after season. Suffice to say, 95% of seed sources comes from previously saved seed of non-improved landraces.[105] It is customary practice for peasant to select best progenies from the population for next year seeding.

The absence of well-developed infrastructure and the weak purchasing power of most peasants in Ethiopia make the widespread use of the exotically developed hybrid seeds difficult task to undertake. According to the FAO/WFP Crop and Food Supply assessment the use of hybrid maize among farmers in 1999 did not go beyond 1,450 MT, abysmally low even by African standard.[106] A decade later the use of hybrid maize by farmers quadrupled and the land covered by improved cereal crops increased to 322,819 hectares a mere 3.5 percent of the total land occupied by cereal crops.[107] Hybrid maize production in 2010/11 was estimated at 19,300 MT.[108] However, the wide scale use of improved seeds by ordinary cultivators can be short changed if complementary inputs are not forthcoming.

Almost all exotic including locally developed hybrid cultivars are highly responsive to inputs especially fertilizer in order to demonstrate their full potential. They also require an optimum amount of moisture to thrive and give maximum yield per hectare. Obviously, this poses a serious challenge to a subsistence farmer whose plan of action always revolves around whether enough moisture is available during a particular cropping season. Thus, insufficient moisture or in some peculiar cases the excess of it, render fertilizer use precarious. Survey conducted among peasants in some of the marginal rainfall areas of the country indicates that eighty percent of peasants base their cropping pattern not on market consideration, availability of labor, oxen power, availability of seed, degree of weed and pest infestation risks but on rainfall prospect or conditions.[109] In many regions of Ethiopia in particular the north, moisture appears to be the primary production-limiting factor.

---

105 Ibid.

106 FAO/WFP, Crop & Food Supply Assessment (1999).

107 Abebe Atilaw and Lijalem Korbu, 2010. Recent Development in Seed Systems of Ethiopia. (Ethiopian Institute of Agricultural Research, Debre Zeit Research Center, 2010), 17.

108 The 2005-2010 Annual Report of the Ministry of Agriculture and Rural Development, (Addis Ababa: 2010).

109 Tilahun Mulat, Indigenous risk management strategies of small farmers in the central Rift valley of Ehiopia. In food security, nutrition and poverty alleviation in Ethiopia: problems and prospects. (eds.) Mulat Demeke, Wolday Amha, Simeon Ehui and Tesfaye Zegeye. Proceedings of the Inaugural and First Annual Conference of the Agricultural Econmics Society of Ethiopia. 8-9 June 1995, (Addis Ababa, 1995), 101.

Introduction to improved seed by peasant requires diligent work from the extension agency. This is equally true with the locally developed composite or open pollinated improved seeds that the peasant is virtually ignorant about. In order, to ensure the widespread use of improved seeds by peasants it is incumbent upon the state agencies to provide rigorously tested packages in particular fertilizer. In any case, their chance to wider use by peasants is improved when planted in areas endowed with abundant rainfall or where supplemental irrigation is available. Availability of credit and the means of delivery of seed materials to fields are important factors that need to be addressed as well.[110] Moreover, few appealing innovations such as crafting unique crop physique with versatile characteristics through conventional breeding methods could help peasant adopt new varieties for cultivation. For example in the highlands where grazing lands are shrinking and processed animal feed is almost nonexistent peasant prefer crops that leave good amount of crop residue or with potentially high biomass.

As allude earlier, the majority of Ethiopian peasants are novel to the use of fertilizer and crop protection inputs to enhance the potential of the improved seeds to their fullest maximum. In exceptional cases, the combined use of seed and fertilizer could have its cultural drawbacks in some regions of the country. For example some of the northern peasant view fertilizer as an inhibiting rather than enhancing factor to seedling development. Basically this stems from that fact that application of fertilizer in particular urea in moisture stress fields leave a physiologically induced burning effect, stunt growth followed by complete wilting. After years of unsuccessful trying, Saskakawa-Global 2000 and the government have acknowledged that this approach is not appropriate for the low potential, semi-arid areas. Instead compost as nutrient source and more importantly as soil ameliorating agent with its high water retention capacity was found to be more appropriate to the drought prone areas of Tigrai and perhaps other regions with similar agro-climatic conditions.[111]

The fact remains that improved seeds as such may not have significant effect on yield unless accompanied by adequate moisture and fertilizer. However, to dispel the actual or perceived notion that fertilizer may have an adverse effect on crop growth when applied under moisture stress conditions, it might prove useful to continue with the experimental trials using different fertilizer regimes under a variety of moisture regimes. The erratic and sometimes insufficient rainfall throughout the country makes such exercise indispensible.

---

110 The emphasis put by the government to extend roads and highways throughout the land may eventually benefit farmers adopt the use of seeds and fertilizers easily.

111 Sue Edwards, Arefayne Asmelash, Hailu Araya and Tewolde Berhan Gebre Egziabher. 2007. Impact of Compost Use on Crop Yields in Tigray, Ethiopia. (Rome: Food and Agriculture Organization of the United Nations, 2007), 8.

Plant Protection

Peasants in Ethiopia rarely use chemical based plant protection measures that enable them combat biotic and abiotic factors. Instead they rely on hand weeding and for raw crops such as maize and sorghum inter row cultivation is practiced using oxen driven plows when the seedling is at knee high. In the past when the farming system was stable and free from extraneous factors such as state interferences, wars, drought and food aid, farmers in Eritrea and Tigray plowed the land two to three times a year to combat pests and weeds infestation. Leaving the land fallow for one cropping season was also another technique of letting weeds grow unchecked to be later plowed under before the onset of the rainy season. It is easy to deduce from this argument therefore, that the use of chemicals to combat pests and pathogens was and still is a novelty in Ethiopia. To date, the modest number of commercial farmers including the burgeoning flower farms that begun to expand after the fall of the socialist regime 1990 consume chemicals imported into the country. The imported chemicals are predominantly herbicides with fungicides coming next. The total herbicide consumption between 1995 and 2001 is estimated to be 1495 MT with fungicides and bactericides taking small share. During the same period not more than 700 MT of insecticides was consumed in the country. In 1999 about 385,000 hectares of land were sprayed with pesticide. In a span of a little more than ten years pesticide covered fields expanded from 385,000 hectares in 2000 to 2.2 million hectares in 2011, that is, an increase of seven fold.[112] Today cultivated fields covered by pesticide reached close to three million hectares.[113] This is indeed a feat considering the fact that few decades ago pesticides and herbicide use by peasants was virtually unknown. Nevertheless, the figure indicates that only 25% of the cultivated farms in Ethiopia are covered by crop protection chemicals specifically pesticides.

Invariably, Ethiopian peasants hand weed their fields with frequency of no less than three times per growing season. Such cumbersome activity calls the support of the entire family including neighbors or distant relatives whenever available.[114] In some instance especially for corn and sorghum planted fields farmers employ a single row inter cultivator drawn by draught animals. This operation is carried out on row planted fields but not on the broadcasted ones, as is mostly the case. Peasants rarely plant their crops on row, as there are no row planters one can speak of. This is the calling for AIRIC, the tools and implement generating center to come up with prototype of row

112 CSA, (2010/2011).

113 CSA, (2013/2014).

114 This safety net of family, friends and neighbors network is today in serious jeopardy in Eritrea where a quarter of million young and middle aged men and women are tied up in Eritrea's National Service and equal amount living in refugee camps and scattered throughout the globe.

planters that would ultimately facilitate inter row cultivation of 2 to 3 spikes at least. Some obnoxious weeds such as the parasitic weed, the *Striga hermontica* in the Gash/Humera plain are difficult to control using either animal drawn indigenous cultivator or tractor-mounted tines. It is known that single plant can produce as much as twenty thousand seeds and staying dormant in the soil for nearly 10 years without losing their viability. Striga is formidable weed difficult to control using tractor mounted cultivators let alone hand weeding. The parasitic weed is found spread in Tigray, Wello, Gonder, Gojam, North Shoa and Hrarghe. Similarly, wild oat is another peculiar weed difficult to combat by simple hand weeding or deep tractor cultivation. In some areas of highland Eritrea and northern Shoa, farmers stopped altogether planting species related to oats in favor of oil seeds and broad leaf crops, such as legume. Planting alternate crop does not necessarily eliminate the infestation of fields by oat weeds but ease hand weeding operation by peasants considerably owing to the distinctive morphology of the weed in question.

Weed loss assessment trials conducted in many of the research centers indicate the superiority of chemical application than hand weeding. But for the reasons allude earlier, peasants are not in a position to use pesticides, herbicides and fungicides to combat the disproportionately existing co-habitants such as insect pests, weeds or fungal diseases. Given the financial capability and the limited technical knowhow inherent in peasant farming the adoption of Integrate Pest Management (IPM) practices appear to be preferred option. The practice of IPM entails the intelligent application of pest control methods using combined chemical, biological and cultural inputs.

Agriculture Research

Structured agricultural research activities begun a little over half a century ago. By late 1960s rudimentary agricultural research infrastructure was in place to deal with the diverse agro–climatic conditions that the country was endowed with. Some had sub-centers focusing on specific problem peculiar to the area while the rest specializing on commodities that are common to an area. The format of the experimental designs was more or less similar to that of the international research institutes the world over. Many of the cereal or industrial crops were conventionally home breed and in some cases introduced from international institutes at their early generational stage. These early progenies were mainly put under replicated trials and tested across locations and seasons. Experimental data were often subjected to statistical analysis to enable researchers arrive at conclusions based on the set of hypothesis. The thrust of the tests was to determine whether the materials are drought tolerant, high yielding, pest repellent, or resistant to lodging. Most often, outdoors research are seldom complemented by indoor testing in laboratory facilities or green houses very common practice in many developed countries. Such facilities are at the bare minimum in almost all the research centers of the country. In that case, research activities in Ethiopia may not be strongly guided by clues and hints that preliminary investigation provides in controlled environment such as laboratories or greenhouses. In other words outdoor experimental trials are not complemented with off-field practices instead are left at the mercy of vagaries of nature that could skew experimental results and outcomes.

Among the seven Research Centers established by the Imperial government, the bioscience laboratory at Holletta was relatively well equipped. In this laboratory nutritive value of crops and their chemical composition are determine. Also analyzed in the laboratory are the fertility status of the soil, its structure and morphology. Animal nutrition related laboratory studies were also conducted at Holletta. The remaining six major Research centers had basic laboratory equipment that one may find in pre-university educational facilities in the developed world. During the last decade efforts were made to equip laboratories with the facilities enough to assist researches in their endeavor to determine cultivars with high protein or oil content, stalk strength, disease resistant or grain quality.

When the *Derg* took power in late 1974, the focus was to expand experimental sites in as many regions as possible. This was done in earnest and in a span of 10 years the Research centers and their satellites grew two fold. Despite the opening of new sites across the country practically no improvement were observed right away with respect to personnel, equipment and laboratory facilities. However, during this time the National Animal Health and National Soil Laboratory was established to provide services to users. To support the Villagization and Settlement programs initiated by the government, few more research centers were also added. Inexperienced staff fresh out of agricultural colleges manned almost all of the newly established research centers. Although not exactly an ivory tower in the midst of backward rural setting, nevertheless the centers without exception were more of a repellent than an attractive force for the rural peasant and the newly established settlement farms launched in the 1980s. In the mind of the peasants the research centers had little technical solutions to offer to the enormous problems that the farming community faces. Ironically the doubling of the research institutions and the establishment of some of the vital laboratory facilities occur at the time when a generation of educated elite where either killed and imprisoned by the *Derg* or left the country clandestinely.[115] It was also a period where expansion of higher learning institutions virtually remained stagnant.

There was one basic flaw that made the research activities in Ethiopia somewhat irrelevant to the need of the peasant farmer. For one, most of the

---

115 The reason for this was that the upheaval that followed the overthrow of the monarch consumed many of the agricultural experts to be useful in building socialist Ethiopia. Worst of all the annual number of entries to the only agriculture college available then was as low as 200 with half of them making it to the finish line, the graduation period.

thrust of the research programs was aimed toward tiny but highly mechanized agricultural enterprises operating under ideal conditions. Here I refer, to farming units operating in an adequate moisture environment, with sufficient agricultural equipment, and basic inputs necessary to boast production. The adoption of such research strategy had more to do with blindly following the international research protocols and methods and the overwhelming desire to cater for large-scale commercial farms. Understandably, the drive to establish new research centers, the plan to make it workable and the fund needed to support it partially came from international agencies facilitated by multilateral agreements. Ironically, the proportion of large-scale farms to that of the small-scale peasant farms to date remains miniscule.

Research on Livestock was also limited in scope focusing in feeding regimes with little relevancy to the actual pastoral and peasant production environment. In this regard the feeding experiment trials are mostly based on industrial waste products practically unavailable in the country at the time. There is little doubt that agriculture research in Ethiopia is crop biased and caters less to animal based farm activities.[116] According to the Ethiopian research organization bulletin out of 110 experimental projects executed in 2009, 17 were activities related to livestock investigation. But if one reduces it to simple ratio of crop related research vs. animal husbandry it is about 4 to 1.[117]

Despite the emphasis put on crop experimentation, however, little if at all of the research results have trickled down to benefit the masses of peasants. As if all the production factors are under control, that is, abundant inputs, equipment, reliable moisture regime, most often experimental trials were designed to determine optimum crop planting dates, weeding schedules, harvesting times, and spacing between and among plants. Undeniably, these are important variables that influence output or yield per unit land; nevertheless, they are secondary in importance when viewed against developing a moderately yielding but drought tolerant cultivars of crops highly sought in many of the marginal rainfall areas of the country. Moreover stable varieties that guarantee optimum yield when environment is favorable and resilient to withstanding adverse climatic conditions are more preferable to subsistence farmers in the developing world. The irregular rainfall pattern, the decline in fertility, the absence of appropriate tools and equipment to tame physical hurdles, the out of reach credit opportunity for the peasant, should have been the cue that the

116 Alemu Ghebrewold, (Addis Ababa: IAR Newsletter. Vol. 12, 1997), 13.

117 Ethiopian Agriculture Research Institute. www.eair.gov (Addis Ababa: September 14, 2009).

research agencies take into consideration when designing experiments. Amazingly, in a farming system where the peasants universally use draft animals as farm power and practice broadcasting rather than drill method to establish seedlings, many of the researchers never posed for a moment to question the validity of their assumptions. One can, therefore, surmise that without easy accessibility to modern farm tools and equipment, agricultural research findings alone does not make much of a difference on the lot of the peasant farmer. Thus, many of the research efforts exerted to date reflected more of solving mechanized than of subsistence farming problems. It would be a while before research endeavors geared toward improving the peasant lot through the combination of conservation practices, affordable farm packages, appropriate tools and equipment would be realized.

To expect definitive answers to many of the agricultural problems that are exclusively dependent on the generosity of nature are as unrealistic as predicting that Ethiopia would be a food exporting country in less than a decade. The uncertainties of rain coupled with other factors render many of the research assumptions hollow. In many cases early planting followed by short spell of drought increase the frequency of pest attack. In such a circumstance it is hard to contemplate as to whether planting early as has been customarily practiced for maize, sorghum or pearl millet but risk serious infestation of army worm, cricket and the stalk borer (*Busela fusca*) or delay planting until the onset of the big rain missing the opportunity of planting this important crop species altogether.

Infestation by armyworm species and periodic locust invasion has been the scourge of the Ethiopian peasant for centuries. Thanks to the DLCO, dedicated to alerting and fighting locust swarms before turning menace for Ethiopia and its neighbors since recorded history, the frequency of their occurrences has been reduced to negligible level. Locust swarm has been for ages formidable pest to peasants across Ethiopia that King Sahle Selassie who ruled Shoa in the early 19th century expressed his frustration to European guest that "The locust are destroying the crops and the priests has been unable by their prayers to arrest their progress....have you no medicine to drive them away?."[118]

Other factors working against the rapid response to the agricultural problems of the country is the time consuming 'rituals' involved in handling the results obtained from tedious research work. Until very recently the procedure followed to release proven packages generated at the research centers to potential users had a rule of its own. Testing materials were subjected to rigorous screening from what is termed in agriculture research lingo as preliminary observation to pre national and national yield trials. Understandably, these various steps of research screening take a considerable

118 Harris, Cornwallis, Highlands of Ethiopia (London: Brown, Green and Longmans. Vol 3, 1844), 209.

time. Packages would not be certified for farmers use before being verified under normal growing conditions away from the pampered research center's environment under which the materials were originally developed. In this regard, technology packages will take close to 10 years before become available to users. Certainly, there should have been a better way of hastening their distribution to farmers once their potential is decisively proved. One case in point was the limited distribution of drought tolerant groundnut variety to peasant in mid-1980s while the material ICG 273 was still in its formative stage of testing across locations. Once proved to be essentially better than the standard varieties that the peasant cultivates, the scientists at the Melka Werer Research Center (MWRC) took the risk of introducing the variety to farmers ahead of time. The result was astounding in two ways; the variety made an impression on those marginal rainfall areas of the country and in the eastern escarpments of the Rift valley where sorghum was the dominant crop became an alternatively most favored newly adapted crop. Although, a number of the crop husbandry and breeding programs initiated earlier focused on the development and/or improvement of composite varieties that are suitable to input-less farming system such as those found in northern and central Ethiopia, under severe financial and manpower constraints few hybrid materials were also developed. At one point Gamma Irradiation on Teff (*Eragostis abyssinica*) was also performed to increase yield. I suppose this is done to overcome the genetic drawback of *Teff* to score high seed yield. All being equal, there is no reason to doubt that hybrid materials can give yield as much as five fold to the indigenous varieties. Field experiments of the past decade irrefutably demonstrated that yield of hybrid maize as hybrid tomato and many others exceed the expectation of researchers. In 1994 out of 250 maize populations received from Centro Internacional de Mejoramiento de Maiz y Trigo, (CIMMYT) the average kernel yield was 7 tons. At the same time, 8 hybrid maize cultivars developed at home and some of exotic nature were tested in the southern region to give an astounding seed yield ranging between 7.3 to 9.9 tons.[119] Today cereal yields in Ethiopia stands at 1.6 tons with maize topping them, at 2.0 to 2.2 tons per hectare. This is a modest if not dramatic increase of seed yield, which has always gravitated between 1.1 to 1.4 tons per hectare for nearly a century.

As explained earlier, invariably all hybrid varieties require heavy doses of fertilizer input to fully demonstrate their potential. Given the financial position of the peasant to purchase hybrid seeds and apply chemical inputs to cultivated fields, it is absurd if not utterly far-fetched to expect their widespread use by peasant. The use of hybrid seeds by subsistence farmers is predictably not promising. The reason is twofold; hybrid seeds are too expensive and beyond the means of peasant to acquire and worse the materials lose vigor after a

119 IAR, Newsletter of Agricultural Research (Addis Ababa: Vol. 11 No. 4. 1996).

repeated cultivation to be useful for a subsistence famer as source of seed for subsequent cultivation. After all it has been traditional wisdom for an Ethiopian peasant to select the best progenies from the population for next year seeding while the crop is still standing.

In spite of being crop biased, the research institutions in Ethiopia have yet to come up with characteristically stable and widely adapted cultivars for grain crops. So far few sorghum, maize, and wheat varieties of this trait are under limited distribution. The most outstanding of them all is *dinkmash* a sorghum cultivar and *laketch* an improved wheat varieties, both with exotic gene base. In addition few non-hybrid composite varieties of pulses and oil crops have been screened but tragically most waiting for distribution to cultivators. The same applies for forage crops as well as horticultural crops. As we shall briefly discuss below the means of disseminating information and the strategy necessary to distributing packages of production remains to be difficult hurdle to overcome.

It is clear by now that research findings generated out of the research centers were not always helpful to the small peasant. The reason is that the line of investigation pursued rarely addressed many or some of the production constraint factors. Most of the research efforts are expended on generating innovations that require considerable resources to apply at the peasant level. Thus, by default the beneficiaries of the newly generated innovations have been the few and economically insignificant mechanized farms run by the state and private developers. The efforts of the agricultural scientist would have been more rewarding had the experiments were geared toward solving soil erosion problems, generating minimum package practices augmented by locally improved implement, selection/breeding of draught animals including the unconventional use of donkey and camel whose numbers are known to be in millions. The breeding strategy that has for so long focused in the development of cultivars for well-endowed environment should shift gear to evolving materials with special attributes to drought/pest/disease tolerance, early to mature and with yield response to moderate application of inorganic and organic fertilizers.

It is important to note here, that research activities carried out to date has limited applicability if at all. Throughout the three decades where agriculture research work was pursued enthusiastically some of the packages, in particular seed, did trickled down to few farmers big or small. Admirably, with all the disadvantages that the Ethiopian agriculture is known to inherit, the experimental stations were able to develop few sorghum, maize and oil crop materials adaptable to the marginal conditions. The limited use of cultivars such as ACV6, ACV3, Melkassa 1,2, and 3, Teshale, Gobye, 76T1, Kelafo 74, E, Roba are testimony to the success of many research scientist who labored under adverse conditions in the past few decades. In addition 58 Durum wheat and 44-bread wheat were architected during the past two decades. Although on a limited scale 15 out of the 44 bread wheat cultivars are currently under cultivation. In 2007 two hybrids of maize and one open pollinated variety was released with the cultivar Wenchi topping them in the list.[120] Although, they may be location specific and needing a favorable environment to flourish, the two hybrids of maize with promising performance are BH540 and BH660. All in all about 400 varieties of different crop types were formally released out of which only 80 varieties from 20 crop species are in use today.[121]

The recent restructuring of the research activity into to two complementary entities, that is, federally managed and regionally handled ones may eventually redirect the research effort into the right path to address the chronic problem of food shortages in Ethiopia.

Agricultural Extension

Numerous variables play in unison to ensure successful agriculture production. Among them extension services play crucial role because of their bridging effect. Innovations both materials and ideas generated at experimental sites can find their way to the fields of production through the relentless effort of the extension services. There are numerous ways by which the materials and ideas are transmitted to farmers. One being, through hands-on method and the rest could be through electronic media dissemination, distribution of written materials, by organizing farmer's day, symposiums and seminars to benefit the peasant. In other words the vitality of the agricultural sector of a country is judged by the extent that this branch plays a role in linking technology generating institutions with those actual users, the farming community. In the absence of this important institution hardly a breakthrough is expected in the agriculture sector.

According to estimates put in early 1975 the country had no more than 660 extension agents serving 50,000 farmers out of the total 30 million people engaged in agriculture most of which requiring peasant advisory service. In contrast the extension agents in mid-1960 were about 120 covering the entire farming population estimated to be 25 million at that time. In the days before the revolution, the absence of structural land reform made extension work as irrelevant as one could guess. In a country where the constitution affirms hierarchal rule and the governance was utterly autocratic nothing short of

120 CIMMYT, International Maize and Wheat Improvement News (Batan, Mexico: Vol.6 No. 1 January 2009).

121 1/ George Jacob, Grassroots Seed Multiplication in Ethiopia, Field Exchange, Issue No. 31, (2007), 21. http://fex ennonline.net/31/grassroots.aspex. 2/ Techane Adugna. Birth of the Ethiopian Seed Trade Association Seed Info, (Addis Ababa: Ministry of Agriculture, Ethiopia. Newsletter of the WANA Seed Network - June 2007).

revolution could address those problems pertaining to agriculture.

In the early 1960s under Freedom from Hunger campaign, the FAO together with the World Bank initiated a combination of input testing and information dissemination programs in the Shashemene and Woliata areas of the Southern Region and the Humera Development Project in the North Western part of the country with modest success. Obviously, the extension work undertaken during the *ancien regime* had served the burgeoning commercial farms more than the small-scale farmer. In fact the importance of extension system to farm productivity was not recognized until the intervention of Swedish support program in the central highlands known as CADU in late 1960s. In a sense it is because of this project and the effect it had on the small farmer at Chilalo district that the MoA was persuaded to embrace an Extension Program Implementation Department (EPID) within it.

The essence of the program envisaged by EPID was that each package will serve 10,000 farmers and each will have 1-supervisor, 5 development agents, five marketing or trading agents. Based on this strategy the extension agency begun its work in earnest when all of sudden its mission was diluted by a wave of 70,000 university and high school campaigners sent out by the post-revolutionary government to the rural areas to "inspire, organize and educate the masses". Because of the land reform proclamation and its trappings, the work of EPID was scaled down and its pilot project suspended. During the 7 odd years that EPID worked independently it has produced valuable data that are still useful today.[122] EPID, which had limited area of operation before the revolution was superseded by Minimum Package Project (MPP I) with the objective of covering the entire farming population in the 14 Administrative Regions of the country.[123] The MPP was mandated to reach peasant through the supply of inputs and the extension of credit opportunities. This program took its course up to 1978 when its life was extended for yet another six years. However this may be, the extension program, which came to dominate the outreach work in Ethiopia for over a decade, was the Peasant Agriculture Development Project (PADEP). Unlike MPP, which was set out to cover the entire geography of Ethiopia, this project systematically classified the country into 8 major agro ecologic zones based on moisture, relief, and population density. The mantra of PADEP was to emphasis on training and visiting peasants, commonly known as T&V.

With the overthrow of the imperial regime the extension activities increased modestly, at least in the traditional farming areas of the highland. Reports by the MoA indicates that, 4,000 development agents were made to

122 Participatory demonstration and training extension program and credit system was the hallmark of EPID.

123 James McCann, People of the Plow: An Agricultural History of Ethiopia, 1800-1990, (University of Wisconsin Press, 1995), 245.

serve 75 % of the 400 *weredas* including the ones in Eritrea and Tigrai where fierce civil war was raging at the time. Under the PADEP program the focus was on the SC members as opposed to the PCs. This new strategy, that is, focusing on the service co-ops more than the commune farms is assumed to be influenced by donor agencies in particular the World Bank, which supported the new initiative. The core program of PADEP was ostensibly to Train and Visit (T&V) members to improve their production skills. Unfortunately, the duty of the extension agents at the time was aimed not only on the technical aspect of agriculture production but also on other duties ranging from home economics to organizing and enticing farmers to enroll in service associations and to explain the merits of PCs, the Commune. Many also took part in political indoctrination and recruitment of local militias to fight the war in the north and east. The effort of the government to expand extension activity as genuine as it may appear was laid bare for the world to see when two powerful episode of drought and numerous minor ones triggered starvation throughout the land. In actuality, the famine of 1984/85 dwarfed that of Wello which became the *raison d'être* for the overthrow of the Imperial Government 1974.

Gone is the policy of the *Derg* that aimed at transforming the Ethiopian agriculture through socialist path of development which later was to be substituted by the policy of mixed economy, implying the marriage of two seemingly incompatible economic systems. The attempt to marry both systems, that is capitalist and socialist mode of production, so as to revamp the economy that almost gone aground is by itself an admission by the *Derg* that the development path charted decade ago was not working. Few years after coming to power, the EPRDF government followed suit but also launched an economic program widely known as ADLI, which is made to focus on agriculture and agricultural based economic activities. Imbedded within the ADLI program was the new agriculture extension system broadly known as Participatory Demonstration and Training Extension System, (PADETES). The system is primarily designed in *show* and *tell* format whereby peasant is engaged in workable technologies generated at the research centers. It is clear that the new extension system adopted by the MoA was influenced by lessons learnt from measures undertaken by private and non-government agencies that took matters upon themselves to disseminate agricultural innovations to farmers.

Furthermore, in an effort to maximize the effect of the services rendered by the state apparatus, areas long recognized as high production potential were once again targeted as priority by the new extension system introduced in mid-1990s. What makes the new system unique is that it also included some areas with chronic food shortages and with relatively limited potential areas in Tigrai and others. Obviously, this well-meaning policy was aimed at benefiting peasant in the marginal rainfall area long neglected by successive regimes. As a result, the number of beneficiaries under the new ADLI strategy jumped from less than 50,000 in the mid-1990 to close to 3 million by the beginning of the new millennium. Unlike PADEP, which focused exclusively on potentially productive areas, PADETES classified the area of operation into three zones based on moisture availability, giving equal emphasis to high, medium and low rainfall areas. This time the long neglected pastoral mode of production was considered equally important as crop based agriculture to deserve attention. According to the five-year development plan of 2010, the projection is that at least 14.6 million peasants will receive extension services by 2015.[124] In real term this amounts to less than 20% of the total farming population. By the end of the 1990s, the number of extension agents reached an impressive number of 14,000 and inputs along other technical packages such as fertilizers, and plant protection measures increased considerably. Given other priorities or festering problems such as the issue of dealing with national stability, maintaining border security coupled with the fundamental problem of less diversified economic base and much lower outputs in those sectors one can surmise that the EPRDF government has performed superbly well.

Complimenting the efforts exerted by the government are other non-governmental agencies notably Sasakawa Global 2000 and the Irish Self-Help program who also are involved in extension work in the Rift valley areas, in Tigrai and in the central highlands. It was these combined efforts that impressed the government to prematurely declare in the fall of 1996 that Ethiopia was in the verge of becoming food self-sufficient.[125] As a testimony to its success the government exported insignificant but symbolic amount of the supposedly excess grains to an African nation. Unfortunately, the euphoria was short lived when in late summer of 1997 the Disaster Prevention and Preparedness Commission launched an appeal for assistance to meet an overall shortfall of 154,000 metric tons in relief food assistance.[126] Later in the year the grain shortfall was upgraded to double of the previous figure and people needing aid relief to 3.4 million people.[127] Unfortunately, the hit and miss talk by officials regarding food self-sufficiency has been repeated more than once to warrant paradigm shift aimed at genuine food security for Ethiopia and the adjacent countries.[128]

---

124 Five Year Development Transformation Plan 2011 – 2015, Ministry of finance FDRE (2010).

125 "We estimate that we will have a surplus of between one-half million and 1 million tons of maize, roughly the equivalent of the grain shortfall we had in 1984 and 1985," Meles said. Michelle Riley, Emory Report, Vol. 49, No. 23. (March 3, 1997)

126 UNDP Monthly Situation Report for Ethiopia (August 1997).

127 Stephen Devereux, Food Insecurity in Ethiopia. A discussion paper for DFID DS (Sussex: October 2000).

128 Since agricultural growth has been in double digit figures for the past two years we might achieve that (food self-sufficiency) hopefully by 2007 may not need food aid. Interview of PM Meles Zenawi with BBC Talking Point. (January 2005).

Lately the policy of EPRDF government has been changed into disseminating without delay whenever production packages become available to the peasant. It considered this as an essential and equally important factor as the timely arrival of rain is to the peasant. Nevertheless, the action taken so far has not translated into qualitative results. This is meant to indicate that the innovation was neither adequately introduced nor widely distributed to benefit the farming community. This lend to the fact that food production in Ethiopia is precarious to give comfort. Should not the annual food aid request by the government proof enough to imply that neither research nor extension have come to the aid of the peasantry to change his/her life qualitatively? Any argument to the contrary will simply prolong Ethiopia's road to recovery. For example, food request for 2.8 million victims of drought placed in early 2011 to offset shortages caused by poor rains in the *Meher* and *Belg* season in pocket areas of the south is indicative enough that the country is still struggling to achieve the objectives of food security. Conservatively, the grain shortfall for that particular year was put at 234,017 MT.[129] The problem of food insecurity and the perpetual need of food aid are compounded not only by the absence of production packages but even if they exist by the lack of mechanism to convey them to the needy at a faster rate. To circumvent the problem of easy access to innovation by farmers the government has recently embarked upon a commendable rural road building program, which is certainly to have an impact on production, and marketing of agricultural commodities soon.[130]

As indicated earlier, peasants in Ethiopia adopted few of the packages generated at the research centers. Some exceptions is the partial distribution of improved wheat varieties, the sporadic use of pest control measures by small-scale cotton farmers and large estates, the adoption of basic irrigation practices in the Awash, the Omo and Wabe Shebelle basins. It is important to note here that the dissemination of many of the packages was augmented by the relentless efforts of non-government agencies, and individuals within the research and extension realm. However, the weakness of well-structured institutions with the mandate to relay innovations as well as to upgrade or at times withdraw redundant technologies including released varieties of crops or cultural practices makes individual contributions futile.

The Case in point is the story of a potentially high yielding wheat cultivar, *Dasheen* that few years earlier proved to be superior to many candidate varieties

---

129 DAG, Ethiopia Thematic Area/Rural Eco. Development and Food Security. 2011 Humanitarian Requirements, (February 2011).

130 By the end of 2010 the total road constructed in Ethiopia reached 103,689 km of which 11,616 km was asphalted, 38,875 km was all weather gravel road while rural road networks is estimated to be 53,198 km. In 1991 Ethiopia could boast a mere 20,000 paved and unpaved roads. Transformation National Journal No. 6; Ginbot – Sene 2005 GC (2013 AD). An interview with the Executive Director of the Ethiopian Roads Authority.

and therefore approved for release to farmers in the Amhara highlands. Unfortunately, few years down the road that is by the end of the 1990s, the variety succumbed to a devastating yellow rust disease-exposing peasant to more hardship. Unable to replace *Dashen* by the next in line variety as quickly as possible, has left a serious production gap to be frantically filled by peasant themselves through other market channels in 1995-97. Despite the conscious effort by the EPRDF government to link research and extension in the service of the producers, there still remains a huge gap between information dissemination and distribution on one hand and the peasant on the other. Admirably there are few wheat, sorghum, maize and oil crop materials already available at hand. However in order to ensure their popularization among users it is imperative that they are available in sufficient quantity. In this regard the role of the extension department together with agency responsible of multiplying and packaging the new innovations become crucial. To circumvent the chronic problem of seed production, packaging and distribution the role of the private sector becomes indispensible.

Whatever economic reform measures were taken by the socialist government or its successor, the gap between technology generation and dissemination remained as wide as ever. As a result the Ethiopian food inventory has reached at the lower end and the specter of famine continues to loom over sections of the country. For Ethiopia to come out from such a precarious condition it needs to devise bold measures such as it has prepared to implement today. In the long run, there is no doubt that some quantitative change may be in the offing especially now when the ratio of development agents to the peasants requiring advisory service seemed to have reached to a discernible level. If the ambitious plan to graduate 55,000 development agencies by the year 2008/2009 is fully realized, one cannot be indifference to the change that this will bring to the life of the peasantry.[131] True enough by 2008 the number of extension agents reached 67,000 covering almost all *weredas* of the country.[132]

To solve the problem rooted in structural deficiency, severe financial and manpower shortages, the government is emphasizing on an approach where cultivators are grouped into a system known 1 to 5 where one early adopter will take the responsibility of transferring knowledge and convincing others follow suit. This is supposed to be achieved through the use of demonstration plots laid out with recommended inputs supported by the best management practices generated at the Farmer Training Centers (FTC).

---

131 Commercialization of Ethiopian agriculture: Extension service from input supplier to knowledge broker and facilitator Improving Productivity and Market Success of Ethiopian Farmers Working Paper No.1.

132 Ken Ohashi, Sustaining Growth: Ethiopessimism? World Bank Country Director for the Ethiopia and the Sudan, (World Bank, 2008).

The FTC that is located in each administrative zone of the country is an instrument designed to narrow down the gap between the generation of technology and delivery. The mandates of the centers are to transfer innovations and knowledge, help develop skill, and provide technical support. With the number of the extension agents soaring to thousands, they could also get involved in other aspect of the peasant life such as in non-farm activities that could supplement farm income. [133] At present there are about 30 Agricultural Technical and Vocational Education and Training Colleges responsible for preparing development agents to deal with the challenge of extension work. Moreover over ten thousand FTC is believed to exist in Ethiopia today.

Water an Abundant Resource: hydropower and irrigation

Water is no more considered as an inexhaustible commodity. The phenomenon of water shortage is nowhere acutely felt than in many nations of Africa whose lack of both capital and the knowhow to exploit water to the benefit of their citizens exacerbated the situation further. On the other hand the highly developed global North with its modern technological capability has been harvesting water from both surface and subsurface stratum for close to a century. As the rate of water extraction increases to satisfy demand for household, irrigation and recreational uses, so was the capital required to meet this insatiable demand. However, it took the drying up of water wells in what was once fertile, albeit semi-arid California's Central Valley for ordinary people to realize the gravity of water shortages. This phenomenon is not confined to parts of North America alone but seen widely occurring in central Asia, Central part of Africa and Ethiopia itself.[134]

Experts long ago have warned that unless a judicious way of utilizing water is devised, life in the planet earth could be in Jeopardy. It is also ominously predicated that wars in the twenty first century will be fought on water not oil.[135] Currently many countries including the USA, China, India and Mexico are experiencing water shortages especially for agriculture use. In this regard Ethiopia and Egypt do not see eye to eye on the issue of water apportionment

---

133 The author as private consultant designed Farmer Training Center for the Gambella Administrative Region in 1997.

134 *Alla*, a strip of horticultural farmland in southern Eritrea used to extract water from borehole of less than six meter deep in the first half of the twentieth century. Today, most of the boreholes are dry and those with financial means to sink boreholes go deeper more than 100 meters to hit water source.

135 Ismail Serageldin, Water Wars? (World Policy Journal, December 2009).

from the waters of the Nile River[136]. Nevertheless, both countries including the Sudan have signed a tripartite agreement to use the upper tributary of the Blue Nile for hydropower purposes that will ultimately benefit all. Despite the outwardly cordial relations between the Nile Valley states with Ethiopia as a vanguard and Egypt the beneficiary of the Nile river (water) since the dawn of history, the apportion of water from the Nile remains contentious and could be the underlining cause for destabilization in the region.

At the time when human advancement in science made it possible to convert saline water into useful commodity, Ethiopia is still at the initial stage of harnessing its surface water potential for good use. The country has 12 major river basins and an equal amount of inland lakes and water bodies. The perception widely held by many that Ethiopia is an arid country devoid of water is therefore, superfluous. Water, as much as 123 Billion Cubic meters is discharged annually from the high plateau draining toward the eastern and western basins to support life in Ethiopia and beyond.[137] During the peak flow, that is, from mid-July to mid-August, the Abbay, the Omo, the Tekeze and the Baro rivers alone discharge about 15 Billion cubic meter of water.[138] Life among the pastoral population would have been unbearable without those rivers which from time to time overflow their banks to create lash green land in what is sometimes categorized as horrid climatic zone. But by far the major beneficiaries of this water resource are Egypt and the Sudan, which receives larger portion of their waters source from the Blue Nile.

The catchments area of the Abay is over 300,000 KM^2 assuring a continuous supply of water to Egypt and the Sudan. Both countries have a combined 5 million hectares of land under cultivation mostly using the Abay waters. In contrast the land under strict irrigation regime in Ethiopia excluding the limited large scale irrigation schemes have reached about 166,384 hectare.[139] In Ethiopia the potential for irrigated agriculture is immense and may reach about 5.1 million hectares. Good reason for the EPRDF government to be a leading proponent among the Nile basin countries for the equitable share of the river. It has persistently requested the revision of the old treaties and riparian laws that left the country in a very disadvantageous position. The demand for the rightful share from the river which until now has been

136 Despite diplomatic niceties between the two countries, Egypt will use its political leverage to dissuade Ethiopia from having an upper hand on the apportionment of water from the Nile. It is rumored that Egypt is back to its old game of political maneuver to destabilize the country by supporting Eritrea and Ethiopian dissidents numbering in thousands.

137 Gebre Tsadik Degefu, The NILE: Historical, Legal and Developmental Perspective (Canada: TRAFFORD Publishing, 2003), 184.

138 Sileshi Bekele, Challenges and opportunities in Capacity Building for water resources, R&D. MoWR/EARO/IWMI/ILRI/ Workshop Ethiopia, (2002).

139 CSA, (2013/2014).

practically neglected by successive governments will continue to resonate among the Nile basin countries numbering ten.

Until recently Ethiopia's draw from the Abay was very low. It is mind boggling to contemplate for a country, which contributes about 80% of the Nile water to make use only 1% of the total whole. The River Nile is the main and only source of water for the great Gezira agricultural scheme in the Sudan located above the southeast confluence of the Blue and White Nile. The scheme has, 900,000 hectares of land under irrigation. Most certainly, Egypt would cease to exist as a nation without the Nile. No wonder why early historians among them Herodotus called Egypt as a gift of Nile. It is no secret that the two downstream countries make full use of the water thunderously pouring out form the mountains and valleys of South West and North West Ethiopia. The long held view that the River Nile is the sole property of Egypt and by association the Sudan is gradually eroding. Ethiopia as a significant contributor of water to the Nile River had played a positive role among the riparian countries to claim what was rightfully hers.

To the credit of the EPRDF government today many of the Ethiopian river basins have their master plan completed and still few others under study. The largest and widest Abay catchment is one of them. In the past years the Baro-Akoba master plan earlier prepared by soviet experts in the 1980s was revised and upgraded by Ethiopian water experts to include both agricultural and industrial development. So far an in-house reconnaissance survey for the Wabe Shebelle and Genale Dawa River basins in the east are in the pipeline. An in depth study is required to employ it for multipurpose projects including irrigation and fishery farms, hydroelectric power and recreation. In the early 1970s the Imperial Government had commissioned a reconnaissance study supported by French technical group on the Wabe Shebelle River shared with the Somali Republic. Moreover, a study to integrate the use of scanty rains together with the flush waters created by infrequent torrential rains is in progress in the Dankali and Aysha dry river basins as well. The initiative of preparing master plan for many of the basins and water outlets has continued steadily.

Although most of the rivers are perennial in nature, nevertheless, the discharge is reduced significantly soon after the rainy season stops abruptly. The flow of the water is at its lowest beginning November till the onset of the big rains in June. Some like the Tekeze and the Mereb-Gash Rivers are so dry with no or little surface water in their bellies during certain months of the year that farmers are forced to dig shallow boreholes on the riverbed to extract water for irrigation. The dam at Tekeze is expected to change all this by regulating water flow to downstream cultivators all year round. Until very recently, Ethiopia had one major dam in the Awash and smaller version at Fincha and diversion weir at Tana Beles to ensure water supply for cultivation all year round. However, in the past decade, an elaborate master plan study was carried out for the Abay Basin, the Tekeze, and the Omo rivers. Based on the study, therefore, the Tekeze hydroelectric and irrigation dam was added to the one at Koka, Fincha and the Gilgel Gibe on the Omo River resulting in much larger dams. Moreover, with the completion of Gibe III, the Abbay Dam and others at Kesem, Tendho, Gode, Kora Dobe, Beles, and Gojeb in the coming 7 to 10 years, the power generating capacity of the country will increase four to five folds. Thus, the electrical supply power of Ethiopia is hoped to increase from its current level of 2100 megawatts to 10,000 megawatts by 2015.[140] No doubt the availability of cheap Energy will make agriculture enterprise flourish by adding values to the produce, which in the past remained off limit to many developing countries relegated to supplying unprocessed agricultural commodities.[141]

With the completion of the Grand Ethiopian Renaissance Dam (GERD) currently under construction below the Blue Nile Gorge, the electric generating power of the country will be envy to many countries lacking the source of such cheap, renewable, environmentally friendly and secure income generating enterprise. Since the beginning of time, and to the astonishment of potamologists, the Abay begun to take its current northern course by gently bending away from its assumed southerly direction to create one of the world's greatest civilizations, Egypt. In that case, Ethiopia has for the first time begun to claim its fair share by building a dam across the Abay. The dam will have some impact on agriculture but will generate 6,000 MW power enough to export to Sudan and perhaps adjacent countries. It should be noted here that when the current government assumed power in 1991 the electric generating capacity of the country was mere 750 megawatts. Chronic energy shortages and frequent outages, notwithstanding, energy output in Ethiopia have increased three fold. It is undisputable fact that the EPRDF has done more than any government before it to quantify and develop the water resource that the country is abundantly endowed with.

Despite its unintended consequences, irrigation water as a primary source for cultivation and/or supplementary to rain fed farming, remains to be the only viable option for Ethiopia to become a food secured nation. It is also considered as one of the safety nets, which lead to increase in labor absorption in agriculture. However, it requires modest investment to commission thousands of earthen dams to catch run off and conserve water. To stabilize the watershed areas that is currently contributing to serious silting not only to

---

140 Five Year Development and Transformation Plan 2010-2015, Ministry of Finance, FDRE, (2010).

141 Marc and Nick Francis, (BLACK GOLD Film Documentary. Picture released in 2006) Watching Black Gold puts into perspective the skewed relationship that exist between the suppliers of primary commodity, the coffee farmer which in effect is excluded from reaping the fruit of his labor and the big corporations and the coffee dealers who siphon the surplus value.

the dams in Ethiopia but also to those downstream dams in Meroe and the Aswan. The Koka Dam built in early 1950s is currently working at less than 60% capacity because of silting problem. Diversion weirs to siphon portion of water from a command position and let gravity wet the relatively low laying lands is not only economically sensible but also doable.

Given Ethiopia's chronic food shortage there is no better choice for it than to encourage proliferation small-scale irrigation schemes tailored to meet the need of the majority of food producers, the peasantry. Exploitation of surface water from small rivers and rivulets diverted through low cost structures is not cost prohibitive. In other words, it is within the means of the government and the community at large to invest on this venture. The sand dams currently in use in the Borona area of southern Ethiopia have one and only one purpose that is, to conserve water within the dry river bed for future use.

In Ethiopia, modern irrigation practices did not come before the 1950s. Although, narratives left by European adventurers in the 16th and 17th century suggests that there might have been an elaborate system of irrigation works on the valleys and slopes of Northern and central Abyssinia.[142] The meticulous and keen observer, Alvarez member of the Portuguese embassy to the court of Emperor Libna Dingel in the early 16th century saw an irrigation system where 'one crop is reaped and the other planted; rich country'[143]. Most of the irrigated areas are today confined in the Awash, the Omo and the Wabe Shebelle basins and other scattered river basins with little to no contribution to food security. Crops raised under irrigation are mostly industrial with cotton occupying the number one place. There is no serious attempt to double crop or practice relay agriculture in those lands, unfortunately. Unlike rain fed agriculture, irrigation being a continuous and repetitive practice it requires an utmost readiness to maximize its use. To engage in irrigation practices it is essential that material inputs together with skilled manpower be in place. Above all it requires

142 1/ William Cromwell Harris, The highlands of Aethiopia: The highlands of Aethiopia described, during eighteen months' residence of a British embassy at the Christian court of Shoa, vol. 3. (London: Brown, Green, and Longmans, 1844), 269, 2/ Walter Chicele Plowden. Travels in Abyssinia and The Galla Country with an account of the mission to Ras Ali, (London: Chichile Trevor, Longmans Green and Co. 1868), 5 and 136; 3/ George Visconte Valentine, Analysis of New Works of Voyages and Travels; Voyages and travels to India, Ceylon, the Red Sea, Abyssinia and Egypt in 1802, 1803, 1804, 1805 and 1806. 1810, (London: Printed for Richard Phillise, 1810), 274.

143 Francisco Alvares, The Prester John of the Indies; A true relation of the lands of the Prester John, being the narrative of the Portuguese embassy to Ethiopia 1520. The Translations of Lord Stanley of Alderley (1881). Additional material by G. F. Beckingham and G. W. B. Huntingford. (London: Published for the Hakluyt Society, University Press, Vol. 1 1961), 255.

unshakable will power in the part of the state to prioritize actions and provide directions to make modern irrigation practices acceptable in hitherto rain dependent culture.

In contrast, the South Asian nations could stand as perfect example where water is treated with reverence and recognized as a source of life to be utilized aggressively for the benefit of farmers. Although the Asian continent claims sixty percent of the world irrigated land in the turn of the century, the Indian sub-continent pride itself of having to manage about 82 million of the total 190 millions of irrigated lands in Asia.[144] Since the downfall of the the *derg* in 1991, there was steady increase in peasant managed irrigated farm. The slow progress of peasant managed irrigated farms can be attributed to the novelty of it but more significantly to the lack of inputs required to run successful irrigated enterprise. The Five Years Growth and Transformation Plan (FYGTP) set targets to develop large and small-scale irrigation in all regions of the country. The Plan, projects that within 5 years beginning 2011 close to half a million hectare of land will be identified, their survey and design completed. The plan also envisages the construction of irrigation schemes of up to 390,000 of which 166,384 hectare was already in use by 2013.[145] The direction chosen by the current government to utilize water resource as a vehicle for national development is encouraging. As a sign of its commitment to the development of water resource, the government elevated the status of the institution from Authority to ministry level.

Irrigation practices require capital and skill to be profitable. Wisely practiced irrigation systems are hard to find, even in the most developed nations such as Israel, the Netherlands and the USA. Nevertheless, irrigation being considered as more or less a controlled practice of water delivery to cultivated fields; it is by far more dependable than rainfed system of agriculture. However, as attractive as it may appear it has also its downside where once productive land can be turned into complete barren and wasteland by excessive water or the lack of it. One of the regions of the world most often cited as victim of injudicious irrigation practices is ancient Mesopotamia the land where humans consciously began to domesticate and eventually cultivate crops.

To achieve food security and perhaps realize the objectives stated in the ADLI, it is likely that the exploitation of water will continue at a rate faster than it is today, hopefully in more judicious way. Despite the claim that there is not a statistically significant trend in mean rainfall in any season in Ethiopia between 1960 and 2006, empirical observation suggests that rain coverage

---

144 Randolph Barker and François Molle. 2004. Evolution of Irrigation in South and Southeast Asia Comprehensive Assessment of Water Management in Agriculture. Research Report No. 5. International Water Management Institute, (2004), 13.

145 Summary of mid-term progress report for the Council of Ministers Information, Ministry of Water and Energy, (Addis Ababa: April 2013).

and intensity has been diminishing in many parts of Ethiopia and Eritrea.[146] Many villages and small towns are suffering from water shortages. Water springs; streams and underground wells, which has been the source of drinking water for both animals and humans, have gone dry subjecting citizens into unprecedented insecure livelihood. Exploitation of water therefore, becomes essential not only to help extricate the countryside from such an abject poverty, but also as a matter of right, to make water available for both human and animal consumption. Despite the injection of foreign aid and the efforts put by the government to alleviate water shortages in northern and eastern part of Ethiopia, potable water remains an elusive commodity not easily accessible to many Ethiopians in both the rural and urban areas.

But conservation of water is more than physically flexing ones muscle to face the challenge. It requires methodic planning all supported by engineering ingenuity and social considerations. Few of the ill planned and ill managed earthen dams built to supply water to many rural towns in Eritrea have failed to meet their expectation. Out of 300 dams located in Eritrea in 2003 only a third had been known to operate efficiently. Some of the dams rarely hold water because of faulty design that fail to consider the extent of the catchments areas that are supposed to serve as a funnel to collecting runoff water. The bursting of the Hazemo Dam in the early 2000 in southern Eritrea is a good indication of poorly designed water holding structure that failed to contain water far less than its intended capacity.

In Ethiopia water harvesting and conserving techniques are in its infancy. It goes without saying, where moisture deficiency is as obvious as in Ethiopia, it is vital to invest on the development of water resources. Undoubtedly, such measures could propel economic growth in all sectors. Water is not only a commodity that can be easily converted into energy but by itself is energy. Water in its purest form gives life to all living things. Water does keep animal and plant alike in rigidity so as to function adeptly. It also provides as conduit to everything that plants and animals require for sustenance. Regardless of our social status or level of development water for human beings is as essential as the air we breathe in. It is therefore imperative that water is conserved by all means necessary to ensure continuity of life. To this end it is paramount to build water storage structures of multiple purposes. Thus, water harvesting need to be regarded as an activity critical in terms of providing water for domestic use, livestock, and development of irrigation at the homestead level.

Despite nature's gift to Ethiopia and the immense water bodies to exploit, until recently it has paid little attention to harnessing its water potential. Prior to late 1950 there was no institution exclusively dedicated to dealing with all aspects of water development. The Water Resources Development Authority

146 ALM Country Level Climate Data Summary, (2009) http://www.adaptation-learning.net/climate-data/ethiopia-country-level-climate-data-summary

established few years before the advent of the 1974 Revolution, acted as a mere repository for water related broad subjects. Some of the earlier works on the development of surface water was conducted through bilateral assistance. The nature of those studies was confined to preliminary assessments of the water resource that had ultimately led to the drafting of master plans for the basins in much later years. In the early 1960s, the US Bureau of Reclamation and the German engineering firm, Lahmeyer had carried out feasibility studies on the largest water basins of the country namely, the Abay, the Awash, the Wabe Shebelle and the Gilgel Gibe. Two decades later, an Indian firm, Water and Power Consulting services, also prepared a preliminary master plan for water resource development for the whole country.[147] To the credit of the EPRDF government, today Ethiopia has detailed master plan for the major river basins that include the Abay, the Tekeze, the Omo-Gibe, and the Baro-Akobo basins. Thus, human and material resources' permitting the country is poised to developing some 2.5 million hectares of irrigated lands and at the same time generating around 20,000 MW power in the 2020s.

There is little doubt in the mind of the students of Ethiopian agriculture that water and soil conservation practices remained to be the single most challenging task to date. The fact that an estimated 42 ton/hectare/year of soil and 100 million M^3 of water are lost annually to erosion and runoff respectively tells a grim reality for one to easily conclude that the problem is profound. It is indeed a formidable task to arrest and consequently reverse the situation. Already there is a clear indication of soil infertility, bareness and complete removal of topsoil in many areas of the plateau. A recent estimate put the status of the Ethiopian soil as 50 percent significantly eroded and 25 percent seriously eroded.[148] In light of this incontrovertible fact, it is important that experimental stations located in both mid and high altitude areas of the country intensify research work on water and soil conservation practices. In order to tackle the problem of such magnitude, obviously the involvement of the community becomes indispensible.[149]

A year to the great famine of 1985, the *Derg* with the help of the World Bank, the UN and other multilateral organizations established the Arba Minch

---

147 WAPCOS, Preliminary Water Resources Development Master Plan for Ethiopia. Final Report. Volumes I to X. Prepared for EVDSA, (Addis Ababa: 1990).

148 Alemneh Dejene, Towards Sustainable Agriculture and Rural Development in the Ethiopian Highlands. Proceedings of the Technical Workshop on Improving the Natural Resources Base and Rural Well-being, 25 – 27 November 2003, Environment and Natural Resources Working Paper No. 17. (Bahir Dar, Ethiopia: FAO Corporate Document Repository, 2003).

149 Currently, the government is undertaking massive conservation work in the four-major regional states (Amhara, Tigray, Oromiya, and SNNPR) and this has now become an annual event. Farmers are providing up to 60 days of free labor for conservation works every year and the government is coordinating such activities.

Water Technology Institute (AWTI) located in the southern tip of the Ethiopian Rift valley. What makes this institute unique is that it deals exclusively with many development aspects of water as related to plants, livestock, and the rural communities' dire need of potable water. Based on its mission, AWTI is mandated to train skilled and qualified water engineers to meet the challenge of water resources development in the country. The curriculum includes the training of civil, mechanical & electrical engineers that will complement with civil construction undertakings in small & large-scale agricultural-led industries. Four years after it opened its gate the institute admitted 2653 students and graduated 905 in degree, advanced and intermediary diplomas. [150] The Water Institute, which upgraded itself to university level in 2003, has conferred engineering degrees to 2,860 students by the end 2015.[151] With the will power and the resources that the country can muster, no doubt it could serve as a beehive to the nations of the Horn of Africa and especially Eritrea known as one of the moisture deficit country. Many of the public universities established after the ouster of the *derg* have engineering syllabus that are complementary to the Water Technology University in the southern part of the country.

Population Factor

Unchecked population growth coupled with continuously diminishing resources has contributed to the slow progress of the country to the extent that a sizeable number of its primary food producers, the peasant has themselves became food aid recipients. Since the early 1970 the number of peasants who habitually depended on food aid for survival has increased regularly from few thousands into millions today. Unfortunately, in many occasions the northern region including Eritrea has been in the receiving end of aid while at the same time population exploding exponentially. Decades of plea for food aid followed by pledges from donors has unquestionably numbed the feeling of sections of the Ethiopian peasant more than the political leadership. As food shortages begun to deepen, the size of the population continued to increase unabated. Neither self-motivated action by peasant nor long term vision by the leaders of the day could stop the population surge from merely 24 million in 1960 to what is today about 90 million people. In the past half century population has increased three hundred percent while food production remained stagnant or at most outpaced by population growth.

---

150 Arba Minch Water Technology Institute. http://www.reocities.com/awti_alumni/About_AWTI.htm# enrollments.

151 ArbaMinch University http://www.amu.edu.et/index.php?option=com_content&view=article&id=544:28th-convocation-contribute-in-nation-building-graduates-told&catid=1:amu&Itemid=244

The staggering birth rate of 2.8% per annum did not only outstrip food production to unimaginable level but also left a horrible burden on tiny minority of able bodied people between the ages of 15 and 50 years to support nearly 44% of the very young below 14 years of age and the elderly, the incapacitated and the infirm sections of the population. This is typical age-dependency-effect-syndrome were few producer are responsible for the maintenance of the majority of the population. More tragically, almost half of the 53% of age structure between 15 and 64 years were at one time, either directly or indirectly involved in the civil wars that raged for close to five decades.

Although pure speculation, one cannot help but deduce that the intervention by food aid donors to fill the gap left by production shortfalls might have inadvertently but artificially kept the ratio of population to food resource in somewhat balance. In addition, the man to food production ratio could have been further skewed had Ethiopia been not ravaged by disease among which malaria and recently AIDS/HIV which took its toll among the population.

In the Ethiopian context population may have a drawback effect and could in fact perpetuate underdevelopment in a most shocking way. This is how it played against steady economic growth and secure livelihood. A typical example to cite is the wanton destruction of forested land to build shelter for human and livestock as well as firewood without meeting the civic or moral obligation to replace what is consumed. Obviously such reckless human intervention encouraged soil erosion to unimaginable level as the case in Ethiopia is today. Ravines and valleys devoid of vegetative cover have low water retention capacity triggering springs and streams to go dry and forcing women and children to trek long distances to fetch water for domestic use. Tragically, substituting animal waste to compensate shortages of firewood has further deprived the soil from being naturally replenished by nutrients resulting in very poor harvest. Use of cow dung and crop residue as a source of energy instead of organic nutrient is blamed for the loss of at least 700,000 tons of grain/year.[152] This is equivalent if not more than the food aid requested by the combined countries of the horn of Africa.

Thus, unless and until some technological breakthrough is in the offing it is frivolous to entertain population as a valuable resource for development. To the contrary uncontrolled population growth could be liability more than asset. Study after study show that people with inadequate supply of basic necessities of life such as food, water, and elementary hygiene are unable to fight diseases causing to their early demise. It is not by accident, therefore, that the life expectancy of many African countries barely reaches 60 years. Most often

152 J. Bojo and Cassells, D., Land degradation and Rehabilitation in Ethiopia: a reassessment. AFTES Working paper # 17, (World Bank, Washington, DC: 1995).

people are unwilling to curb their family size for economic and religious reasons. It is also firmly established that population growth is positively correlated with education, wealth and life style.[153] The less fortunate, the uneducated and with simple life style the more people tend to multiply. Thus, large number of population with limited resource base and traditional means of production tends to over utilize meager resources such as wood, the soil, the grass and whatever surface water is available to nothing.

153 Thomas W. Merrick, Population and Poverty: New Views on an Old Controversy. International Family Planning Perspectives. Vol. 28, No. 1. (2002).

V

POLICY: PAST AND PRESENT

The agricultural sector remains our Achilles heel and source of vulnerability.
—Prime Minister Melees Zenawi,
Few years before his untimely death

An Overview of Past Agriculture Policies

Policies are formulated to guide efforts aimed at achieving meaningful national development goals. In general societies produce goods and services as much as their resource endowment permits but also suggests that a country's institutions and policies adopted are decisive for its economic performance.[1] What is regarded, as relatively successful policy is the one that is flexible but not dogmatic, people friendly and not narrowly based and exclusionary in its vision. Moreover, adopting a workable formula to existing problems will undoubtedly release people's energy with hitherto unseen courage to change, modify or redo in the event the policy fails to yield the desired result. When formulating policy, it is essential that people are persuaded to participate either directly or through their traditional representatives who understand the existing problems better than the most learned persons with the

1 Mancur Jr. Olson, 1996. Big Bills Left on the Sidewalk: why some nations are rich and others poor. Journal of Economic Perspectives. Vol. 10 No.2, (1996), 3 -24.

capacity to design theoretical models that may or may not work. Nevertheless, policy is guiding tool aimed at achieving a set of goals put by the state.[2]

In essence sound policies that are based on the socio-economic realities of a country will have a greater chance of succeeding than the one based on unrealistic expectations. In the case of developing countries like Ethiopia, the external relations that they consciously cultivate could also have a direct bearing on the formulation of policies and the realization of the goals set in motion. This notwithstanding, policies need to focus on the development of both the material and human resources with major emphasis on the later. The reason is that unfettered human spirit will release forces that are capable of generating new ideas and novel things for the good of the society. Without tapping the human resource potential no breakthrough is expected in the field of cultural enrichment, individual self-expression and enlightenment. History shows that it is the curiosity inherent in human beings that is the driving force for inventing, creating and adopting new things all at the same time. One of the primary causes of Africa's underdevelopment stems from edging rather than focusing on developing, cultivating and nurturing its human resources. Needless to say that innate human resourcefulness play decisive role in the exploitation of material resources for society's well-being.

Invariably, institutions that fail to place people's needs; desires and aspirations as the center core of their policies rarely succeed. The significance of policy as an essential guide to steady development is therefore, not debatable. Based on this argument perhaps, one can find a rational explanation to the existing disparity in wealth among nations. In the case of Ethiopia the sluggish performance of the agriculture sector among other things can be attributed to the absence of well-articulated policy to tackle problems related to agriculture production, conservation, and consumption. Obviously, the inability to articulate policies supported by relatively stronger institutions emanates from the rigid socio-political system that Ethiopia inherited from its archaic feudal system encased by an impermeable shell of the most conservative institution of all, the Ethiopian Orthodox church. One can argue, therefore, that right until the abolishing of monarchy there was no consistent policy of any sort targeting agriculture as such to address the visibly stagnant life of the countryside.

The wake-up call to Ethiopia's economic ills and its place in Africa and the world at large came rather late. It took for the leaders of the abortive coup of 1960 to remind Ethiopia that despite its claim of being independent for longer years than many countries in the world has nothing impressive to show in terms material and human development. The poignant statement issued by the young rebel officers almost five decades ago still reverberates today but in

---

2 Sherri Torjman. 2005. What is Policy? Ottawa: The Caledon Institute of Social Policy. PDF at: http://www.caledoninst.org/Publications/ PDF/544ENG.pdf.

a louder tone. It was this incident, a mere footnote in the long history of Ethiopia together with the persistent agitation of the students following the coup, that the monarchy begun to respond Imperiously to the life and death questions of landlessness and unemployment almost at the end of its life.[3]

The Aftermath of the Coup d'état'

The clarion call of the December 1960 rebels and their accomplice, the tiny student body was the equitable distribution of land and the installation of good governance. These demands kept the Ethiopian students busy throughout the 1960s and into the first half of the 1970s. Their student activism was bolstered by embracing leftist ideology which brought them closer to the international student movement whose demand at the time was peace, justice and equitable share of world resource.

After the brief interregnum where fascist Italy was ignominiously defeated by joint Patriotic-British military action, the one-government institution expected to act expeditiously on agriculture reform was the parliament. This body far from championing agriculture reform as a necessary measure to development, instead it stood on the way of reform up to the day it was shamefully locked out of its premises in April 1974 by surging popular revolts. Incidentally, the land reform bill tabled in late 1960s was left unattended for five years and to the chagrin of the young technocrats, who labored very hard to make the draft a reality, was scornfully criticized of its contents. The parliament packed with the landed gentry from head to toe had wrongly perceived that any measure that advocates an equitable land reform or that encourages the promotion of improved technology and affirm justice for the tenant farmer was viewed as inimical to its interest. The land reform bill astutely supported by the USAID, the World Bank and the Swedish bilateral initiative was neither an antidote against agriculture stagnation nor a blue print to a long-term agriculture based national development. It was simply a test of will on whether the *ancien regime* is genuinely prepared to take a new step toward a meaningful socio-economic reform to radically transform the archaic land tenure system that fetter the countryside for centuries.

Prior to 1974, if ever there was policy to guide agriculture development, it was mainly designed to maintain the status quo where a bunch of feudal lords, their associates and a tiny section of the civil servant benefited most.[4] It is quite

3 Ethiopian Radio Broadcast of the 1960 coup d'état 'Ethiopia's history has been renowned for 3.000 years but ...all other spheres of life have seen no changes. There has been no progress whatsoever.'

4 Micheal Stahl, Ethiopia: political contradiction in agriculture development. (Uppsula: Publications of the Political Association No 67, 1975), 63-64. Between 1942 and 1967 land grant unprecedented in modern Ethiopian history was made to reward ex-patriots, returnee exiles, members of the armed forces, the Territorial

clear that, unlike the peasant who given the opportunity was willing to accept any change even if the change is not substantial, the landowners whose share from the fruit of the land was about 50% remained short sighted and insular to consider any broad based national development endeavor. The system they represented cared less of the wellbeing of the land not to mention the peasant folks upon whose shoulder and back lived luxuriously.

The *Derg*, which assumed power in the wake of the popular revolt in the spring of 1974, was the antithesis of the imperial government in every respect. It begun tackling land issues that for centuries plagued the country and showed readiness to address social and cultural inequalities that the old Abyssinian state consciously overlook. Land for a long while was concentrated in the hands of few by excluding the majority. Therefore, it became urgent for the new socialist government to declare that ownership of huge tract of land by individuals is a thing of the past and warned the landowning class from challenging the new edict. Soon after, a policy governing land was formulated granting every person wishing to pursue farming a maximum of 10 hectares. It also forbade renting or mortgaging land and nor allowed the hiring of person to work on the land. The good intention of the *Derg*, notwithstanding, the consequence of the sweeping measures taken to implement the policy proved to be fatal further throwing the country into whirlpool of instability and food insecurity.

The longstanding problem of land degradation, low agricultural output and the festering problem of nationalities and their aspiration for self-determination required a bold measure to address. Unfortunately, the *Derg* has partially touched these nagging problems that Ethiopians were forced to live for centuries. Many of the burning issues that kept Ethiopia at the lower stage of development were left for EPRDF to juggle, some with profound success while many still remain unresolved. From the start, the government that supplanted the *Derg* realized that land is inextricably linked with the recognition of the rights of the various nationalities that make the country. To that effect with a blink of an eye it demolished the unitary system of governance by granting nationalities the right to administer their own affairs.[5] In other wards it endorsed a federal structure where the various nationalities could exercise their right to use their language, exploit their resources and enter into voluntary alliance with others within the federal system.

The measure taken by the new government did not only instill confidence on those perceived subservient to the Abyssinian state but also allow them to

---

Army, and civil servants. The lowest of the low in the category received one *gasha* equivalent to 40 hectares or 99 acres.

5 The right of nationalities and the equality of religion in Ethiopia were enshrined in the 1987 socialist Constitution. The *Derg* short of instituting federal system in the country encouraged nationalities to use their own languages and express their cultural heritage in various national forums. The 1987 Constitution Articles 1, 2, 3, 5, and 35.

make use of the land of their ancestors within the broad guidelines put by the center. Here, the new government policy on land did not diverge dramatically from the policies espoused by the previous governments including the Imperial government where land belonged to the state. However, the new policy adapted by the EPRDF was open to many changes short of auctioning land to the highest bidder. Undeterred by the danger of secession in particular by elements in the Somali Administrative region which for long harbored the sentiment of greater Somali at the expense of Ethiopia or the resource rich Oromia Region with its advocates for independent Oromo state, the government went as far as giving the option of nationalities to secede from the union but in orderly fashion.

According to Getaneh (2002) Ethiopia begun strategizing an economic development path in late-1940. Since the 1950s three policy regimes are recognized. The first as promoted by the imperial regime was market economy based on the infusion of foreign capital with its focus on urban industrial model. In this case the impact of the model will be felt more on urban than rural. In contrast the policy of *Derg* was to follow socialist road of development based on purely centrally planned economy. Here, the commanding height of the economy including the agriculture sector was highly influenced by the intrusion of the state. The third and the last to policy to appear after the takeover of the country by EPRDF was to design market oriented economy with the state taking the lead in major capital layouts and at the same time making the agriculture sector the cornerstone of its policy.[6]

The *Ancien Regime* and the Agriculture policy

Agriculture in post fascist Ethiopia was not given the priority it deserved despite the sector was the backbone of the economy. The land owning class possessed huge tract of lands ranging from 50 hectares to perhaps 100,000 hectares in all the tenancy dominated Southern, Western and parts of the Eastern region. In this regard, one might guess, the royal family and their closest associates, the hereditary nobles stand out to benefit most from the introduction of modern agriculture innovations leading to agro based industry. Sadly this was not meant to be. In all those regions were tenancy was the rule more than the exception; hardly does one discern the presence of modern technical innovations, tools and equipment that tend to promote socio-economic change. In some countries where the big landowners dominate the social relations, it is not uncommon to witness some sort of mechanization accompanied by modern agricultural inputs and even agro based industries doting the landscape. In Ethiopia until the dawn of the present century the countryside was frozen in time with tools and equipment unchanged and

6 Mehari Getaneh, Agricultural Development Models Reviewed.First International Conference of Ethiopian Economic Association, Addis Ababa: EEA, 2002), 3.

outlook still traditional. Undoubtedly, introduction of modern novelties in otherwise traditional societies would have had an impact in the construction of modern agriculture enterprises in the country. This in turn would have helped in achieving food self-sufficiency earlier than later. One can argue, therefore, there was no or little policy guided administrative efforts on land or the infusion of production enhancing technologies in rural Ethiopia for decades before and after the restoration of the monarchy 1941.

Obviously, one cannot expect an agriculture policy to be drafted by landowners who were dead set to stand against any meaningful reform. They are almost the opposite of the Meji dynasty in Japan who pushed for land reform and laid foundation for the modernization of the Japanese agriculture in the 1860s. It appears that there is more to being a landowner in Ethiopia than to think and act progressively for one's own good and perhaps for the country's well-being. The Ethiopian feudal lords the majority of whom were traditional and illiterate were quiet satisfied with siphoning surplus from tenants, commodities such as grains, honey, goat and raw cotton. Sadly, they have neither the drive nor the ability to crystallize market-oriented economy with all its capitalist superfluities.

Unfortunately, the Imperial Government lost the will to completely break away from the centuries old practices of overseeing the land on behalf of the powerful few that has no desire or intention in initiating ground breaking policies to spur agriculture based national development.[7] The stagnation that led to the downfall of the imperial regime was long in the making for the ruling class to notice. At the end, the goose that laid the golden egg was expired beyond resuscitation affecting the landowner fundamentally. How else does one explain the string of famines that took place between 1956 to the widespread northern famine of 1974 without a single far-reaching initiative to avert what is an avoidable incidences. Agriculture, unlike many economic sectors in Ethiopia was undervalued for quite a long time to contribute to the overall national development. This, despite it was and remains to be the mainstay of the population. The end result was massive degradation of land amounting to one million cubic meter of soil being washed every year, starvation in a massive scale and finally a leader less revolution that toppled the pompous monarchy and all those associated with it. Ironically, budget allocated for agriculture remained the lowest throughout the second half of the past century.

A comparative study on the national budget for the periods extending from the early 1950s to late 1960s by Bekelle and Chole (1969) showed that the

---

7 Emperor Haile Selassie ruled Ethiopia as reagent and later as the sovereign for over half a century. In the first quarter of the 20th century Emperor Menelik II and His daughter Empress Zewditu presided over a realm less settled, bewildered and unsure of its place in a rapidly advancing world trying to encroach its impregnable image in both economic and militarily terms.

annual fund allocated to the MoA rarely exceeded 5% of the total budget.[8] The TFYP which run between 1968 and 1973 shortchanged the agriculture sector by nine fold giving it only 10% of the total whole. Yet the increase to the sector by the new development plan was double or 100% when compared with Second Five Year Plan (SFYP), which run from 1962 to 1967. Unfortunately, the peasant sector received only 10% of the total budget allocated to agriculture development, leaving the lion's share to commercial farming.[9]

The *Derg*'s agriculture policy

Decrees, regulations and Practices

The military that soon filled the void created by the February revolution of 1974 came with vengeance to reverse the sliding course caused by feudal mode of development. Although, the popular uprising was not purely rural instigated, nevertheless, the question of land together with the prevailing economic stagnation of the country was high in the agenda of the rioting mobs throughout the months preceding the dethroning of Emperor Haile Selassie in early September. It is also important to note that barely a year had passed since the death of hundreds of thousands of starved people in the north for anyone to forget.[10] It is not by accident therefore, that the regime's first revolutionary act was to nationalize rural as well as urban lands in 1975. Although hailed as a landmark measure by revolutionaries at home and abroad at the time, soon after, the program begun to experience snag in the implementation phase because it lacked clear cut methodology and backup plan should in case the need arise to revise the entire reform program.

Unable to articulate agricultural policy relevant to the socio-economic realities of Ethiopia, the military regime rushed to issuing barrages of decrees that were later found out to be reckless, socially irresponsible and economically disrupting and totalitarian at most. Without doubt the decree to nationalize rural lands was announced in haste. The timeline between its accession to power and the pronouncements made on land reform was a little more than six months. One would have hoped that more time is expended in dissecting the anatomy of Ethiopian feudal system using numerous tools including the Marxist one to arrive at workable reform genre. Ethiopian feudalism has through time evolved into a system where mores, traditions, religious beliefs and rural ethics fused together to prevent one making a clear distinction

8 Asseffa Bekele and Eshetu Chole, A Profile of the Ethiopian Economy, (Addis Ababa, Oxford Press: 1969), 48.

9 John Cohen and Weintraub. 1975. Land and research in Imperial Ethiopia, (1975), 8.

10 A young English report Jonathan Dembelby showed a grizzly pictures of starving people while the Emperor was celebrating his 83 birthday at a gala.

between the 'oppressor' and the 'oppressed' at least quantatively. Invariably, such historical development, unique to Ethiopia begs a home grown solution to satisfactorily deal with it. It is without doubt that the experience of the Russian Revolution influenced the decision making process of the *Derg* and its diehard Marxists to deal with the land issue. In as far as agriculture production is concerned, it is this fateful decision that send Ethiopia deeper into socio-economic crisis of the 1970s and 1980s resulting into periodic famine and unabated land degradation.

Tragically, the tiny and feeble left wing groups were utterly divided to voice their misgivings on the reform for better or worse. While sections of the left group were quick to support the radical reform long advocated by the student movement, few rejected it out rightly not for its lack of symbolic effect or substantive significance but dogmatically arguing that it was not in the nature of the military to institute far reaching changes. The antagonists openly insisted that the task of carrying the goal of the revolution is through the involvement of civil group in the sphere of local and national administration not the other way round. The intransigence of the opposition group annoyed the army who are well aware that 'power comes out of the barrel of the gun' and dismissed their criticism as empty talk. Those who joined hands with the government in this long awaited journey of land reform did so in the hope that they will be able to outsmart their rivals in the left wing camp by sharing power with the Military. As events showed later, they also were hatching plot to send the military back to the barracks and perhaps introduced a reform much closer to the Soviet system.[11] Nevertheless, none but the land gentry objected to the reform unleashed. Certainly, from the standpoint of its vested interest in keeping the status quo unchanged, the measure taken was unsettling to the landowning class.

The intersectional strife that soon followed and the inability of the enlightened to formulate strategy that would benefits the former serfs cost the Ethiopian people dearly. Among the few who dared to argue against the land reform format just announced by the *Derg* and its left wing ally was the newly appointed Minister of Agriculture whose insistence on moderation landed him in trouble.[12] Earlier the short-lived government of Premier Endalkachew, that

11 The *Dergh* ruthlessly hunted members of the MEISON party its old allies against EPRP. In the course of seek and destroy operation many radicals among them those who took part in the so-called 'Paris Uprising' of 1968 perished never to raise again.

12 Dr. Dagnachew Yirgu a noted scientist advised the reformist not to go overboard by nationalizing productive modern private farms run by nationals and expatriates. He insisted that instead of going forward with blanket nationalization of land to distribute land according to the need. True to his social background he remained the champion of the peasant serfs until his mysterious death early in the revolutionary period.

came soon after the forced resignation of the imperial appointed cabinet in mid-1974 came up with similar pronouncement but was out rightly rejected as a lukewarm reform that does not fundamentally alter the existing land tenure system.[13] In short, when the land reform came into effect the peasantry was caught by surprise and the countryside thrown into chaos. Here is the reason why.

Without proper guidance and directives from the top, the implementation of the land reform program was left to the discretion of the already shattered but potentially dangerous bureaucracy with vested interest in maintaining the status quo. What made the process more complicated was the dispatch of 70,000 students from urban setting to the countryside to assist in the execution of the land reform. As it turned out, hostility soon erupted between groups of different political tendencies especially between the students and the government in one instance and among the left on another. No doubt at this time the countryside was turned into hotbed of contention between various forces vying for power. Caught in the middle, the peasantry was left to ponder when the urban rooted madness would end to resume the pathetically wretched life that they are so accustomed to.

Unlike the *ancien regime*, which was not particularly keen in issuing frequent pronouncements, the *Derg*, was obsessed in doing just that. Soon after seizing power and throughout the 1970s and 1980s numerous decrees were issued to regulate production and land use, to set new land tax, and to formulate and reformulate peasant associations, on settlement and movement of people and trade. However, none seem to have a long-term effect on agriculture improvement and few were contradictory to each other to register any visible change in the life of the peasantry. With the *Derg* in power there was one major and few minor Save Our Lives distress calls by the RRC to warrant a quick action by the food aid donors.

The famine of 1985 was a clear manifestation of the failure of the agricultural policy pursued by the *Derg*. None of the measures taken so far were able to mitigate the chronic food shortages that the country has experienced for long. During the socialist reign, the agriculture policy in Ethiopia could best be described as a caricature of those Euro-Asian socialist countries that had socio-economic structures remotely resembling that of a traditional and predominately agrarian Ethiopia. Some of the unique features that distinguished the Ethiopian farming community from their counterparts in Indochina and Eastern Europe were that of the subsistence livelihood that they led. Rough estimate put by various sources in the 1980s suggest that between 12 – 15% of the peasant produced grains reach the market, indicating that the Ethiopian farmer still existed at a subsistence level to supply the market

13 ETHIOPIAN Herald, (Addis Ababa: April 19, 1974).

adequately.[14] Even then the subsistence farming system that kept the peasantry within a narrow margin of existence for many centuries has in recent times begun to break down exposing the peasant and his family to further hardship. Still today many are unable to carry themselves from one cropping season to the other. Coupled with the outdated method of cultivation inherent in the system and his inability to manipulate water for his own benefit put him in a very disadvantageous position to mimic any other agriculture systems with ease. Thus the misguided policy relentlessly pursued by the *Derg* to transform the peasantry who has barely come out from the stifling and parasitic system of feudalism cost the nation irretrievable opportunity time and also perpetuated the long standing image that Ethiopia being the poster child of famine, poverty and ignorance.

What were the most regressive elements embedded in the policy that the socialist regime was blamed most? They include among other things the failure to address the much-needed structural changes with the people as the center of it all. Most often development strategies are leadership driven with no or little participation from the people themselves. Thus, when policies are formulated they rarely reflect people's immediate needs but that of the armchair formulator who has no inkling of what is best for the peasant farmer. In league with the consultants both foreign and local, policies are quickly put into action in the hope that they work as planned. It is unfortunate, however, that projects conceived and implemented within the general guidelines of the stated policy are rarely monitored or evaluated.

The *Derg* in particular was less receptive to useful suggestions much less constructive criticism to amend or improve policies already in motion. It took 15 years for the government to partially admit that the Euro-Asian socialist road of development so aggressively pursued was inappropriate for Ethiopia and as a way out devised hybrid economic policy with the government controlling the commanding heights of the economy and the private sector involved in consumer manufacturing and import export ventures. However, the new policy failed to implicate the agriculture sector broadly. Despite the government's willingness to restructure agricultural and farm price guidelines in early 1988 neither this hopeful sign nor the mixed economy policy launched in 1989 could prevent the continuous decline of agricultural output.

Up to the period where the mixed economy was declared the government focused on three major agricultural activities that are genuinely perceived to be the way out of the quagmire. These include the reorganization of the countryside into communes, SCs and large mechanized estates. In order of priority more attention was to be paid to the growth of communes, the PCs and the state farms. Furthermore, to alleviate overcrowding in the highlands,

---

14 Ajit Kumar Ghose, Transforming feudal agriculture: agrarian change in Ethiopia since 1974, (Journal of development Studies. 22 No 1 1985), 127 - 49.

the *Derg* needed to activate settlement and villigization program that in all fairness was in the pipeline way before it came to the scene.[15]

The Beneficiaries of the Land reform

The land reform as such was never the primary target of criticism at least by the peasantry and by the majority of the Ethiopian left. After all who is to shed tear on the land tenure system that enslaved the peasant masses and sunk the country into an abyss of misery? As allude earlier the relationship that existed between the peasantry and the overlord was no better than the one that existed in pre-industrial Europe and worst. The peasant who in some cases, was required to hand over half of his produce to the landlord on top of being obligated to provide *corvee* and other services of ceremonial importance to him was in constant fear that he may run the risk of forfeiting the continuous use of the land if refuse to toe the line. The reform, therefore, was overdue and necessary if Ethiopia has to break away from the vicious cycle of poverty.

The paradox lays in the arbitrary application of the reform, which failed to take into account the size and the fertility status of the land allocated to each *Kebelle*, its proximity to potential market places, and the formula by which production-enhancing factors are equitably shared. Moreover, the lack of recognition and the utter disregard by the government to the basic instinct of human beings to grow and prosper was troubling. Undoubtedly, actions taken throughout the *Derg's* reign stifled the initiative of the peasant to amount something. Regrettably, the blanket nationalization of land once hailed as the hallmark of the revolution was no longer so but an impediment to transforming agriculture in Ethiopia in a profound way. The rigid but biased government policy on dealing with the various categories of farms, namely, the peasant, the cooperatives and the state owned were harmful. The ironic of it all was that behind the rhetoric of equality and justice that the land reform was alleged to have brought to the peasantry, the rigid control imposed by the state and its relationship with the individual remained unaltered. The reform instituted, therefore, was less than far reaching if not cosmetic to have any direct bearing on the life of the peasantry. In effect the peasants similar to their forefathers continued to be indebted to the central power that used all means to extract surplus including human as cannon fodders in the civil war that raged since the 1960s.

According to the land reform decree of 1975 each peasant association comprising of 200 households was allowed to operate within an arbitrarily defined boundaries of 800 hectares. Although, the land reform decree put a

15 The Third Five Year Plan (TFYP) of 1968 emphasized on the need to relocate people from the high-density area of the North to the lowlands, mistakenly regarded as fertile. Finally the MLRA was given the mandate to oversee the resettlement of peasants in the early 1970.

ceiling of 10 hectare that each peasant is permitted to own within the *kebelle* allotted land in reality all they can own was a hectare at most. Thus, each beneficiary of the land reform was shortchanged. Moreover, the figure was uniformly applied to all without the slightest regard to the variability in soil fertility administered by each *Kebelle*. This in effect implies that many PAs would have to settle for less fertile land with land classification of category IV and perhaps with rainfall of marginal character while some was fortunate enough to sit on productive and easily accessible land.[16] The negative consequence of such simplistic but also unfair land allocation practices is that it did not only foster unequal development within the PAs but also encouraged land fragmentation to accommodate the young coming of age. Nevertheless, the fragmentation of land was universally true for all PAs regardless of land quality or rain dependability. Be that as it may, those PAs operating on productive lands with sufficient moisture to raise variety of crops fared better than the rest. At the end of the day, the peasant farmer regardless of his fortune in the land allocation process was left stranded and unable to lead a decent living let alone move vertically.

Pursuant to the land reform act of 1975, the *Derg* without due consideration or technical justifications nationalized all lands including the nascent but small commercial farms owned by local interest. As proved later this was a serious mistake if not reckless act that ultimately deprived individuals to act independently. In fact it stifled their initiative to flourish resulting in production depression throughout the 1980s. The negative consequences of the blanket nationalization of land, in particular as applied to those indigenous enterprises which begun to emerge in late 1960s was felt very early in the course of the revolution. Production begun to slump and cultivated farms reverted to fallow conditions or reduced in size. These semi-mechanized farms were also an income-generating source to many rural communities. The Setit Humera farms, once the source of foreign revenue earner for the nation went down from its previous size of 300,000 ha to below 10,000 hectares by the time the *Derg* was ousted from power[17]. Similarly, the rapid increase of Haricot bean cultivation in the upper Rift Valley in late 1960s dwindled into nothing more than few scattered farms managed by farmers whose financial capability to invest on farm inputs was totally lacking. In this regard the only exception was the cultivation of cotton in the Awash valley, which increased phenomenally in both acreage and total output dwarfing those of the Baratollo Cotton

16 Douglas Helms, Readings in the History of the Soil Conservation Service, (Washington, DC: Soil Conservation Service, 1992), 60-73.

17 Getahun Bikora, Oil Seed Production and Marketing Situation in Ethiopia (Addis Ababa: Ministry of Trade, Nov 2013). 8-12. Within a span of a little over a decade Ethiopia become the 4th largest sesame seed producing country in the world and the oil crop's contribution to foreign exchange earnings is next to coffee. In 2012 the country produced close to 400,000 MT of the seed.

Plantation in the Lower Gash basin of Eritrea. Sadly enough once nab from the bud these commercial farms never recovered to their pre-revolutionary level to make any meaningful contribution to the local economy or the state coffers.

The most amazing aspect of this period was the spontaneous development of irrigated cotton cultivation in the Middle and Lower Awash Valley by agriculture trained young individuals. To his credit the Emperor paid special attention to these farms and the pioneers who managed them successfully until his ungraceful exit from the scene. In contrast the *Derg* hastily expropriated those farmlands without due process for compensation. Add insult to injury, the government transformed the owners to salaried staff right there and then. The growth of the indigenous commercial farms in the 1960s in a span of short time was not only inspirational to many daring individuals who took up modern farming as an honorable profession but as trail blazers set perfect example that modern agriculture could be a base for national development.

In place of the nationalized farms the government imposed a new management structures similar to that of European socialist system which in effect were bureaucratic in their *modus operandi* but still with relatively sufficient resources to make some difference. In contrast the state farms in Ethiopia lacked the necessary production enhancing inputs and technical knowhow to allow the farms operate at a profitable margin. This is despite getting major share of the country's agriculture inputs and assets such as 80% credit, 82% fertilizer and 73% of improved seeds. Years after their expropriation, many of the farms were pitiful and financially bankrupt.[18]. In 1982 alone the rainfed sesame farms located in the Northwest of the country absorbed millions of birr from the state treasury and used hundreds of thousands of unpaid man-days with nothing to show after the harvest. During these horrible years the human cost was incalculable to say the least. Tens of thousands of peasants from the Amhara highlands were forcefully recruited and sent to the malaria infested region to assist in manual weeding and harvesting operations of sesame without adequate preparation for shelter, food and medicine.[19] Except for the brief attempt to revive the farms, which were once modest foreign revenue earners for the country, large-scale operation was temporally suspended leaving a token few thousands of hectares only to be managed by the State Farm Agency. To confound matters further, the North and Northwest were the hot bed of contention between the Northern Liberation Fronts and the central

18 Girma Kebede, The State and Development in Ethiopia, (1992), 8-9.

19 Wosen Yefru, "The African Challenge to Philosophical Paradigm: The Need for a Paradigm Shift in the Social, Economic, and Political Development of Africa," (*Journal of Black Studies,* vol. 30, issue 3, Jan 2000), 362. Yifru argues some 250,000 peasants died in the agricultural experiment of relocating people to new settlements and to establish producer cooperatives and state farms as well as in forced labor.

government for nearly two decades.

The same is true with the once profitable private farms in the Rift Valley and the Awash basin were cost of operation was exorbitantly higher than any state run farms elsewhere in the country. Unlike rainfed cultivation, these irrigated farms demanded much more attention and required regimented but repetitive practices to yield any sort of profit. Moreover, cultivation of cotton under irrigation is almost impossible without the use of pesticide to control a host of sucking and chewing insects that cotton plant is known to attract. As a result not only was the cost climbing but also the degradation of soil caused by the injudicious use of irrigation practices continued unabated. Despite the manifestation of early signs that the state operated farms were in dire technical and financial troubles the farms expanded exponentially from their original size of 88,200 hectares in 1975 to almost 300,000 in mid-1980s.[20] Be that as it may, the state run farms was regarded by the regime as prestige items and the reaffirmation of socialist resolve rather than profit making enterprises for anybody to emulate.

True to its ideology, the *Derg* was determined to push with projects that give legitimacy to its left wing revolutionary position in the socialist communities the world over. To that effect it set out to establish a Chinese style commune system known as cooperative farms. Three stages were laid out before achieving the ultimate level of cooperatives status where everything including household items was to be managed and owned communally. The hope was that if farmers pull their resources in the first two stages of cooperative existence the chance of them extricating from the cycle of poverty would be greater. Furthermore, by their actions they could be an example where others PAs could follow. To that effect they were much favored by the regime to receive inputs denied to other peasant groups namely the PAs and SCs who constituted the majority and with limited capacity better placed to contribute to the food self-sufficiency. When all is said and done by 1987 there were about to 2,820 PC with a membership of 215,141 of which 255 were registered as full-fledged communes. In contrast the PAs had an approximate membership of five million households.[21]

Until their demise in 1992, the cooperatives together with the State farms received hundreds of millions of Birr worth inputs from the Agricultural Input Supply Corporation (AISCO) and an open credit line facilities from the AIDB. Despite the unusual pampering by government agencies, the cooperatives were unable to stand on their own two feet. In the course of their existence the cooperatives breed a parasitical hierarchy out of the bureaucracy created to

20 Edmond J. Keller. 1991. Revolutionary Ethiopia. From Empire to Peoples Republic, (Midland Book Edition, 1991), 256.

21 Eshetu Mohe. 1988. The Ethiopian Economy. (International Institute for Relief and Development. Food for the Hungry International. Individual households. 1988), 2.

facilitate administrative functions. The biased policy pursued by the *Derg* for over a decade in favor of farmers and farms committed to contributing a mere 4% of the country's total agricultural output is mind boggling.

It is very instructive to show that during its decade and half stay in power, the *Derg* gave a blind eye to the peasant sector by writing it off and not considering it as a pivotal player in the overall drive for food self-sufficiency. Between 1980 and 1984 the *Derg* consistently under budgeted this category of the agriculture sector to less than 4% of the total national budget.[22] This practice of underfunding continued throughout the 1980s except for the early years of the revolution, that is, before it was being sucked into whirlpool of civil wars that ultimately consumed it. In 1975, the *Derg* boldly apportioned 33% of the total capital expenditure to agriculture improvement followed by an additional 8% in 1976.[23]

Resettlement and its Consequence

Like the land reform, which in theory remained to be the single most important measure taken since the 1974 revolution, so was the human settlement program initiated by the government in mid-1985. In principle, the settlement program was launched with the sole purpose of releasing human pressure from the overcrowded and unproductive areas of the country to the relatively fertile lands situated below the plateau. It is also important to note that the characteristics of agricultural marginality was and is more pronounced in the north than in other naturally endowed areas of the West and Southwest of the country. Although human settlement was part of the government program and an item in the 'to do' list, nevertheless the hasty way of its execution can only be described as a backlash to the famine looming over the country in 1984-1985. By some estimate the cruel famine claimed over a million lives. It is therefore, plausible to suggest that the settlement campaign was not well thought of a scheme but a bravado response to the famine looming over the country by the regime, which has just lavishly celebrated the founding congress of its party, WPE. Suffice is to say, that the settlement program was poorly executed to enable settlers be self-sufficient in few years. Like the land reform, however, the methods employed to relocate people from their domicile situated hundreds of kilometers away to the low and mid-altitude areas of the country was so inhumane that it is better left to those who lived to tell the tale.

Given the recurrent drought and poor agricultural productivity in many regions of the country, it could well have been in the mind of the rulers that sometime in the future, the transplanting of people from their origin of birth to fertile areas is inevitable. After all, Article 18 of the Land Reform

22 IMF International Finance and Statistics. (August 1992)

23 Harjinder Singh, Agricultural Problems in Ethiopia. P. 17 (Delhi-110007-India: Gian Publishing House, 1987), 17.

Proclamation of 1975 did provide perfect justification for the massive movement of people that took place at a later date. But the idea of organized resettlement of the land hungry peasants did not start with the *Derg* but the imperial government. Earlier in 1972, the MLRA activated a project aimed at resettling 60,000 landless farmers.[24] However, because of the uncertainty caused by the mass uprising in 1974, the project was temporarily shelved in order to give time for the new government to deal with other major pressing issues specifically the civil war and the Somali invasion of Ethiopian territories in 1977. Without the privy eye of the western world in 1979 close to 50,000 families were resettled in locations long believed to be favorable for pursuing productive agricultural activities.[25]

Judging by the ferocity and the speed of its application, however, one cannot help but speculate that the settlement effort set in motion was conceived out of desperation. It is not surprising that, the program was announced at the height of the famine of 1985 where hundreds of thousands of peasant folks trekked to temporary feeding centers established throughout the north and the central region of the country to avoid death hovering across the horizon. It is relevant to mention here that those unable to make the long journey to the feeding centers simply perished in their villages. Like those left behind, the weak, the old and the young who made it to the feeding shelters also met similar fate. At the height of the famine the death toll at the Batti camp of Northern highland was 50 persons per day. In retrospect, it was this famine, that by all accounts the severest in the 20th century, that catalyzed the government's scramble to finding an 'everlasting' solution to the chronic problem of food insecurity. How humiliating it must have been for the 'communist' regime to continuously seek food aid from the western nations in particular the United States of Ronald Reagan known to be unfriendly to the *Derg*. Thus, the *Derg's* nihilistic approach to addressing the problem of food shortages was to launch the ill-timed and ill-prepared settlement program that inadvertently affected all facets of life in the country albeit, for short time.

At this time the state propaganda apparatus went in full force to convince the prospective settler in particular and the people in general that Ethiopians must take full advantage of the vastly uninhabited and/or relatively fertile land situated in the mid and low altitude regions of the country. In order to realize the lofty objective of 'settling our unfortunate compatriots' in the new lands, people have to make sacrifices including the diversion of resources to support the settlement program. Individual and group contribution of materials including money to famine relief and toward the establishment of settlement projects was touted as time honored traditional Ethiopian practice. As a result

24 Kurt Jansson, Michel Harris and Angela Penrose. 1987. The Ethiopian Famine (London and New Jersey: Zed Books, 1987).

25 Ibid.

basic consumer goods became scarce overnight bringing the whole country in par with Eritrea, which was the victim of it all since the intensification of the civil war in mid-1970s.

Reminiscent of the 1975 DTCC campaign which had propelled the entire secondary school and university students to the countryside ostensibly to assist in the implementation of the land reform, this time as well the government ordered all university students together with their instructors to join hands in an effort to build shelters and hygienic facilities in the designated places of settlement for the prospective settlers. In what the *Derg* termed as a 'payback time', it reminded the students to serve the masses who through hard toil contributed to the relatively privileged status that they enjoy. In an unprecedented move all campuses were shut, classes suspended and scientific experiments in the institution of higher learning left unattended all for the reason of implicating in the devious program of mass dislocation the very elements of the society who would otherwise be a credible witness to its bravado fueled crimes. The weekly held Marxism education forum which in effect remained a mandatory ritual to attend for all workers in government institutions, factories and the peasant associations was transformed into a resource generating platform for the benefit of the settlement projects. Thus, short of declaring moratorium on the wars raged in the north, the government fully mobilized all the resources available toward the success of the settlement program.

In a span of 6 months close to 600,000 peasants from Tigrai, Wello and the densely populated areas of Wolliata, Hadia and Kambatta were moved to the Barro/Akobo and the Blue Nile Basin of Asossa and Pawe, Gambella and Didessa. Except for few sparsely distributed pockets of farm lands the greater part of the proposed settlement areas are inhabited by nomadic people. The proposed settlement sites are flat to undulating savannah grasslands mixed with shrubs and in the case of Gambella with dense woodlands. In addition few of the prospective settlers from the north were scattered among traditional farmers in selected areas of Wellega districts known to have a dependable rainfall and fertile soils. Despite earlier misgivings that minor cultural differences between the host Oromos and the Tigraian guest farmers would trigger ethnic conflicts, the settlers quickly adapted to their new environments with tremendous resilience and stay put for a very long time. This is attributed to the cautious but welcoming attitude of the host farmers and more so to the Psychological acceptance of the new comers to a region that have similar farming systems as theirs. For many new settlers it was also comforting to reside in areas where accessibility to local market was easier and transport facilities relatively abundant.

The settlement schemes elsewhere did not fare well and in many cases completely failed. The primary reason one can cite is that the government lacked the means to make the settlements livable. For one the Ethiopian people were not convinced that a nation at war with itself and with others can pull such a gigantic project without putting strains on their lives and the lives of those very people on whose name the project is conceived and carried out. At the time when the people were asked to make a sacrifice for a non-workable scheme and yet the government trying to prove to detractors, both nationally and internationally, that it could muster all the local resource to make it succeed, is an adventurous measure lacking honesty. Moreover, in sections of the society, in particular the urban poor and the working class where the effect of drought was equally felt, it was cynical in the part of the government to attempt sell such diabolic scheme hatched by the deranged few. In this case, the urban poor had nothing to offer but watch silently as events unfold.

Be that as it may, the effect of the famine was real for any sober person to look at it lightly. Even, after the much-needed relief was in place, hundreds died in the temporary shelters built for the settlers in waiting. In a nutshell, for the settlement project to succeed it had neither the resources nor the backing of the western nations who ironically were maintaining the hungry and the poor in the feeding centers across the northern plateau. This is in contradistinction to the settlement program launched by the EPRDF in 2003 that had full financial backing from the international organizations. The ultimate aim of the current program is to resettle 2 million peasants from the stressed areas of the country to the relatively fertile lands at the cost of 220 million USD. The outcome of the voluntary resettlement carried by the current government is yet to be evaluated but believed to have a mixed result with low human cost.[26]

Here is the summary of the root cause of the human settlement failure:

1. Farmers even those on the verge of starvation were reluctant to leave their birth places for unknown lands*
2. In an effort to meet the quotas assigned by the government, zealot cadres randomly selected people and scattered them to different settlement areas without the slightest regard to the family unit. This involuntary separation of family members increased the aura of desperation followed by grim resignation throughout the settlement areas. It took few planting seasons for the settlement authorities to sort out their mistake.
3. Despite the genuine effort exerted by the students, their instructors and the staff of the various government ministries to construct makeshift camps out of the local materials there was no enough shelter to accommodate each family at least in the first few months of the campaign. Understandably, this has contributed to the many deaths reported soon after their arrival.
4. Hygienic conditions were so deplorable that parasitic ailments such as diarrhea and dysentery, coughing was rampant especially among the young, the old and breast feeding mothers throughout the settlements.

---

26 New York Times, In Ethiopia, Promised Land of Broken Promises (5/30/2004).

5. In a country where every 35,000 – 40,000 citizen is attended by one health professional, one cannot expect elementary health matters be addressed without making stress on the nation's health status. Nevertheless, each settlement areas had one health center with the capacity to dispense medicine, dress wound and refer patients to the nearest hospitals some as far as 100 kilometers away from the settlements. Almost all the settlements areas were located in the malaria infested low laying regions causing an epidemic among the plateau dwellers, which incidentally constitute 80% of the settler population.
6. Food was scarce throughout the period of 1986 to 1989. Unable to feed themselves settlers were continuously issued food aid donated from the international community.
7. The implantation of rural people whose lifetime travel experiences is limited to one or two day travel distance by foot from his/her immediate environment is bound to send waves of shocks, fear and apprehension. In such circumstance it is difficult to expect a gregarious pioneering spirit among the settlers. All were transported hundreds of kilometers far from an area that they can easily identify with and too unfamiliar place devoid of their ancestor's legacy. It was heart breaking to see people consumed with homesick day in day out.[27]
8. Technically, there was little information at hand on the type and the behavior of the soils of the designated areas of settlement. In that case the peasant may encounter abiotic elements that are unfamiliar to him. Managing weeds and other out of place plants can be a daunting task. Moreover, there was no well-established technical information generating institution at or adjacent to the settlements. Extrapolation of existing agricultural data generated from experimental stations located long distance away could not be as helpful as one expects. For example the settlers from Tigrai and Wello were no matches to the challenge posed by the heavy vertisolic soils of Gambella, which turns into waterlogged condition upon receiving rain, and a V shaped hard cracked soil when dry. Accustomed to type of crops of marginal character, settlers had difficulty managing familiar crop species of variant nature or an entirely new species all together unknown to them. After fifty years of experimentation on crop adaptability, the low laying lands of Ethiopia were found to be suitable to growing cotton, sesame, sorghum, rice and many others unfamiliar species in the plateau. Despite some efforts made by the government and non-government organizations to provide livestock to the settlers for some obvious reason it did not succeed. In later years there was modest livestock trading spearheaded by young settlers who totally abandoned cultivation and instead specialized in limited livestock fatting practices for homebound slaughtering.

---

27 It was common for settlers to run away from the settlement villages in the hope of reaching their ancestral areas. Since strict patrolling was in effect their hitchhiking effort was aborted before reaching their destination.

Villagization

Among few of the pet projects that the government was able to execute without much eye brows raised by its detractors abroad and by the silent majority inside was the villagization program that started soon after the uproar on the forced settlement died down. Arguably, it is one of the stealthiest operations that the *Derg* has undertaken for a long time. This massive movement of people affected a little over 12 million peasants. As expected few zealots hankering for attention in the party headquarters conceived the program and pushed their agenda to finally be embraced by the government. The objective was to cluster all scattered households in a given area into one compact village or hamlet.[28] The rationale was that by pulling the entire scattered household into well-defined sub-municipal areas the government with the support of some willing NGOs would be able to supply provisions and social services quickly and economically. In ideal setting this may sound reasonable. But for a nation that had surely bled to the bone because of the misguided internal policies, civil wars and unfavorable external relations, it was an adventure that played on the sensibility of the Ethiopians. Many at the time knew that a government stretched to its limits could only pay lip service to the program already in motion. But it went ahead and in a span of six months one of the most important and populous province, Harar went from devillagized region to a land with hundreds of closely neat corrugated zinc roofed villages. Rough estimate put of people subjected to forced villagization at 12.2 million.[29]

It was projected that by the end of the 1990 fifteen million peasants will be crammed into villages provided with the basic necessities of life and some amenities for the families to enjoy. A master plan was also drawn to set up cottage industries as a means of transition from purely subsistence agricultural practices to advanced agro-industrial activities. In other words, it was intended to propel the country from its current rural based subsistence economy to agriculture based semi-industrial country. Certainly many of the promises made by the regime were not fulfilled. Except for few, conveniently located near major highways, that seemed to enjoy hand dug latrines built by the villagers themselves and an understaffed and undersupplied clinics provided by the government, many of the villages resembled more of a mini feeding centers common in the north than naturally evolved villages of the distant past. Farmers who lived humbly but spaciously in the past found themselves crammed into environmentally oppressive spaces and unhygienic conditions.

28 Although he may not be the actual architect of the viligization program, a cabinet minister by the name Kasa Gebre orchestrated the need of it on national scale when giving tour to Mengestu Haile Mariam, the president of the country to the model new villages that he created in the Hararghe region.

29 Teffera Haile Selassie. 1997. The Ethiopian revolution 1974-1991. (From Monarchial Autocracy to Military Oligarchy, 1997), 270.

Most villages lack hand pumped water supply and therefore, have to trek for a considerable distant to fetch water from streams and springs. Carrying pail of water on ones back or head is a daunting task few in the urban areas of Ethiopia can contemplate. Those fortunate few with hand dug water wells in their villages have to take turns and wait in line to fill their earthen pots or plastic cans bound for home. In short, stench during summer and extremely muddy condition during winter made the newly constituted villages quiet inhospitable.

The negative effect of the villagization policy was felt more in Harar than anywhere else in the country. It is important to underline here that in Harar as in many southern districts of the country domiciles are found randomly scattered across the tableland, the ravines, and plains. In some cases the dwellings are as far as 100 – 200 meters apart from each other. This natural setup lends for each homestead to look and behave as a unit with a distinct characteristics of its own. The homestead, therefore, was a minute part of the total whole surrounded by hedge or shrubs sprinkled with castor plants and consisting of family dwelling with animal enclosure within it. Many of the dwellings had what is called a backyard garden with medicinal, horticultural, and stimulant crops growing within and around it. In recent years this has become an indispensable source of supplementary income to many families. Thus, the cost of abandoning an established homestead in favor of a mechanically constructed village was the deconstruction of a safety net built years ago for the survival of the subsistence farmer. Environmentally speaking, it had destroyed a micro ecosystem that the peasant and his family have enjoyed for generations.

Given the fact that Ethiopia has been food insecure country for many decades to put it mildly, there is indeed a compelling reason to take drastic measures such as the one taken by the *Derg*, namely the vilagization of the Ethiopian countryside in the east, south and west. By the same token, one cannot sensibly argue against moving people from marginalized to naturally well-endowed areas of the country if it's for the benefit of the cultivator. Moreover if providing health services, education and other social amenities are in mind, thus, bringing closer all scattered settlements is indeed a socially responsible thing to do. Given the dire economic and political situation that Ethiopia was facing at the time, however, the measures taken were ill advised. In short, it is simplistic to justify the launching of project of such magnitude on the account of "desperate time needs desperate measures". This is tantamount to defying reason. A government fighting internal wars with many armed groups and unfriendly with its neighbors in the East and west was in no position to embark upon such gigantic project. Above all it lacked the resources necessary to implement the project with less pain. In a nutshell, the agriculture policy that the *Derg* pursued brought more horror than relief to the people, in particular the peasantry in the 1980s. No doubt the villagization program that started in mid-1980s was disruptive, socially insensitive, and harmful to the fragile ecosystem and most certainly dangerous to the well-being of the citizens. Frankly, it was senseless policy that worked against the very goals set; achieving food self-sufficiency.

In retrospect the reason why the government failed to address the chronic food problem inherent in the agriculture system was because of its fixation both in rhetoric and action to the socialist mode of production championed by the Eastern European nations. One can argue that the rigid adherence to the socialist ideology and the blanket application of centrally designed economic policy cost Ethiopia an opportunity time to do better. The lack of proper planning supported by concrete physical and social studies exacerbated food shortages. Nothing whatsoever, substitutes the imperative of sound technical planning and careful consideration of the social implications. Unfortunately, the agricultural policy of the *Derg* to transform Ethiopia into food self-sufficient with healthy and educated population was misplaced.

As time revealed the food situation far from improving become precarious by the year. The calamities that befall the country in 1984/85 is just one of the few episodes that send jolt to make the government think for the first time to change direction, hence for the adoption of mixed economy.[30] It is however fair to state that it is not only the misdirected agriculture policy that was the primary cause of Ethiopia's failure to sustained agriculture growth but also the indifference of the Western nations together with the vagaries of nature that is to blame. By the time the *Derg* was driven out of office the food aid to Ethiopia was estimated at 912,000 tons of cereal and non-cereal foodstuff.[31]

The Grain Bank

Other project that the *Derg* has diligently pursued during its decade and half in power was the stockpiling of grain reserve ostensibly, to be used during lean times. Unfortunately, for some regions of Ethiopia crop failure was so common that it was crucial that extra food be stashed to fight famine whenever it occurred. Thus, building strategic food reserve through local purchase or procurement from abroad is perfectly sound policy to pursue. In the aftermath of the 1972-73 droughts, therefore, the *Derg's* first order of business was to restructure the Ethiopian Grain Corporation established in the early 1960s. During this period of uncertainty where grain market volatility was rampant, the *Derg's* only option, at least in theory was to procure locally produced crops from farmers at fixed prices for future use. Thus, Agriculture Marketing Corporation (AMC) as it came to be known was established 1976 and had one

30 The new economic reform Perestroika announced in July 1987 must have influenced the *Derg* to change course.

31 WFP, 1990, 1991, 1992, 1993 FOOD AID REVIEWS Development (food for work) and Emergency aid in Millions of dollars (1993).

objective, that is, to regulate supply and stabilize food prices in urban areas.

However, with the widespread rural instability and urban upheaval that gripped the nation soon after the declaration of land reform, production of crops begun to slump further defeating the objectives set by the ruling military government. In 1976 where cereal production was roughly at 4.7 million MT went down to 3.9 million MT in 1978. In subsequent years, that is, until the period where the country was hit by the intractable drought of 1984/85, grain output begun to improve slowly increasing by an average of 100,000 MT per year. Despite the efforts made, grain production remained at standstill; 4.9 million MT in 1986. In absolute terms the population growth rate increased at steady 2.8% per annum, while grain production depressingly leveled to prerevolutionary period, which in 1974 was 4.3 million MT. In short the per capita food production declined to an alarming level. This is clear indication that exactly a decade after the land reform was announced agriculture in Ethiopia did not show much progress.

It was under this circumstance that the AMC continued to enforce supply quotas on the PAs, the State Farms and the retail grain dealers throughout the country. By virtue of their accessibility and constitution as an appendage of the state, the primary clients of AMC were the state farms whose grain yield per hectare was tragically as low as that of the peasant farmer. This is despite their favored status as beneficiaries to the government provided agricultural inputs. Unable to hide from the enforcers, the grain trader was also victim of the senseless quota system who by law were required to supply 50% and in some case a bit more of their purchase to the AMC at price slightly higher than that of the farm gate prices.

Understandably, the bulk of the procurement came from the PAs, as they constituted the majority and known to be responsible for over 92% of grain production in the country. In 1981/82 alone the provinces of Gojam, Arssi and southern and central Shewa contributed about three-quarterof the national average purchased by the AMC. This being the case not all farmers in these or other provinces was affected by the stringent measures adopted by the purchasing agency. Due to its inefficiency and because of the agent's inability to access the PAs, many farmers escaped the obligatory supply of grains to AMC. A study conducted in late 1980s showed that 51% of the household sampled expressed that they had never paid quota imposed at all since the agency begun to implement the procurement program in 1978.[32] There is no surprise here the agency could not reach the nook and cranny of the country to extract the 'surplus' because the road network at the time was rudimentary at best. Nevertheless, the AMC continued to pose as Damocles' sword ready to strike at any time.

---

32 Patrick Webb, Braun von Joachim, Yohannes Yisehac, Famine in Ethiopia: Of Coping Failure at National and Household Levels, (IFPRI, 1992), 75.

From the finding, one can deduce that it is not the aggressive approach of the agency as such that was the problem but the absurdities that followed its shadows. For almost a decade it was impossible to conduct not only inter but intra provincial trade of grains no matter how small without one landing into trouble. Practically at every large and small town, checkpoints were installed to confiscate unauthorized transport of grains without mercy.[33] In addition some PAs with added advantage of being near large towns were deprived of the opportunity of selling grains in particular *Teff* (*Eragrostis abysinica*) which fetched higher price at all times. The price of this most favored indigenous grain was 410 Birr per ton at farm gate price while in the open market it fetched as much as 1500 Birr in the early 1980s.

Although not widespread some PAs were infrequently asked to supply the AMC, types of crop not indigenous to their areas. For reasons still unclear peasants in many cereal-growing areas were asked to supply oilseeds or legumes to meet the quota set by the agency. Thus, to avoid the wreath of the agency, peasants were forced to sell their produce in order to buy the 'quota crop' demanded upon them. This bizarre way of enforcing government regulations has caused more confusion than good to the agency that it soon dropped the practice. The absurdities did not stop there. Peasants harvest (sickle cut) their crop in broad day light and thresh by night to avoid the watch full eye of the agency or its surrogates, the cadres and the militias of the village. In many cases they can get away by simply pleading to the representative that this year's harvest was lower than last year. In some cases inter personal rivalry or feuds render their effort fruitless putting them into more debt. In all honesty the agency never acted in a spectacularly ruthless manner to extract already set quota in time and in a desirable quantity. The reason is that it was not equipped to forecast and monitor harvest. It also lacked the necessary infrastructure to enforce the quota system. Moreover, it never had competent staff or adequate transport facilities or even environmentally controlled warehouses to effectively store the produce.

Aside from the callous policy toward the peasant sector, the AMC's comical posture as a bogeyman unmistakably played a part in depressing production. Based on FAO compiled statistics, the average cereal production for 16 years was 5.3 million MT. This, when compared with the figure produced in 1974 (4.3 million MT) it is not that impressive. In short, this amounts to a mere increase of less than one million metric tons for a nation that has doubled its population from 26 million to 55 million by the end of the 1980s. When the *Derg* relinquished power in 1991 the agriculture growth rate stood at negative 2.5%.[34] No doubt; Ethiopia was in much worst condition than at any time in

---

33 Ironically this was also practiced albeit for a brief period of time in the Eritrean Liberation Fronts controlled areas of the highland.

34 World Bank, World Development Report 1992; Trends in Developing Economies

its history

The EPRDF: the generation after

Since the EPRDF came to power, food situation has not improved drastically but managed efficiently by an elaborate system of relief providing mechanisms. This include the strengthening of an early warning system were information is adequately relayed from the outlying areas and analyzed to help asses the overall food situation in the country. In addition, some warehouses were made available in strategically located areas of the country to facilitate rapid response. Therefore, without much change in the overall food production business the government was able to control events before they occur.

Nearly two decades after the exit of the *Derg*, the US Famine Early Warning System Network predicted that by the year 2007/08 Millions of Ethiopians will need food aid to survive.[35] In retrospect the predication proved to be true when people needing food in Ethiopia in 2009 reached up to 13.7 million[36]. The forecast was a bit inaccurate but credible enough to cast serious doubts on the infrequent pronouncement by state officials that achieving food self-sufficiency is the horizon. Nevertheless, one can surmise that in the past decade millions of citizens could have suffered seriously if production gap was not met by donations from the EU and the US. In the meantime, ensuring an adequate food supply for a fifth of the population remains a formidable task to the policy makers of the country. Be that as it may, there were about 7.5 million chronically food-insecure Ethiopians receiving assistance through a largely U.S.-funded program known as the Productive Safety Net. Ironically, the figure is equal to the one estimated by the RRC of people needing food aid a quarter of a century ago.[37] In October 2009, yet another plea for an emergency aid was made for an estimated 6.2 million people suffering from crop failure.[38] According to the recent published document in year 2012 close to 3.2 million people were in dire need of food aid to the tune of 366 thousand metric tons to survive.[39] This figure is by far lower than the one shown in previous years. To meet the challenge the government in this instance has purchased 135

1991, 1992, (The IMF International financial statistics, Aug 1992).

35 FEWSN, IRIN News.org. UN office for the coordination of Human (October 17 2005).

36 Peter Heinlein, Ethiopia's Meles accuses donors off exaggerating food crisis (Voice of America (VOA), 14 Oct 2009).

37 Relief and Rehabilitation Commission. Current Situation & Future Prospect. (Addis Ababa, April 1985), 6.

38 Alemseghed Negash. Ground Report, (Internet Blog 14, October 2009).

39 Humanitarian Requirement Document (HRD) (Addis Ababa: Published by joint Ministry of Agriculture and donor agencies, January, 2012).

thousands metric tons of grains from abroad.[40] It is tragic that the 'Food Aid Industry' is more active today than it has been two decades ago.

Unfortunately, this is a challenge too familiar to successive governments who most often became superbly adept to cope up with problems related to food shortages, principally by organizing a very efficient food aid soliciting state machinery. In this case the *Derg* and with more finesse the EPRDF kept the flow of food aid uninterrupted for the last forty years. As pointed out earlier, the socialist government's response to the chronic food shortage that made Ethiopia the poster child for famine was to organize cooperatives and promote state farms as well as to forcefully villigize the countryside. Unfortunately, this did not work as planned. Therefore, it is no secret that many of the *Derg's* agricultural policies did not work to the extent that the EPRDF government could build upon towards fulfilling the goal of Freedom from Hunger Campaign launched by UNFAO and Ethiopia over half century ago.

Regardless, Ethiopia under the EPRDF administration is still dependent on food aid to meet production gaps. According to FAO Media center, Ethiopia is one of the 22 countries in the world whose food supply status is exceedingly low to be included in the list of nations living in 'protracted crises.[41] The United State government alone provided one billion dollar worth of foodstuff in 2008.[42] Unlike its predecessor, the current government has tried to address major economic problems in particular those related to agriculture through the supply of inputs and by providing technical assistance to increase food production.[43] Be that as it may Ethiopia remains one among the largest recipients of humanitarian aid in the world.[44]

The economic reform announced after the removal of the *derg* from power unequivocally put the agriculture sector as driving force for industrializing the country. The strategy of ADLI was, therefore, to negate many of the policies pursued by the *Derg*. To that effect the EPRDF took numerous measures to mitigate food insecurity. In other words, the new government made it is primary goal to revitalize agriculture by lifting certain restrictions put by the defunct government. It also provided the peasant with provisions and technical support. Among the significant measures taken by the government was to dismantle the AMC which many believed that its heavy handedness in the past

---

40 Reuters news agency (July 10, 2012).

41 Media Center. In State of Food Insecurity in the World, (October 6, 2010), 13.

42 Peter Heinlein, (VOA, 11 October 2009).

43 Ken Ohashi, Sustaining Growth: Ethiopessimism? World Bank Country Director for the Ethiopia and the Sudan, (World Bank, 2008) He reports that the number of DAs has risen from about 15,000 in 2000 to about 67,000 2009. That is a significant institutional capability build up.

44 1/ The USAID supplied 274,750 tons of food aid to Ethiopia in 2014 www.gov/ethiopia/food. 2/ World Food Program supported some 6.5 million disadvantaged citizens with food aid in 2014. www.wfp.org/countries/ethiopia

did more harm than good to grain producers. For good or bad, it allowed the market to play a role by encouraging free flow of grains without much restriction. This measure clearly demonstrated that the peasant is no more obligated to sell his produce at a fixed price. He was also allowed to cultivate crops of his choosing not what the state dictates. At this point they began to distribute fertilizers, sometimes at a subsidized price where private and parastatal took part in the endeavor. More significantly, some of the lands previously operated by the state farms were revert back to their original owners, the pastoral and in some cases the peasant, to be managed by or leased to private developers who were eager to operate the mechanized farms on profitable base. Pursuant to the long term objective set by the government, it continued disposing the state run farms where in 2012 alone put seven state farms worth over 2 billion birr in the market.[45]

Significantly, the land reform law with its basic tenets untouched has also undergone with some sort of overhauling. This does not imply that the fundamental provisions of the proclamation 31 of 1975 and Article 40 sub section 3 of the 1995 constitution, which kept land under, state ownership is changed. The new laws like the one before it categorically forbade land sale or even exchange for the same kind. Nevertheless, the stricter rule against renting or leasing land was relaxed and employment in peasant-operated farms was permitted. Moreover, the government also begun to issue Together with land certification the bit and pieces………. certificate of usage, which enable farmers, some flexibility in utilizing land assigned to them. Significantly, the certificate will empower peasants transfer usufruct right of the land to their siblings. Indeed there are clear evidences showing that land certification have an obvious advantage over other arrangements.[46]

However it remains to be seen whether banks will honor the certificate of 'usage' as collateral to lend money for land improvement. Assuming that banks would be willing partners in this venture, one question arises, would peasants in their part be convinced to engage in long-term production enhancing or conservation measures on land that is not 'technically' theirs. But since the right to use the land is spelled clearly in the constitution and the land certification allows 'deed' transfer to sibling, there is little or no reason why peasants would not incline to invest in long-term land and water improvement ventures. Without doubt peasants and farmers in Ethiopia are closer to the market today than any other time in their history. Given the current environment prevailing in the country it is likely that the path of the peasants, the banks and the market will cross to give respite to the peasant in his long journey of overcoming food

45 Reuter News Agency, (3/29/2012).

46 Stefan Dercon, Ruth Vargas Hill and Andrew Zeiten, In Search of Strategy: rethinking agriculture-led growth in Ethiopia. (Oxford, UK: 2009), 5.

insecurity so future governments could aim higher for sustainable development.

Despite cereal production gone up by 40% during the past decade and milk production is an all-time high, one can argue that there is little qualitative change apparent in the Ethiopian agriculture.[47] This is to underline the fact that the per capita food production in Ethiopia has decline significantly to make a difference in peasant life. In light of this, food aid sent to the country has not decreased substantially. There is however, some quantitative change discernible in Ethiopia today. One being the price incentives offered to peasants to produce more and at qualities acceptable to the consumer at home and abroad is exemplary. In addition, production of potentially exportable crop went from their lowest minimum to rival coffee, which at one time was 80% foreign revenue earner for the country.

But there is more that the ruling party in Ethiopia can proudly show to skeptics. For example the increase of peasant managed small scale irrigated farms, the expansion of floriculture and horticulture farms in unprecedented scale, the modest increase of cereal yield per unit land, the expansion of medium and large scale farms in the Humera/Angereb as well as in the Baro Akobo and Omo region, the conservation of water through the construction of many multipurpose small, medium and large dams, the increase of student enrollment from its previous 3.3 million in 1990/91 to almost 22 million in 2012, the expansion of higher institutions by tenfold, the network of roads from their modest 20,000 kilometer in the first part of 1990s to about 150.000 kilometer in a mere two decades is but to mention few. There is no doubt that the projected 5,000-kilometer rail system currently under study will give a boast to economic growth with agriculture as a base.[48] The eight railway corridors radiating from the center to all corners of the country will not only bring goods and services where they are needed but also help in the integration process of peoples in a profound way.

Thus, the change in the horizon is admirable not to be overlooked. Despite this tremendous achievement, however, Ethiopia still remains undernourished and almost always at the mercy of the vagaries of nature. According to the interactive hunger map drawn by the WFP the undernourishment rate is estimated at 40%, the second highest in the world.[49] Overall, the state of agriculture in Ethiopia has been lingering between life and death that some equate the country to a bottomless pit conjuring God's hand to extricate it from its predicament. Thus, for deliverance as if the country is waiting for the fulfillment of the biblical prophecy 'Ethiopia shall soon stretch

47 CSA, (2009) The estimated milk production in the year 2008/09 is 2.76 billion liter from cattle and 162 million liter from Camels.

48 Ethiopian Railway Corporation. www.erc.gov.et

49 http://www.wfp.org/ 2013.

out her hand to God'.[50] God's helping hands, notwithstanding, Ethiopia's own resolve matters more to produce food and industrial crops in sufficient quantity.

The new regulation charted on PA also encouraged cultivators to engage solely in activities related to agriculture production and focus on solving their immediate social problems. Gone are the days when peasant associations were regarded as an appendage to the government to be used as cannon fodders in a war or worst their efforts redirected to support wars and other government initiatives irrelevant to their immediate needs.[51] Moreover, the land certification process underway may infuse sense of security among peasants to invest in land improvement activities long neglected in Ethiopia. With credit unions and banks honoring the certificate of usufruct issued by regional governments, the Ethiopian peasant could easily transform himself from a subsistent peasant to a farmer-producing surplus for the wider market. Through time, the certification of land could be a helpful instrument in drawing appropriate land use planning, in the rational management of land and water and also seal the fate of land litigation practices common during the imperial era for now and ever.

Considering the magnitude of the problem existing in Ethiopia, these measures of success are nothing but a drop in a bucket. Still the situation lend to the fact that at the rate in which farm inputs are distributed and farm credits dispensed it may take a while for Ethiopia to declare hunger as things of the past. Thus, to accelerate the utilization of farm inputs by peasants for steady growth, it is crucial that peasants continue receive a remunerative price for their produce. The more they employ farm inputs the more they produce for the market, hence for the absolute need to subsidize farm inputs. Without doubt, this will empower peasants to employ improved seeds and use other production enhancing materials as well as encourage local entrepreneurs to thinker with innovations including searching markets for agricultural supplies in their raw or processed form. Perhaps one should give time for the Ethiopian commodity market, a recent phenomenon to appear in the agriculture scene to mature into becoming an important stabilizing factor for commodities in the local market and a conduit for foreign destined industrial crops.[52]

---

50 Psalm 68:31.

51 Many in the opposing camps however, accuse the government of using the peasant associations for political gains in local and national election. Given the support of the rural population to the ruling EPRDF one cannot completely dismiss the claim.

52 The credit of establishing the Ethiopian Commodity Market goes to Dr Eleni Gabre Madhin former World Bank economist and researcher at the International Food Policy Research Institute (IFPRI), who in 2008, took the challenge of establishing the ECX.

Given the low level of economic development that Ethiopia had to experience throughout the 20th century, it is only fair to conclude that it will take time for the efforts underway to mature and yield tangible results. The next section will outline policy imperatives on key issues of agriculture and related topics that is aimed at breaking the vicious circle of rural poverty in order to avoid massive human disaster resulting from food shortages. There is no question in my mind that unless Ethiopia and the surrounding states are open for a periodic appraisal of policies and the introduction of new ones that are exclusively focused on the production of food and more food, by comparison the starvation of 1972, 1985, and the few minor ones that occur between the 1990 and 2010 are kids play.

Be that as it may, the policy put by the Ethiopian government to revitalize the countryside leaves one with an impression that indeed the country may at last come out from the quagmire that it find itself for more than a century. Resource permitting, the focus that the policy makers put on the small farmer will eventually transform the country from being a food aid recipient to a self-sufficient one. However, the production of food and industrial commodities is not the end by themselves unless accompanied by an all-encompassing national development program that first and foremost aim at slowing down population growth. Currently the birth rate stands at 2.8% and population is expected to triple within the next 40 years. It is frightening to notice that the population of Ethiopia has quadrupled from 20 million in 1950 to what it is today close to 90 million. It is imperative, therefore, mass based campaign accompanied by incentives (for having less children) such as is practiced in China and in few other countries of the world are instituted.

It only takes a quick glance at the government report to appreciate the importance of balancing population growth with the actual and projected food availability in the country. For example in 2005 the aggregate food production of Ethiopia was estmated at 18.08 Million tons against 39.50 million tons of food produced in 2010.[53] At the same time within 5 years the number of population went up from an estimated 74 million to 84 million. Notwithstanding, traditions, religious beliefs and intra-ethnic rivalry to boast numbers to score political points, uncontrolled population growth could be the undoing of all efforts put so far to transform Ethiopia into self-sustaining middle income nation that it should be. Thus, despite the tremendous achievement in food production the country continues to solicit or import food more than ever. In this regard one cannot help but reiterate that the capita food production has decline to give comfort for the people, the government and the partners in food aid business.

---

53 Five Year Development and Transformation Plan 2011 – 2015 Ministry of Finance, FDRE (2010).

A little before the FYGTP (2010-2015) came into effect, the government drew an ambitious plan to reduce poverty to a manageable level by devoting considerable resources from its coffers. With additional support from the US and the European Union, it drafted a program of action to ameliorate poverty in the country. According to the government white paper, this 'pro-poor growth' sustainable development and poverty reduction program was estimated to have cost the government 9.9 billion Birr in 2003 to 2005. Although Ethiopia ranks the lowest in the Human Development Index surpassing only Eritrea and 12 other desperately poor nations, the World Bank testifies that it is making a remarkable progress in key areas of development.[54] The *African Development Indicators 2006 Report* published by the Bank also places Ethiopia in the top category for policies of social inclusion and equity. The country is also praised for its exceptional performance in the domain of structural policies and public sector management and institutions.[55] It is difficult to make an issue with either the Banks' report or the EPRDFs periodic claim that Ethiopia's overall condition is better today than at any other time of its history. Earlier in the decade some observed that the will and commitment of the government to overcome the Malthusian trap in rural areas have improved moderately.[56] By government estimate people living under poverty level in Ethiopia is about 29% and mostly of rural background.[57] However, the World Bank put the poverty level at 39 percent.[58] Whichever figure is correct, the projection is that to reduce poverty by at least 7 to 8% from its current level by the end of 2015.

Policy imperative

In this section I will try to outline or perhaps modestly suggest the wide policy measures that are likely to work in the context of Africa's underdevelopment and more specifically in Ethiopia's current food production predicament. Broadly speaking, the agriculture policy of the current government except for its ill-advised move to go ahead and rent huge tracts of land to international corporations leaves little room for severe criticism. However, given the chronic food shortages that people face today it is imperative that the policies pursued by the current government and those before it is examined. Food, the basic

54 hdr.undp.org/en/statistics/

55 Business in Africa. Online (Nov, 10 2006).

56 Mulat Demeke and Bekele Hunde, Population Pressure and Intensification of Agricultural Production in Ethiopia: The Will and Cost of Overcoming Malthusian Trap, (Addis Ababa: EEA, January 3 - 5, 2003).

57 Five Year Development and Transformation Plan 2011 – 2015, Discussion paper for all regions, Ministry of Finance, FDRE (2010).

58 World Bank, **World** Development **Report—**Data & Research—econ.worldbank.org/...8093~piPK:477627~theSitePK:477624,00.html

human requirement for sustenance and perpetuation of the species has been in short supply for many Ethiopians close to a half-century.

Given this fact, therefore, what more compelling reason does one require not to put all efforts national and individual than to combat food insecurity by exploring an alternative development path that calls for designing a self-sustaining agricultural development course? Lack of participatory approach in the design and execution of projects are the cause of these failures. In this regard past government effort failed to include, organize and lead the populous toward self-sustaining food production paths?

Cursory look at post *Derg* policy and navigating through some piece meal activities and community driven economic measures taken so far one observes that the EPRDF government appeared to be exceptionally conscious of its responsibilities. For one, it began to identify some of the bottlenecks deserving immediate attention. In this regard among the salient ones are the attempt to curb population growth, to systematize the early warning systems and to prepare workable methods by which emergency food aid is delivered without delay and the exploration of an alternative venue for engaging the productive force of the population in sectors other than agriculture.[59] Moreover with community cooperation it devised a strategy where economic gain to peasants is guaranteed while still caring for the ecosystem to flourish instead destroyed as practiced by residents in Sheka Zone South Ethiopia.[60] Today many farm localities are encouraged to protect and rejuvenate as well as utilize judiciously forest products and other resources falling within their administrative jurisdiction. This is a sound policy indeed where farmers are allowed to enjoy the fruit of their labor but also by virtue of the usufruct right bestowed on them to be treated as stakeholders in any future commercial farm establishment, national or otherwise.

In early 20th century when Afewerk Gebreyesus indict the Ethiopian war lords and their lesser chiefs 'living luxuriously pre-occupied with drinking and eating like a fattened cattle ready for Easter slaughter; this is at the expense of the poor and disadvantaged", he meant to underline the fact that as cog in a wheel is essential to moving a cart so would the actions taken by men and women in the helm for societal development.[61] His frustration and anger toward the ruling class is, therefore, well placed. As history shows, leadership

---

59 Off-farm activities and urban-based economic empowerment is in the rise. The author was heartened to observe cobblestone pedestrian paths in some major cities of the country and community policing of natural habitat with monetary gains to villages. The hundreds of thousands of people involved in the informal market in cities and towns must and should be taken as complimentary to the larger market promoted by the government as well.

60 Here communities are given access to all the bounties that forest can give in return for their protection and custodial care of the forest itself.

61 Afwerk G.J. , Guide du voyageur en Abyssinie (Rome: 1908), 182.

at any level and at an opportune time can be instrumental in setting pace for national development. In this regard the Japanese experience can be cited as an example, where the feudo-agrarian culture had made tremendous stride in the past 150 years with the Emperors at the helm. Link between the agrarian reform, trade and manufacturing helped propel Japan to heights in economic and military sphere for close to a century. It is not by accident therefore, that over fifty years ago, one of the prolific Ethiopian writer Kebede Michael inspired by the rapid progress of Japan had to write a book in his native tongue titled '*How Japan Developed*'. His underlining message to the ruling class was that it is possible for them as was for the Meji dynasty in Japan in the 19th century to lead a revolution without necessarily committing class suicide. Indeed the world is witness the speed by which the South East Asian nations pull themselves from mainly agrarian system of development to what they are today, highly industrial first nations with full capacity to play in international economic and political theatre. Conversely, successive government policies in Ethiopia have been regressive in nature than being progressive. That is why highlighting the drawback of many of the *Derg's* policies not to mention the feudal government before it becomes relevant. Then and only then can one suggest a defensible thesis upon which development policies ought to be built?

With all honesty any attempt to reverse the stagnation process set in motion at least two centuries earlier is by itself a Herculean task. This pessimistic assessment is based largely on the state of the Ethiopian agriculture in the past century. In this regard, among the discouraging factors are the geometric expansion of the population, the spread of malaria and AIDS pandemics, the alarmingly high rate of soil erosion, the helplessness to harness water potential, the lack of appropriate technology to complement the back breaking farm chore, the absence of energy power, and above all peace and security that continued to elude modern day Ethiopia and by association the Horn Region. Ironically, its within the second half of the past century that many nations in the world, the majority of whom are in Asia and some in South America begun to make headway in all fields of endeavor including science. To the contrary Ethiopia's movement toward modernity was snail pace making a superficial dent on the existing socio-economic structure without effecting a substantial change in the lives of its citizens. This however, does not negate the fact that even at this eleventh hour, people friendly and sound agriculture policy that put people as the central driving force cannot be formulated and implemented as well.

Policies are borne in response to challenges faced by society. The primary objective of drafting a policy is therefore, to ensure the well-being of the society at large. Thus, success is guaranteed if and only if the needs and aspirations of the members to whom the policy is drafted for is concretely reflected in the formulation of the policy in question. Unfortunately, this truism is either undervalued or given lip service when drafting past policies in Ethiopia.

Frequently, policies including development strategies are leadership driven with little popular participation. There are countless cases one can cite as an example to make the point clear. In the past health clinics were built according to a blue print drawn by bureaucrats and not upon a concrete study and/or request of the inhabitants on whose name the facilities are established. At times schools and hospitals are built with no likelihood of their immediate use but in anticipation that they could be needed in the future or simply to adorn provincial and district towns in the hope of scoring political advantage or due to outright favoritism. Many are built without providing adequate health or educational provisions and the staff to manage them. It is not uncommon for magnificent edifice or humble buildings to be left idle or used for unrelated purposes.[62] Frequently, one come across well intended criticism by nomads whose immediate need as it were was the establishment of watering points for their livestock in the range lands but instead faced with an option of incorporating date palm planting in their opportunistic life style. Conversely, while construction of strategically located bridges could be more useful to farming communities in a given district, at times, the promotion of livestock marketing outlets or building children playground take precedence[63].

Many countries in Africa and in particular the state of Eritrea is not immune from such sin. The Eritrean leadership delude by its rural based development undertakings during the armed struggle is notoriously known for its do it alone attitude. It is allergic and cocky to any views expressed by expert and adamant to doing things its way no matter how primitive and inappropriate. Developmental programs are almost exclusively formulated in isolation and tragically subservient to the whim of recalcitrant guerilla fighters turned bureaucrats. The feat brilliantly demonstrated during the armed struggle, albeit, in monastic settings, was far from being replicated in independent Eritrea. If it did it was with disastrous consequence. No doubt the leadership of the onetime ragtag liberation army is responsible for commissioning projects that are

62 *ADISS FORTUNE Vol 7. # 336, (October 8, 2006).* The Ethiopian Academy of Financial Studies (EAFS) built on a 165, 000sqm plot of land at a cost of 120 million birr left idle 2 years after its completion in 2003. It was designed to enroll 500 students at a time. Likewise, a pharmaceutical establishment built in the city of Keren, Eritrea had to remain idle for many years before it was put to good use. To begin with the town had neither dependable electric power nor sustainable water source to produce the required product.

63 Petros Tesfagiorgis, Diary of EFD-UK: a continuation (Awate.com, July 2007) An Eritrean-American who is an ardent supporter of the ruling PFDJ went to Eritrea to celebrate the 16th anniversary of independence said, "I went to Massawa and I was shocked to find it a ghost town. No movement at all. I drove to Gahtelai and I took the new road via Solomuna and Serejeka to Asmara. The new road is of high standard but there was no cars using it. It was mostly enjoyed by camels and monkeys".

irrelevant, resource consuming and environmentally unsustainable. One typical example of some of the senseless projects arrogantly pushed by the government in Eritrea is the huge micro dam build in the Hazemo plain that busted soon after its completion, severely eroding the soil down below.[64] Sadly enough no follow-up study was undertaken to explore the untimely burst of the huge micro dam. Moreover, there is no economic justification whatsoever to build a major airport at the port city of Massawa located a mere 50 kilometers distance as a crow flies from another international airport in Asmara. There should be other rational justification to invest on airstrips other than the foolish notion of upgrading the seaport at the Red sea coast to the status of free port. Not a novel idea indeed, where you already have a bunch of well-tended ports in the Red Sea rim and Indian Ocean coast. Wait until the huge and state of art freeport is completed in 2020 at Lemu, Kenya to serve East African countries among them the land locked but with potentially large economy, Ethiopia.[65]

Agriculture Sustainability: The Target

Food insecurity has been and remains to be the most formidable foe for the Ethiopian leadership.[66] The downfall of the Imperial government in early 1970s and the crumbling of the socialist regime after it, the *Derg*, can all be traced to the explosive issues related to food and shelter. Indeed the undoing of many regimes in Africa is associated with lack of basic resources to provide their citizens. The current government's bold move to tackling food shortages through an investment that focuses in agriculture and natural resources are commendable. With stable political climate and some perseverance from the beneficiaries there is little doubt that many of the long term investments including the grass root initiated projects will impact life as we never knew it before. For start the 10 billion Birr that it had spend on agriculture revitalization between 2002/3 and 2004/5 was an impressive sum. The commitment went beyond Band-Aid measures to include investment on livestock development and the exploitation of natural resources as well as in expanding agricultural technical and vocational training to aid in the process.

The use of farm inputs in particular fertilizer, seed and pesticides have quintupled since the EPRDF took power. Today over half of the 12 million

64 Out of the 300 micro dams built since 1978, a mere 120 were able to store or seem to storing water on regular bases in 2003. The remaining micro dams are either silted, busted, or because of faulty design unable to store sufficient runoff water.

65 Kenya is building the largest free port on the eastern coast of Africa at Lamu. This port will serve not only Kenya, but also the nations of Ethiopia, South Sudan, Uganda, Rwanda, Burundi, east coast of the Democratic Republic of Congo and a few other nations within central Africa and to which this will be a conventional port. http://lamuportagency.com/Port%20Area.html

66 Prime Minster Melees Zenawi at various occasions lamented that food insecurity is the Achilles heals of his government and Ethiopia.

cultivated farms in Ethiopia are under fertilizer regime and 20% are sprayed with pest control chemicals. Considering the magnitude of the problem that Ethiopia faces in the agriculture sector, this significant measure of success is nothing but a drop in a bucket. However, to a student of Ethiopian agriculture or keen observer of Ethiopia's economic development this can be considered as a remarkable achievement. It is important not to lose sight that the greater portion of the 20th century in particular the last three-quarters of it was an opportunity time lost for Ethiopia. Thus, to accelerate the utilization of farm inputs by peasants, it is imperative that special efforts are exerted to empower peasants within the PA to produce seeds and other production enhancing materials and encourage local entrepreneurs to tinker with innovations including searching markets for agricultural supplies but also in the packaging and processing of agricultural commodities for internal consumption. In this connection the emerging Seed Producers Cooperatives (SPC) must be encouraged and technically supported to complement the production of seed by the nascent private seed industry. Energy permitting many essentially consumable product such as edible oil, confectionaries, inland fish farming and animal feed can have an impact on the well-being of the rural community.

Currently, the focus of the government is to make the small-scale farmer namely, the peasant to be self-reliant while being provided with modest technical support for his own good. The government through its action has reassured the peasant that it will refrain from the heavy handedness that the state has traditionally practiced over them. Definitely this would disallow the state to engage the peasant above and beyond what citizenry obligations demands. The peasant in Ethiopia has for centuries been victim of wars, exploitation in the form of heavy taxation and *corvee* involving physical labor. Because the efforts so far expended by the government to transform the peasant into self-sufficient and resource permitting into surplus producing farmer is encouraging. Indeed if Ethiopia is to recover quickly from its prostrated position and fulfill the undying dream of becoming self-sufficient in food and agricultural commodities, the producer must be engaged in every turn of the way for the simple reason that he constitutes the greater majority among the population of the country. Therefore, it is only logical to focus more on the peasants and the role he/she could play in fulfilling the objectives of ADLI formulated when the EPRDF assumed power in the early 1990s. It is illustrative to state that, in 2013/2014 *maher* cropping season the share of the commercial farms to food production was about 5%. In other words, while the peasant holdings produced 25,153,662 MT of all types of crops the commercial farms reaped a mere 1,191,995 MT.[67] This clearly shows the importance of smallholding in Ethiopia and will remain so for decades to come.

67 CSA (2013/2014).

It is extremely dangerous otherwise, to expect large agricultural corporations of foreign origin to act and think in sync with the nationalist goals. Those objectives being agriculture led industrial development but also secure the supply of food to millions of Ethiopians in dire need. The heavy flow of capital generated from abroad not dispensed through traditional banking system but hedge funds, equipped with high tech as a means of production and geared toward maximizing profit anywhere and everywhere should be a matter of concern. The argument that foods produce by any means necessary is really indefensible and in the long run unsustainable. It is therefore essential to exercise prudency from recklessly opening the country to foreign-based large corporations that may or may not have the interest of the country at heart. Once again, the government should not weaver from continuously formulating policy that recognizes the peasant as a central figure for achieving the long talked issue of food self-sufficiency and as a means of producing industrial and cash crop to aid in raising foreign reserve. Nonetheless, the government in its effort to achieve secure livelihood for the peasant masses has outlined policy for sustainable development, which focuses on the following sphere:

1) Raise the number of Agricultural Technical and Vocational Training (TVET) personnel from 30,000 to 270,000
2) Inject capital for the development of the Livestock sector
3) Enhance natural resource management actions to a level where it makes a difference
4) Strengthen the food security measures in rain deficit areas of the country.
5) Channel the meager resource to raise productivity in potential high yielding areas of the country

In addition it has established bureaucratic structures such as PADEP, the Animal Resources Marketing Department (ARMD) and so many other agencies and commissions to deal with specific agricultural problems.

The Land vs. policy

Through various official channels the EPRDF government has made it clear that, the land reform program of 1975 cannot and will not be fundamentally altered without the risk of inviting rural mass dislocation and far worst starvation in rural as well as in urban areas. However, it remained receptive to some ideas that call for its modification and went as far as allowing 'Land for Rent" and permitting hired labor to work on the land. The land reform act of 1975 was devoid of such concessions. Furthermore, it supported the notion of land registration and certifications that begun sometime in early 2000s in selected Administrative Regions of the country. The government also contends that wholesale privatization of land as passionately advocated by some experts and few opposition parties today will deprive the peasant masses the means to lead even a subsistence life.[68] Nevertheless, the claim by many that neither the land reform championed by the *Derg* nor its modified form as it exist today has benefited the peasant in a tangible form is somewhat an accurate assessment.

The point held by many observers is that the state of agriculture in Ethiopia has shown no or little improvement since the implementation of the land reform in 1975 or its modified form since the 1990s to spare sections of the Ethiopian society from food aid dependency. Given the fact that, Ethiopia since 1958 has been the recipient of massive food aid from the United States of America and later the European Community made the argument for privatization plausible. It is, therefore, unfair to dismiss it as simply demagoguery. Food insecurity in Ethiopia and Eritrea is more apparent today than it has been during the past regimes. Thus, the call for more innovative policy regarding land use is indeed in order, especially if the need to unlock the mystery of perpetual food insecurity and lack of agriculture productivity is in mind. However this may sound reasonable, few including the most notable student of rural Ethiopia objects to the complete privatization of land but advocates for what they termed as 'associated ownership', which is the combination of freehold and community owned. This, in a sense bestows the peasant full rights over the land but subject to state approval.[69]

On the other hand it is unrealistic to advocate in unequivocal term that lands currently under nominal peasant ownership be left to savage market forces. The state of agriculture in Ethiopia is such that, it will take minor vagaries of nature let alone unwarranted human interference to throw it off balance. Who will deny that the agriculture sector is dominated by close to 70 million small holding subsistent peasants who hardly produce surplus grain? The agricultural lands, in their most fragmented state have accommodated close to 80% of the population to unwisely encourage peasants sell, mortgage or exchange their land for cash. Tinkering with the land at the moment increases the danger of triggering a revolutionary upsurge of the 1974 magnitude. Undoubtedly, if land is allowed to be a disposable commodity it will leave behind hundreds of thousands or perhaps millions of citizens victimized by famine, pestilence and the general sense of insecurity. In such circumstance, it is only reasonable to envision a scenario where the government has to resort purchasing grains and/or appealing to international donation to avert an impending catastrophe. There are indications that peasants are not particularly keen on the notion of unrestricted freehold and perhaps in support of the

68 BBC's Stephen Sacker, BBC Interviewer (July 5, 2005) In his interview Hailu Shawl, an opposition figure underlines the absolute imperative of land privatization for better agricultural performance.

69 Desalegn Rahmato, UN office for the Coordinating of Human Affairs. IRINnews.org (Oct. 9 2006).

current policy, which forbid the sale and mortgaging of land to the highest bidder.[70]

The current policy on agricultural land under PA holdings is ironclad to make peasants worry losing their livelihood due to land sale or mortgaging. The government has repeatedly made it clear that it will not entertain land sales in any form or shape. On the other hand the antagonist continues to believe that unless market value is put on land, productivity will not increase and Ethiopia's nightmare will persist for eternity. However, these elements will find solace in what the government is bent in doing in the river basins and vast plains hitherto, occupied by pastoral and agro-pastoral populations of the country. With the government's active support and the guidance of the Ethiopian Investment office, foreign corporate entities has been scouring the country side with the intention of investing in large scale agriculture, that may have little to do with helping Ethiopia achieving food self-sufficiency. The investors, especially those engaged in irrigated culture will have their priority set in growing crops ranging from industrial to horticulture including floriculture. Some could be enticed to grow sesame and rice that may or may not have an economic spinoff in as far as stimulating the local economy is concerned.[71] Worst, 35% of the land earmarked globally for foreign investment is slated to produce raw materials for agro-fuel, very amusing to entertain in the Ethiopian context.[72]

By all accounts the land grab, which gathered momentum after the 2008 global food crisis is a symptom of the general malice that gripped developing nations such as Ethiopia for long time. Unable to exploit their vast resources, and enticed by idle cash from abroad many of the developing countries would eventually fall victim to the 'United Fruit Syndrome' affecting Central America. Thus, resorting to the extreme measure of leasing lands to those strangers with huge capital and technical knowhow at their disposal could have unintended consequence very difficult to rectify. In reality it is the equitable distribution of land and the massive involvement of the state to provide goods and services that will enhance food production. The danger of converting one's own land

---

70 Berhanu Nega, B. Adenew and S. Gebre Selassie, Current land policy issues in Ethiopia. Land reform-land settlement and cooperatives, (FAO Corporate Document Repository, 2003).

71 Ermias Amelga, interview with World focus radio. BlogtalkradioEnterprenurship in Ethiopia. (October 29, 2009) Karuturi an Indian Company where it is believed to have leased 300,000 hectares.

72 Olivier de Schutter. The United Nations Special Rapporteur on the right to food United Nations A/65/281 General Assembly Sixty-fifth session, 11 August 2010. Original: English 10-48230 (E) 150910 Item 69 (b) of the provisional agenda Promotion and protection of human rights: human rights questions including alternative approaches for improving the effective enjoyment of human rights and fundamental freedoms.

into speculative asset will only benefit the foreign companies who will come out winners by putting their money in safe investment, that is, into a commodity that will certainly appreciate not depreciate.

The government policy on leasing large tracts of land is not going to be helpful to the majority of Ethiopians who are engaged in subsistence agriculture. To date, about 1.6 million hectares of land are either contracted or in the process of being given out to foreign land developers in the country.[73]. Worldwide, between 15 and 20 million hectares of land are put at the disposal of foreign entrepreneurs whose liquid asset is so huge to leave it idle.[74] According to UN agencies this is about the size of all the farmlands in France.[75]

If the objective is to increase food production and accelerate economic growth, the government is best advised to rely on its emerging business class and technical entrepreneurs no matter how arduous the path may appear. It is no secret that Africa's unending pang arises from its outward oriented commodity export that are not even covered in the General Agreement on Tariffs and Trade (GATT) to make the produce competitive in the Euro-American market. If the United States sponsored African Growth and Opportunity Act (AGOA) was designed to redress injustices embedded in the GATT, it is not doing its job because it is viewed as political gimmick more than commercial and trading tool to allow Africa participate in the market. It is paternalistic with gamut of conditions attached to it to be useful to a peasant or garment manufacturer. While maintaining standards are perquisite others such as forbidding child labor, respecting human rights, providing educational opportunities, improved health care, intellectual property, fight against graft are also mentioned as an important requirement for eligibility.[76]

A paradigm shift is necessary where millions of small-scale farmers perhaps in association with local and expatriate elements are involved toward developing medium scale commercial farms. Such is the Indian agricultural landscape where millions of satellite farms revolve around factories of appropriate size to convert the raw commodity into processed agricultural product. The peasant farms in and around Zuwai are placed in a better position

---

73 REUTERS, (JULY 29, 2010) Ethiopia set aside lands for foreign investment "The government has verified and delineated 1.6 million hectares of virgin land suitable for large-scale commercial farming in different parts of the country," Esayas Kebede, Director of the recently formed Agricultural Investment Support told Reuters.

74 (IFPRI, 2009) Land grabbing by Foreign Investors in Developing Countries. www.ifpri.org/sites/defaults/files/publications/bp013all.pdf

75 (UN PRESS RELEASE GENEVA and BRUSSELS, 11 June 2010) The UN Special Rapporteur on the Right to food recommends principles and measures to discipline "land grabbing" *Large-scale land acquisitions.*

76 AGOA track Info, (Business Daily Kenya November 2, 2009).

to be modeled as proto type where commercial enterprises and peasant farms are operating in symbiosis for the benefit of all. In this regard the policy pursued with regard to these huge prime lands waiting to be developed by highly mechanized agriculture enterprises sustained by dependable irrigation water is ill advised.

Therefore, placing the peasant farmer at the center of the land reform debate becomes vital to finding solution to some nagging problems in the current land tenure system of the country. It is important to understand that peasants are better equipped to tackle land fragmentation more efficiently than any government agency. The solution to this acute problem could as well come from the members of the PA or the corporate body in a village who would decide that further parceling of land to accommodate new members coming of age is untenable. In this regard Tigrai should be cited as an example where land parceling is outlawed to accommodate new members of the corporate village.

It is encouraging to note, however, recently the Federal government has approved measures pertaining land certification that the regional governments have initiated. Regarding land certification, the Federal government devolved power to the regional governments to enact laws governing land use including the transferability of land and taxation. In response the regional government in Tigrai declared an end to land redistribution to accommodate new comers. This is indeed a well come move (in a situation where the majority of peasants are hard pressed to make a living out of the diminutive land size. As it stands now the size of land in Tigrai and many districts in the Southern region have reached a point of diminishing return. Considering the number of landless household in some areas of Tigrai which grew by 140% in the past decade, the measures taking by the Regional government is quiet a relief to those on the verge of becoming landless.[77] Similar laws are passed in Oromia and Amhara regions where scarcity of land is the rule than the exception.

The Oromia regional government has boldly issued directives to the extent of allowing land rentals to proceed in grand scale. However the effectiveness of any measures taken by the regional governments will be judged by the extent they empower the peasants to be partners in any commercial farming enterprises. A system not necessarily structured uniformly across locations but applied appropriately for each region is a key to raising productivity. It is important to underline here that if the Federal government allow matter to take its own course then it would be for the first time in the history of Ethiopia that the center has willfully conceded power, albeit limited, to the peripheries to administer the land.

Finally, the option to allow limited privatization of land should be left to

---

77 Berhanu Ghebremedhin ILRi; John Pender, IFPRI and Simeon Ehui ILRI, Land Tenure in Tigrai. International Conference on the Ethiopian economy, (UN conference Center Addis Ababa. January 2003).

the discretion of localities, districts and regions to use it whenever applicable. Although it is most appropriate to leave the formulation of broad land policy at the hand of the central government, nevertheless providing ample space to localities to choose what is appropriate for their specific condition is wise and also necessary. Thus, the decision to decentralize administration of land and give local governments the power to administer it is not only historically significant but also a means of tackling specific problems of land development instead of applying blanket rules and regulations governing land.

Harnessing Water Resource

No government to date had worked tirelessly toward the development of water resources in Ethiopia than the EPRDF led government. Unperturbed by thinly veiled threats from the downstream neighbors who had benefited from the waters of the Nile for millennia, the government had undertaken a vast water work studies including the drawing up of a master plan for the major river basins of the country.[78] The implementation of policy on water development has just begun but will require massive capital to translate the plan into action. Currently work on massive water holding structures are underway at the Gibe, the Tekeze and the Abbay. Water Conservation work is also underway in major tributaries of the Awash River at the Kebena and the Didessa of the Blue Nile. When the Blue Nile Development project is fully realized it is hoped that in the first phase over 200,000 hectare of land will come under irrigation. This amounts to 10% of the total land estimated to be cultivable in the basin. In addition, a modest Hydropower work is completed at Tana-Belese, a major diversion waterway from the Blue Nile, and at the Tekeze and the Omo-Gibe basin with combined power generating capacity of 3070 MW. According to the detailed master plan study conducted in the 1990s, resource-permitting Ethiopia can build 200 dams with the conserving capacity of 624 million M^3 of water.[79] Yet the largest dam to ever have been contemplated is the GERD Dam started in mid-2011 located below the Blue Nile gorge close to the Sudanese border. The dam when completed in 2016 will have the generating capacity of 6000 MWh power and is expected to hold 63 billion M^3 of water, the second largest in Africa.

These large-scale water works including the massive structure at the Omo-Gibe river basin is obviously important. They are the primary source of energy and can command hundreds of thousands of hectare of cultivable lands. In the long run, though, they are not as vital as small dams, water reservoirs, and water ponds built with community participation. For one they are easily manageable

78 To its credit the EPRDF government draw a detailed master plan on many and revised/ updated on few of the following river basins; Abay, Baro Akobo, Omo-Gibe, Genale-Dawa, Wabeshebelle, Setit/Tekeze/Atbara and Awash).

79 WALL STREET JOURNAL (08/22/1997).

with affordable cost to run. They also confer a sense of ownership on the farming community to protect and develop it further. Because of their relative size and the materials used to build these modest structures it is not hard for one to imagine that they will require minimum technical inputs to maintain and run. But more significantly, the participatory approach adapted when building the water storage facilities will increase their chance of being replicated elsewhere and also help knowledge proliferate unhindered. Small dams are also known to enhance food sustainability and guarantee a continuous drinking water for human and livestock. They also help recharge the groundwater table all throughout where they are located, further facilitating the harnessing of it at later stage. Moreover, the impact on the environment will be minimum to attract public attention by concerned groups at home and abroad.

To date there is no concrete evidence to support that small dam cause serious environmental damage. But bigger dams do. For example, the Gilgel Gibe III project is likely to affect the lives of the inhabitants whose dependency on livestock rearing, hunting and gathering is obvious as the damming of the Awash six decades ago. The impact on the environment is therefore more pronounced and with disastrous effect to humans as well as to fauna and flora that Ethiopia is renowned for. Such was the case when the Aswan High Dam was built in Egypt in 1965, where entire villages were inundated to make room for Lake Nasser with ~~a~~ bigger water storage capacity than the Renaissance Dam. To date over eighty million people are displaced worldwide to make room for huge man-made lakes, which in turn becomes the primary source of methane gas emission to add to the environmental degradation[80]. In addition one of the most spectacular human heritages, the Abu Simbel built during the New Kingdom in 13th BCE was saved from complete submergence by painstakingly removing the edifice to higher ground above the artificial lake.

No one in Ethiopia, least the persons involved in policy making are oblivious of the fact that water conservation is the key to achieving food sustainability. The *Derg* in its 17 years of power have built few small earthen dams and numerous community water ponds. In Eritrea alone it commissioned more than 200 small dams and water ponds before its ignominious defeat by the armed insurgents. The EPRDF keenly aware of the importance of water to meet crop and human needs has initiated numerous water conservation projects that have yet to make an impact on the lives of the peasants and pastoral people. In Tigray, one of Ethiopia's water deficit regions the local administration is reportedly known to have built about 65 multi-purpose small-scale dams. Indeed this is a good initiative that would partially help eventually

80 Barlow Maude, Blue Covenant: the global water crisis and the coming battle for the right to water. (The New Press, 2007), 22.

solve the chronic problem of moisture deficiency in the region.[81] Similarly efforts are underway in the Amhara Administrative region where up to mid-2010 about 11,414 trapezoidal plastic-lined ponds were constructed for human and livestock use.[82] Indeed this is a worthwhile effort by the local governments that should not be dismissed as insignificant measure.

Precipitation in Ethiopia ranges from less than 200 mm in the eastern lowlands to above 2000 in the south and west of the country. The arid regions of the northeast and east may receive rain between 200 to 500 mm and sometimes more. The western plateau, however, receives generous rainfall ranging between 1500 to 2000 mm. Yet despite more than sufficient rainfall during most of the *Kirmet* season so far the attempt to make use of this nature provided gift is less than satisfactory. If properly conserved, water from the rains could be more than enough to sustain human life and support livestock for considerable period of time. Thus, the key to solving water deficiency in Ethiopia lays on the degree to which the government is committed to conserve water for multiple uses.

Irrigation practices; then, now and future

Despite allusion by earlier travelers to Ethiopia on the irrigation works that they had witnessed on the plateau, there were practically no wide spread irrigation practices one can boast about by the late 20th century.[83] A foreign resident Plowden, who lived in Ethiopia at the close of what is generally recognized, the era of warlords (*zemene mesafint* 1769 -1855), wrote that irrigation 'in Ethiopia *is*

81 Kifle Woldearegay, Surface water harvesting using small-scale dams in northern Ethiopia: Challenges and opportunities Workshop on "Challenges in Application of Integrated Water Resources Management", (Sana'a University, Yemen, Water and Environment Center, March 15-16, 2010).

82 Eguavoen, I.; Derib, S.D.; Deneke, T.T.; McCartney, M.; Otto, B.A. and Billa, S.S. 2012, Digging, damming or diverting? Small-scale irrigation in the Blue Nile basin, (Water Alternatives, www.water-alternatives.org 5(3), 2012), 678-699.

83 1/ At the plains of Yeha we saw irrigated field of wheat, barley, beans, pulses, peas (which they have all year round) garlic (very big), onions, garden rue, (and around the houses), much mustard; in the water channels much water snips and water cresses. Francisco Alvares, The Prester John of the Indies; A true relation of the lands of the Prester John, being the narrative of the Portuguese embassy to Ethiopia 1520. The Translations of Lord Stanley of Alderley (1881). Additional material by G. F. Beckingham and G. W. B. Huntingford. (London: Published for the Hakluyt Society, University Press, 1961), 141. 2/ Theodor Bent, The Sacred City of the Aetopians. (Longmans, Green and Company, 1896), 142. "Like F. Alvares "before us" describes Bent who travelled to Ethiopia about 4 centuries after the Portuguese embassy "we were struck by the unusual fertility of the neighborhood of Yeha and the conduits of water which they irrigated about their fields".

practiced wherever necessary and the numerous rivulets rendering it an easy task'.[84] Irrigation may not have been widespread practices in the pre or during the era of Axumites but there are evidences showing that cultivated fields might have been supplemented with runoff water and with limited small-scale water diversion coupled with terracing techniques to conserve moisture in the fields.[85] Given the amenability of climate then, it is hard to imagine that irrigation was used as a sole source of moisture for raising crops.[86] A curious look at the ancient water reservoir at Mai Shum in Axum and the Safra cistern at Kohaito not to mention the still useable 17th century water pools at Gonder castle one leaves with an impression that water harvesting techniques must have been practiced in the distant past, albeit in a limited scale. Despite the indication that the technique of water manipulation had existed in ancient and medieval Ethiopia, during the past centuries there was no evident of widespread irrigation practices in the highlands today. The exception is that few and insignificant number of Italian settler irrigated farms begun to appear in early 20th century. In the third quarter of the past century, however, irrigation schemes proliferated in the low laying lands of the country where large and small rivers traverse through the wide plains and narrow valleys making it conducive to pursue irrigation practice. This implies that the farms were mostly commercial and operated by intruding foreign capital in the Awash Valley and the Gash Basin in Eritrea. It is not by accident, therefore, that research on irrigation practices and water use were primarily focused toward catering larger commercial farms. Moreover, studies on water application methods, the amount of water required to raise a crop and the intervals of water application to the field were focused on high water consuming and long duration cash crops and not food.

Generally speaking, 16% of the world croplands are under irrigation and may contribute as much as 40% of the total output. While productivity per unit land is 3.6 times higher than rainfed, the monetary value derived from irrigated crops is 6.6 times than that of non-irrigated crop. In Ethiopia, without considering the commercially irrigated farms, those under peasant management are significantly small amounting to less than 2% of the total cultivated land in the country. In this regard it would be a while before irrigation practices could contribute substantially toward food security. Regardless of the danger posed by the injudicious use of irrigation water, the choice for Ethiopia is not to shy away from adopting this valuable agricultural practice but to continue

84 Walter Chicele Plowden, Travels in Abyssinia and The Galla Country with an account of the mission to Ras Ali, (London: Longmans Green and Co. 1868), 136.

85 A. Catherine D'Andrea, Andrea Manzo, Michael J. Harrower and Alicia L. Hawkins, The Pre-Axsumite and Axsumite Settlement of NE Tigrai, Ethiopia, (Journal of Field Archaeology. Vol. 33, 2008), 159.

86 David W. Phillpson, Foundation of An African Civilization, Aksum and Northern Horn 1000BC – AD 1300. (James Curry Publishing, 2012), 66.

perfecting the art of conserving, manipulating and utilizing water in most efficient way to raise crops and offset the effect of the vagaries of nature which cause periodic havoc on the lives of its citizens. According to the FYGTP, the government has projected to increase small-scale irrigated fields by 8% annually. If the current plan is fully realized by 2015 an estimated 90 to 100 thousand hectares more peasant's land will come under irrigation.[87] It is a commendable effort in the part of the government. Today close to 166,000 hectares of lands are under peasant irrigation.[88]

The MWRC located in the Middle Awash basin was established in 1956 with the mandate to conduct irrigation studies for crops suitable to grow in the Rift Valley. The center was so dedicated to finding solutions to numerous water related crop development problems, that in a span of two decades the cultivation of cotton and sugar and few hardy fruits including banana became very successful. The research center was initially supported by the FAO and later on by the World Bank to carry experimental trials in irrigation water use. It also served as a testing site for management of ruminant animals and crossbreeding center for local and exotic cattle breeds. On the side it provided an additional advantage to the crop research scientists' eager to promote generation of breeding materials faster than under rainfed condition. Such technical flexibility is possible only when moisture is not limiting. The downside is that the center has little relevance to the majority of peasant cultivators, which constitutes the backbone of the country's production system.

It was not before the mid in 1980s that a serious measure was taken to redirected the course of the irrigation studies to benefit small farms. The *Derg*, in collaboration with UNDP established the first ever, Water Technology Institute at Arba Minch located at the southern end of the rift valley. This institution was upgraded to a university level by the EPRDF in early 2000 to include many facets of water engineering including small-scale irrigation to benefit peasants. The primary mission of the institute is to create skilled and qualified water engineers vital for the effective utilization of the vast untapped water resources for various aspects of development. The institute is also engaged in water research studies pertinent to small-scale irrigation schemes so as to effectively reflect the conditions existing in the country. In addition, the curriculum is focused on training small scale Irrigation technician who in turn will help promote small-scale Irrigation schemes. An earlier survey indicates that there was approximately 121,493 hectares of land under the category of small scale or peasant irrigated farming practice.[89] In total, since the coming to power by the EPRDF small scale irrigation activities in the country has increase

87 Five Year Development and Transformation Plan 2011 – 2015, Ministry of finance, FDRE (2010) http://www.ata.gov.et/priorities/national-growth-transformation-plan/

88 CSA, (2013/2014).

89 CSA, (2004/2005).

extensively from what was 63,000 hectares in early 1990s to over 166,000 hectares in 2011 including fields under seasonal irrigation regime.[90] The expansion of small and medium scale irrigation projects is likely to continue in the foreseeable future. Currently feasibility study and construction work on 426,097 hectares is underway in many parts of the country.[91]

It is important; therefore, that research on irrigation practices become an integral part in the rain dependent research center in order to enable peasant exploit rainwater even though it is seasonal. It is understood that programs aimed at changing the lives of the people require the participation and even approval of the farming community at large. This will only strengthened the capacity and the confidence of the peasant farmer whose farm practice is centered on mitigating the unpredictable and erratic environmental behavior. It is also a canny way of involving the peasants from the outset to a domain very unfamiliar to them.

Irrigation is an art as well as science upon which the safety net of human survival was built in the past. It is an art because it demands not only reverence to the physical environment with which one interacts but also the ability to internalize the practice of it. Furthermore, it is a science that requires, precision and accuracy to effectively utilize not only the resource available but also to stretch the knowledge drawn from empirical observations and repetitive practices of past human generations. Earlier civilizations in Mesopotamia, the Nile valley, Indies and Ganges flourished partly because of irrigation practices which provided predictably uninterrupted supply of food and basic necessities of life. The life span of many of the past civilizations and the survival of those who built them depended on the careful manipulation water and their judicious interaction with the land. The demise of Mesopotamia, the region of the world most often cited as victim of injudicious irrigation practices begun in the early part of the second millennium BC. Thus, without the proper use of irrigation techniques, the danger of rendering previously productive lands to a barren wasteland is highly probable.[92]

Today instead of abandoning lands affected by excessive salinity, sodicity or water logging conditions, irrigated farms are reclaimed at extremely prohibitive cost. In other words, using time tested engineering techniques the land can be rehabilitated for effective use over and over again but at a considerable cost. Unfortunately, peasants in Ethiopia, unlike their counterparts in Pakistan, or the Egyptian Nile Delta or the Gezira scheme in the Sudan are novice to irrigation works that it could take years before the

90 CSA, (2013/2014).

91 Ministry of Water and Energy. FDRE, (Addis Ababa: 2013), http://www.mowr.gov.et/index.php?pagenum=4.2&pagehgt=1000px&ContentID=60

92 The Gode large scale irrigation in the Wabe Shebele basin of Eastern Ethiopia is a typical example where injudicious irrigation practices can convert once fertile land into barren waste land difficult to reclaim.

practice becomes an integral part of the agriculture system. The danger lays not so much so in their adeptness to irrigation as source of moisture but more in their ignorance to prevent degradation to the already fragile soil system by promoting sheet erosion in the plateau where land gradient is more than 10%. Similar scenario is expected in the lowlands where arid soils are prone to salinity effect. Hence, for the crucial role that the Arba Minch Water Technology Institute and other engineering colleges are expected to play in the future.

The water delivery method to cultivated fields is as important to ensuring food production as water conservation is to guarantying supply of moisture all year round. In this regard the cheapest method of conveying water to small-scale farms and peasant holdings is gravity system where water is diverted from a commanding position to water cultivated fields. Depending on the gradient, drop structures are installed to minimize uncontrollable flow of water and avoid erosion. Once water is diverted from the source, it can either be admitted to cultivated fields through one of the following methods; flood, basin or furrow systems of irrigation. These are collectively called surface water application system. Such a system does not only help raise food and forage crops but also facilitate the recharging of ground water. This technique is also useful to wildly flood natural pasture in the rangelands. Given the low level of technical requirement to operate and the reasonable fund needed to maintain the system, therefore, gravity irrigation is appropriate to peasant farming. Management risks notwithstanding, irrigation remains the best option for Ethiopia blessed with abundant human resources including water.

The disadvantage of this system is that it exposes soil to serious degradation and substantial amount of water are lost due to seepage and evaporation. Since irrigation Water is admitted indiscriminately along the furrows or inundated over the flat basin wastage is unavoidable. Moreover, in order to wet the tail end of the field more than necessary water is provided to the head section of the farm closer to the channel inlet. This system also encourages the infestation of weeds on cultivated fields. To circumvent problems caused by seepage through the water channels many countries line their canal beds with impervious clay, concrete or cover the length and breadth of the canal with plastic sheets. It is estimated that the indicative values of conveyance efficiency is 30% less in earthen than in lined canals.[93]

Almost all the irrigation system in Ethiopia is surface type where water is admitted through furrows or pounded in the basin. In many cases water is withdrawn from source using diesel engine operated pumps. Funds permitting, peasants could construct diversion weirs from the source to extend the exploitation of the river/stream by raising the water level. This technique not only ensures prolonged use of the water from an ebbing river but also allows

93 FAO, Irrigation water management. Irrigation scheduling. (FAO,Corporate Document Repository Training Manual #4, 1989).

one to admit water to the field through gravity system. As indicated earlier, the downside of the gravity commanded surface irrigation system is that it is less reliable simply because of its dependency on the perennial flow of the source. In Eritrea's Gash basin water to irrigated farms is delivered by lowering down water pumps right on the dry river belly to draw water from below the surface a month or two after the big rain formally ended. In the past this practice was common to cultivators in the Humera area as well before the construction of the Tekezze Dam in 2010. Moreover, salinity in soils is more apparent with surface irrigation than with the rest of the irrigation systems.

The alternative to surface method of water application is the Center pivot sprinkler and drip irrigation systems. Resources permitting these systems are relatively environmentally friendlier and in the long run more enduring venture to pursue than any systems known to man. Still, Drip is superior to any irrigation system to date. Its efficiency is universally acknowledged to be 90% against surface irrigation, which is less than 60%. Long term, comparative cost of surface to drip irrigation study conducted in India's large scale farms in 1998 shows that the overall cost for surface irrigation systems was three fold higher than drip while cost for one hectare cm. of water for drip was 25% less than surface.[94] In the long run sprinklers or drip are superior to any systems of irrigation because of the ease of operation, uniform and measured application of water to cultivated fields. Unlike surface irrigation which requires leveled ground for even application of water, sprinklers or drip faucets works very well on rough terrain thus, reducing cost of land leveling. Despite their desirability, pivot and Drip irrigation system are too expensive to adapt by an individual peasant farmer because of the initial cost involved. For the system to operate smoothly, it requires many vital components such as low energy pumps, Pressure regulators, timers, sensors, adapters, thousands of meters long plastic tubes and emitters. Thus, the challenge to install Drip irrigation can only be met by members of the peasant associations willing to muster resource for this purpose.

Although many of the major accessories for drip and sprinkler irrigation are expected to be procured from abroad, the possibility of locally manufacturing some essential parts such as PVC hoses, nozzles and couplings is well within the capacity of the plastic factories in the country. Undoubtedly, using locally manufactured irrigation accessories, will lower prices and reduce initial investment cost drastically and make drip attractive to producers. As water becomes scarcer and likely to be the bone of contention among communities and between countries in the future more of this system will be adapted as a matter of choice.

---

94 Sree Ramulu, Management of Water resources for Agriculture, (New Delhi, New Age International Ltd. Publishers, 1998).

The amount of water required to raise raw crops and horticultural plants by using drip irrigation system is half to a third of the flood system. In other words application efficiency for Drip and sprinkler irrigation methods can go as high as 84 percent and 73 percent respectively.[95] Adapting drip and certainly sprinkler systems may also give some element of flexibility in the farming system by inviting farmers to cultivate wide variety of crops ranging from cereal, to horticultural, and industrial. Water being one of the production limiting factor, this system will permit the cultivation of forage all year round and encourage pastoralist to maintain few but highly productive head of cattle instead of the multitude and unproductive as that. It should be borne in mind that, raising huge number of cattle is one of the safety mechanism devised by people leading a marginal life. Such practices are necessary for the pastoralists in order to avert or withstand production risk factors, example disease and drought episodes. With water and animal feed available all year round it is possible for the cattle rearing population to engage in livestock enterprises that are much more productive, economical and sustaining.

Water works that include irrigation as a source of moisture to crops ought to assume a participatory approach to guarantee relative success. In hindsight, if only the MWRC had geared many of its water related experimental activities half century ago to cater for small-scale farming and moisture deficit areas of Eritrea, Tigrai, Wollo, Hararghe and parts of Shewa and Gondar the outcome in the struggle for food self-sufficiency would have been positive. Unfortunately, the development policy pursued by successive government failed to recognize the importance of the peasant sector to raising productivity for the common good. It is essential; therefore, that attention is paid to peasant needs and efforts must be exerted to help them utilize their resources effectively. The participation of peasants in irrigation and other water and soil conservation works can take many shapes and forms.

1) research planning and execution
2) in situ experiments with slight regard to cost
3) farmer/researcher managed observation posts
4) direct technical support through development agents
5) field days to demonstrate and verify research result

---

95 1/ A.G Smastral, Efficiency of Florida Agriculture System. University of Florida Coop. Ext. Serv. Bulletin # 247(1988). 2/ Faten Al Juneidi & Jad Isaac, Water and Irrigation Management in the Palestinian West Bank, Applied Research Institute, (Jerusalem: (ARIJ), 1999), 5.

The FTCs are perfect vehicle to promote the indispensability of irrigation to community of farmers and introduce water management practices as well.

Despite its unintended and frightful consequences, wisely managed irrigation scheme is the only way out of the intractable problem that Ethiopia faces today.

History of Research/Extension

Structured research activities begun in Ethiopia over half a century ago. Within the few decades of its existence, the research institution had made some strides especially in the development of crop sciences. Today, there are fewer than forty varieties from different crops species in use or in the pipeline waiting for release. Most are superior in their performance and in almost all cases dwarf the indigenous types. Unfortunately, many are also location and season specific instead of being stable, that is, widely adaptable across different agro-ecologic conditions. But more tragically, the absence of well-structured seed production and distribution mechanism within the country made their widespread use quiet restrictive. Among the newly developed crop species cotton stands high. The crop that had for centuries occupied an insignificant position outside the cereal cultivation areas of the plateau suddenly became one vital to the textile mills of the country. Indeed cotton is the one and only crop that have reaped the benefits that research activities bestowed on agriculture in the country. With the exception by which the cotton lint is picked, the crop is almost fully mechanized.

The expansion of cotton cultivation through research was complemented by the rapid and ease distribution of the cultivars among farms that are big and compact. Cotton cultivation still remains under medium and large-scale management. The beneficiary farms are also conveniently located and in almost all cases operated under irrigated conditions. During the rule of the socialist regime, the irrigated cotton farms expanded exponentially along the banks of the Awash River and the southern rift valley of Blate, Arba Minch and Sille. Later the cultivation of cotton was extended to Omrate in the Omo valley and along the banks of the Wabe Shebelle Basin of the Ogaden. It is highly likely that cotton together with sugar cane cultivation will occupy prominent position for some time to come.

Perhaps the second most focused area of research was on horticultural crops in particular citrus and banana. The research which begun in early 1960s had by mid-1980s enabled the Horticulture Development Corporation of the State Farm to export insignificant amount of horticultural crops to neighboring countries and Europe. The collapse of the Horticulture farms in Eritrea resulting from the civil war that engulfed the territory beginning the 1970s gave further impetus to horticulture development in Ethiopia, hence for the rapid increase in research. In any case at the expense of the limited international market prospect for Ethiopian fruits and vegetables in the 1980s, the local market flourished.

Today, expansions in horticultural and floricultural cultivation are going at a steady pace earning about 160 million dollars in mid-2011. By the end of this decade the country is projecting to export 1.4 billion dollar worth of horticultural goods annually to Europe and the Middle East.[96] Most of Ethiopia's export to the outside market is flower amounting to 80% while fruits and vegetables make the rest. Overall, research endeavors of the past two decades on fruit, vegetable and flowers is more than satisfactory. All being equal the country is endowed with diverse agro-climatic condition to make it competitive in exportable agricultural commodities. Taking advantage of the investment opportunity provided by the government a number of local and expatriate investors are vying for more land to develop horticulture enterprises.

However, one cannot say the same when it comes to cereal, pulses and oil crops that are predominantly cultivated by the peasant sector.[97] Although research on cereal and pulse was simultaneously conducted throughout the decades, the yield per unit land for most raw crops and pulses remained unimpressive. For the past five decades the yield of wheat, maize, sorghum and Teff (*Ergrostic abyssinica*) ranged between 0.7 MT/ha for Teff and 1.1 to 1.4 MT/ha for barley and wheat respectively. Although hard to attribute to research alone, today the national yield per hectare for wheat fluctuates between 1.6 to 1.8 tons per hectare an astounding increase of more than half ton per hectare from a decade ago. Similarly the yield per hectare for maize has increased by half a ton from 1.6 to 2.1 MT today. The yield for improved cereal seeds and in particular maize under experimental condition is astonishingly high ranging between 8 and 9 MT per hectare.[98] Under the circumstance, in good years, that is, when nature is merciful and all other production factors are in place, the annual cereal production tops at 13 million MT with food aid request standing at three-quarter of a million MT. While in lean years, food aid request soars to close 10% of the total annual production, which amounts to 1.3 million MT. Today cereal output is an all-time high with 21.6 million MT reaped in the *meher* season of 2013.[99] This indicates that within the past two years cereal production increased by 2.8 million MT[100]. As encouraging as the total cereal production might seem, population size has also increased from 57

96 Ethiopian Horticulture Producers and Exporters Association,(Addis Ababa: Feb 14, 2009) http://www.ehpea.org.et/aboutus%201.htm

97 1/ Reuters, In 2008 the country exported 400,000 MT of oil seeds, pulses and spices worth 368 Million dollars, (February 2009) 2/ Bloomberg Business, Ethiopia's Bank-Secured Sesame Seed-Sales Seen Boosting Exports. In 2012 sesame seed export went up by 20% earning close to half a billion, (December 2012).

98 IAR Newsletter of Agricultural Research Vol. 11 No. 4, (1996).

99 CSA, (2013/2014).

100 CSA, (2012).

million in 1996 to close to ninety million today striping the benefit achieved during the past decade.

The irony of it all is that, a great deal of resources was devoted to the development of cereal, pulses and oil crops more than livestock. Moreover, there are more research personnel involved in breeding and selection of crop species as there are in crop husbandry. And yet research on wheat, maize, sorghum, sesame, rapeseed, and fava bean to mention few have failed to make a dramatic impact on food sustainability. There are numerous reasons for that. As shown in section three, research undertakings were not directly focused on solving production problems facing the peasant population. What needs to be underlined here is that the peasant farmer constitutes the most single vital force in the struggle for food self-sufficiency. The sheer size of it makes the peasant sector important in any agriculture undertakings. The fact remains that agriculture accounts for half of GDP, 60% of exports, and 80% of total employment. In spite of this, research strategies bas been and to a limited extent continued to be conceived, planned and executed beyond the realm of the peasant farmer. True enough, few in the research field were completely oblivious of the fact that the peasant should be the central focus of their activity. Unfortunately, many of the objective reality under which the peasant operates were overlooked and not even acknowledged in the planning process. In the past, peasants are rarely consulted either informally or through structured questioner on setting up agricultural research priorities. Thus, despite their tireless efforts research scientist, most often find themselves on hot water and sometimes harsh criticism and ridicule by their compatriots frustrated by the horrible image that their country have on the world scene.

Notwithstanding the views expressed by many critics, the poor agricultural performance witnessed in the country is directly linked to the weak research infrastructure that does little to support the effort of the farming community to produce enough food for the nation. It is always lamented that the gap between agriculture research and the extension is so wide for the agriculture sector to suffer in Ethiopia. To this end the new government focused on one central issue, the decentralization of research activities, that is, divesting power from the main coordinating body the Ethiopian Institute Agricultural Research (EIAR) to regional research centers. This step was aimed at bringing research and the producers closer. For the past five decades the EIAR, was the hub of all research activities administered centrally. It was this body that reviewed research proposals, follow their implementation and dispense funds for each activity.

For many the action taken to decentralize the research activities is no more than a political gesture, which does not address existing problems. Critics argue that the decision to reorganize agricultural research activities on regional bases is to reinforce the idea of Ethnic driven administrative setup that the government has encouraged since coming to power. In other words it is hard to dispute the fact that, the measure taken so far was in line with the EPRDF's commitment to establishing and strengthening federally autonomous regions based on ethnic constitution. In any case the law of the land granted the autonomous administrative bodies a considerable executive, legislative and judicial authority with the freedom to pursue their separate development strategies. With this as a background, therefore, the government was less hesitant to push the idea for decentralized research activities within each region. Immediately, thereafter a minor friction emerged between those in the regional centers and at the Headquarter, which in previous decades was the ultimate power. Furthermore, many of the senior scientists resented the government action to the extent that some left for the private sector and few look employment elsewhere. Today the scenario has totally changed with thousands of students graduating from agricultural and related colleges to make this as contentious issue.

Decentralization as perceived by the EPRDF was meant to encourage linkage between the research personnel and the peasant farmer at various levels of activities. It is accurate to state that up to this period there was little empathy by the research personnel toward the peasant farmer. Perhaps because of their social make up as the majority are of urban background or became class on their own to be sensitive to the need of the peasant. As a result the producer is unable to achieve the desired objective that is, of transforming the country from food deprived to surplus producing one. However, decentralization on the field of research does not automatically solve the problem of the much-needed rapport between the service provider and the user. To the contrary it may further stretch the meager resources available especially in the field of scientific endeavors. In the short term it could also create loss linkage among the between research personnel in different regions and their activities may foster duplications that the country cannot afford. Until the early 2005 cooperation among the regions and between headquarter was casual. This type of relationship is bound to continue as long as regional interest overrides the rest. Given the country's economic position it is more rational to strengthen few research centers that seem to cover broader agro ecologic zones but investing more on farmer-peasant inspired experimental undertakings at the regional level.

As described briefly in section three, the protocols used to generate production packages are quiet tedious and time consuming. The time spent from preliminary screening of a production input to observing its performance and finally to its certified release may take as many as 9 years. This indeed is an accepted procedure most often followed by the agricultural research institutions worldwide. Other production support inputs such as fertilization of the soil, tillage practices or the screening of pesticides and herbicides for their efficacy may take fewer years than the release of an outstanding crop variety. Given the magnitude of the agriculture problems in Ethiopia and the precarious position of the peasant who toils the land without the use of improved seeds or production support inputs most of the time; it is indeed unwise to strictly adhere to the international protocols of rigorous testing.

On the other hand, the risk involved in the production and the early release of packages to farmers and bypassing the 'normal' procedure of testing and retesting could be enormous. Specially, when one relates the incident of the newly released variety of wheat, *Dashen*, that broke down by yellow rust in the middle of the growing season creating havoc among farmers during one of the rarest best season of the decade. However alarming this may sound, in a circumstance where farmers' lives are practically from hand to mouth, early release of varieties and production packages outweigh the risk. What counts most is the multiplication of many back up cultivars and packaging them for any eventualities. It is important to underline here that, unlike many agrarian based economies where subsistence livelihood for peasants is somewhat assured, in the Ethiopian case even a bumper harvest does not absolutely guarantee food for great many of the peasant household all year round. In other words many if not all peasants in Ethiopia today lead a precarious subsistence level of existence. It is imperative, therefore, that research methodology and the means employed to dispense research outputs are more flexible and reflective to peasant reality.

The protocol followed to release improved seed and the authentication required to recommend production packages for wider use have eased quiet dramatically in the past years. The distribution of improved seeds gained a modest increase from 13,805 MT 1998/99, to 22,000 MT. 2007/08.[101] The use of hybrid maize also went up from 8,700 MT in 2008/2009 to 19,300 MT in 2010/11[102]. This in a way is a testimony that many of the backlogged materials are made to see the light of day. Despite its good intentions and its stated policy of allowing private seed developers to take part in the production and distribution of seed materials, the government have fallen short of its promises. In recent years private seed companies begun to appear but their contribution to the overall supply of seed is less than 10%. The federal and few regional seed enterprise dominate the production and distribution of improved seeds. The current seed requirement is put at close to half a million tons and only less than 10% of the cultivated land are under improved grain seed.[103]

---

101 1/Teshome Adugna, Birth of The Ethiopian Seed Trade Association, Seed Info, (Addis Ababa: Ministry of Agriculture, Ethiopia. Newsletter of the WANA Seed Network - June 2007 *2*/ Marja H. Thijssen, Zewdie Bishaw, Abdurahman Beshir and Walter S. de Boef, Farmers, Seeds and Varieties: Supporting informal seed supply (Wageningen International, 2008).

102 The 2005-2010 Annual Report of the Ministry of Agriculture and Rural Development, (Addis Ababa: 2010).

103 1/ 400,000 to 500,000 Metric tons of the major cereal crops are required to cover the current cultivated lands in the country. However, the amount of improved

No doubt it takes considerable time to introduce new innovations to peasants. Under such circumstance, therefore, it is neither economically wise nor technically justified to limit the selection criteria to one superbly performing variety by excluding many promising ones as candidate for immediate multiplication. Today, with government encouragement researchers are no longer obligated to follow the strict protocol of seed release. Some promising types are multiplied at the research station grounds and approved for limited release giving the producer a room to maneuver and expand his/her choice of planting materials or production packages. Notwithstanding, the problems that the Ethiopian peasant faces every day, it is not wise also to underplay the strict requirement of rigorous testing of planting materials or production enhancing packages and their release. With a bit of common sense and the genuine participation of the peasantry in the selection process, a middle ground that could have a far-reaching effect can be achieved.

Assuming that the research centers were in a position to release multiple varieties from different crop species one after the other, in all honesty, this would be a daunting task for the minuscule Ethiopian Seed Enterprise (ESE) to bravely take the challenge. However, it is possible to extend the current practice of contracting large and small farmers to multiply seeds according to the standards set by the enterprise. The current practice by Self-Help, a European NGO to supply seed through the farmer-contract seed multiplication scheme is truly innovative to emulate.[104] The ESE can provide the necessary technical back up and inputs for the production of up to standard types of seeds. More importantly, the government needs to encourage private seed developers together with the SPC to accelerate the production of improved seeds and replace inferior types.

On average Ethiopia's food deficit is estimated at 1 million MT annually. All being equal, the food gap could only be met, if there is an across board but steady increase of yield by 100kg per hectare per year. This translates into an increase of million-ton food crop planted in 10 million hectares annually. This also suggests that the country could slightly stagger toward food self sufficiency without attracting any attention from benevolent donors. This is of course wishful thinking considering the low level of agriculture development at the

---

seed supply is very small i.e. about 22,000 Metric tons or about 6% of the potential demand for all crops per year. The establishment of a national seed trade association was one essential component lacking in efforts to strengthening the private sector. The Ethiopian Seed Trade Association (ESTA) was established in 2005. *2/* Marja H. Thijssen, Zewdie Bishaw, Abdurahman Beshir and Walter S. de Boef. Farmers, Seeds and Varieties: Supporting informal seed supply, (Wageningen International, 2008). Other estimates put the annual seed requirement to 150,000 MT.

104 George Jacob, Emergency Nutrition Network. Grassroots seed Multiplication in Ethiopia Issue 31 (September 2007) http://fex.ennonline.net/31/grassroots

moment and more significantly of unchecked population growth. Moreover, almost all agricultural activities are rainfall based and erratic as that. The intensity and the pattern of rain distribution is such that it calls for an alternative approach of maximizing rain water or using supplemental moisture through irrigation.

Agriculture research in Ethiopia has so far did not come up with practical solution through tillage practices to conserve moisture nor develop genotypes with resilient characteristics to withstand adverse growing conditions. It is imperative therefore, that resources are expended at local level to explore moisture conservation measures such as minimum tillage, stone hedging, field bunds etc. In this regard the Tigrai and Amhara Credit Associations are exemplary for extending loans to farmers engaged in water and soil conservation practices. Moreover, drought tolerant varieties including the landraces that make the backbone of the peasant seed stock for millennium should be developed or improved and their availability assured as an alternate source of seed supply. More importantly as repository in situ for universal use by mankind. Ethiopia in her own right as one of the center of origin for some and the center of cultivation for many plants, boasts of having the seed gene bank, the Ethiopian Institute of Biodiversity located in Addis Abeba build with the help of the German government decades ago.

Fertilizer: the miracle stuff

Fertilizer is one of the essential inputs sought by farmers throughout the developed world. Its ability to restore the fertility status of soil and by extension boast production is universally recognized. In Ethiopia, despite the common practice of blanket application, that is, without the slightest regard to soil physical and chemical variations, the response by many crop species and in particular cereals to DAP and Urea has been great. In other wards the yield advantage of fertilized fields over the control has always been impressive. Unfortunately, the per capita consumption of fertilizer by farm household is low even by the developing countries standard. Past estimate put the consumption of fertilizer by farmer as insignificant.[105] The main reason for this extremely low level of fertilizer use in the country can be attributed to the weak purchasing power of the peasant and the benign neglect by previous governments to introduce it diligently and widely.

Barely one year after the bumper harvest of 1996/97 was celebrated with fanfare, the government reluctantly lifted fertilizer subsidy in its entirety.[106]. The

---

105 FAO Statistics (2001/02).

106 Former US president Jimmy Carter and Norman Borlaug, one who lead a team of Asian and US scientists to bring the Green Revolution attended a workshop at Economic Commission on Africa (ECA) to herald a new beginning toward food

reason for doing away with the subsidy was as comical as the parading officials who came out one after the other to herald the near end to food insecurity in the country.[107] No doubt, that particular season referred to here was memorable in that close to 10 million MT of cereal was harvested and at the same time some surplus grains were exported in regional market. The basis for the exceptionally good harvest in 1996/97 cropping season was partly due to the rational distribution of rain across the country and good amount of fertilizer use by farmers. Sadly, a year after the subsidy was lifted; grain yield was reduced by an estimated 25% from the previous year.[108] Again this was compounded by poor rains during both the *kirmet* and *belg* season, literally amplifying the effect of fertilizer cut back.

Admittedly, it is indefensible argument to simply rely on the outcome of the back to back seasons where the first one happened to score bumper harvest while the succeeding season ostensibly due to the removal of fertilizer subsidies (and poor rains) showed poor production performance. No matter how one try to look at it there is little doubt that, removal of fertilizer subsidies the year after the bumper harvest must have compounded for the poor harvest of 1997/1998. Indeed production was drastically reduced as the purchasing power of the peasant plummeted to almost half of the previous year.[109] But, it is not completely accurate to justify the abysmal performance of the 1997/98 season simply on the bases of the presence or absence of one input alone without examining the underlining interactions of both biotic and abiotic factors such as pest and weed control and most of all adequate moisture, timely land preparation etc. This is tantamount to oversimplifying the complex agriculture problem in Ethiopia.

It should be born in mind that the famine of 1984/85 that is believed to have claimed close to one million lives and debilitated another 5 to 10 million people was neither abated nor prevented by the record harvested obtained the previous year, that is 1983/84. Agriculture in Ethiopia has no shock proof to withstand any deviation from the norm. In other wards it is prone to climatic change, biotic attack and abiotic infestation and a whole lot of institutional pressures to be influenced by fertilizer alone.

This is not to argue that application of fertilizer to soils in areas receiving

---

self-sufficiency for Ethiopia. Indeed the 1996/97 cropping season was the best ever reaping over 10 million MT of grain altogether.

107 The bumper harvest obtained that year was announced by the then Minister Agriculture Dr. Teketel Forsido in a press conference pledging that the coming year will be as good as the past.

108 Feeding the Future Newsletter The SASAKAWA African Association. Issue # 2 (April 1998).

109 FAO Statistics, (1996 and 1997), Based on FAO statistical data the purchase/consumption of nitrogenous and phosphorus fertilizer in 1996 cropping season was 140 thousand MT compared to 100 thousands the following year.

high to medium rainfall is not beneficial. To the contrary, fertilizer augmented by modest production enhancing inputs such as herbicides and pesticides, optimum time of planting, timely agro-technical operations can give remarkable results. Suffice is to mention that in high production potential areas, harvesting operations and storage capacities not to mention transportation facilities can be daunting task to a subsistence farmer who because of favorable season and the use of fertilizer/seed were able to maximize on their modest investment.[110] However, in some regions of the north in particular in those moisture deficit areas of Eritrea and Tigrai peasants are usually reluctant to use inorganic fertilizer for fear that it may burn the crop at the seedling stage. Apart from this valid concern expressed by some, most farmers are eager to apply fertilizer as a means of boasting production.

Fertilizer use by peasants is steadily increasing perhaps not at a desired pace or at rates recommended by scientist in the country, nevertheless it is much sought commodity by peasants today. It is important therefore, that for a government which made ADLI as a center piece of its policy need to shoulder if not in full but partial cost of the fertilizer to encourage peasants incorporate the novelty in their farming systems. If indeed the argument holds water that the reason for decreased cereal production after the bumper harvest of 1997/98 was due to fertilizer cutback then one cannot help but advocate for increased fertilizer use at a rate affordable by the peasant. It is bad policy to abruptly cut support from the sections of the society that need most. For whatever reason it is premature to cut subsidizes that are proved to be major factors for increase production. In other words it is wise to heed the physical indicators at home than accept prescriptions that are alien and unworkable.[111]

The distribution of inputs needs to be accorded the highest priority in the overall development strategy until such time that the nation is declared self-sufficient. At this point in time it is farfetched to talk beyond attaining food self-sufficiency where citizens in general are adequately fed and children receive the required calories and body building nutrients.

Price notwithstanding, the use of fertilizer by peasants has increased drastically since 2001. The total area covered by fertilizer has increased from 2.8 million hectares in 1997 to 5.6 million hectares in 2009. At present about

---

110 Taye Work the General manager of the Institute of Agriculture Research (IAR) shared an interesting story with the author in 1985, regarding the predicament of peasant associations in the Negele /Shashamane area who reaped bumper harvest due to fertilizer and herbicide use but lacked the means to harvest, package and even store the technology induced gift.

111 Prime Minster Melees Zenawi was one of the vocal advocates of Developmental State where he argued that government need and must participate in facilitating development efforts in the third world. This is diametrically opposed to the neo liberal economic theory peddle to Africans for decades, that is, downplaying the role the State plays in supporting economic development of a nation.

0.7 million MT of fertilizer is consumed by grain farmers.[112] However this may be, the figure does not include areas covered by crops other than grain. Commercial farmers cultivating cash crops, flowers and fruit species consume a great deal of fertilizer. In the past decade or two many high input commercial farms were established in the country to share whatever is available with the peasant farmer.

Today Ethiopia is in the process of overtaking Kenya the leading flower exporting nation in Africa. The floriculture and horticulture farms everywhere are known to share a bulk of the fertilizers consumed always at the expense of food crops. It is also assumed that cotton commercial farms are expanding as the number of textile mills has increased the last few years. Sugarcane one of the most water-consuming crops is currently propagated outside its traditional zones of Wonji and Metahara and is likely to get the lion's share in the fertilizer consumption. In the next five years, Ethiopia is aiming at exporting 1.4 billion dollar worth of horticultural crops annually to Europe and the Middle East. Most are privately owned and intensively managed commercial farms.[113]

Livestock- Resource Extraordinaire

Recent figures indicate that the livestock sector contributes about 15% of national GDP, 40% of agricultural GDP.[114] The bulk of Ethiopia's animal resources are located in the plateau and the escarpments, leaving only one third of the total whole in the lowlands where the mainstay of the population is predominately pastoral and agro-pastoral. In the plateau, animal power plays an important role in land preparation and transportation. Small ruminants are also an essential source of income to the peasantry as well as an important supplementary source of food. While in the lowlands, livestock including goat, sheep and camel constitute the main source of food without which the livestock herder find it difficult to exist. The Ethiopian lowlands are also an important source of draft animals for the peasants. It is estimated that over 20% of the draft animal need by the sedentary population comes from the large pool of livestock found in the lowlands.[115]

Despite the periodic drought that first and foremost impact the animal

112 CSA (2012).

113 Ethiopian Horticulture producers and exporters Association Web site, (Feb 14, 2009).

114 Aleme Asresie and Lemma Zemedu. Contribution of Livestock Sector in Ethiopian Economy: A Review. Advance in Life Science and Thecnology. ISSN 2224-7181 Vol.29, (2015), 79.

115 Ayele Gebre. Pastoral Development and policy orientation in Ethiopia: appraisal of policies and experience. In *Agriculture Policy in Ethiopia. Economic development: scope, issues and prospects.* Proceedings of the 6th annual conference of Agriculture Economic Society. (2003).

population, still the number of cattle and small ruminants stands at staggeringly high to cause anxiety among environmentalists in the country. The livestock population in general went form 38 million in 2004 to about 51 million in 2010. This suggests that a yearly increase of two million animals[116]. It is understood that the sheer number of livestock living on land characterized as less endowed with nature's gift can cause havoc to the environment. The effect of the millions of livestock on grasslands including the scavenging action of goats and camels on shrub trees facilities erosion to an unprecedented scale. A land void of vegetation cover is subjected to rill erosion caused by rain flashes and wind action. In total, the livestock population in Ethiopia comes close to 111 million. This record number of animals put Ethiopia as one of the countries with high per capita share of animal resource in the world and most certainly in Africa. All being equal, it is correct to assume that livestock could play major role in Ethiopia's development.

According to the country's Customs Authority, in mid-2006 alone the country exported close to 75 million dollar worth of livestock and livestock products.[117] Given the amount of livestock available in the country, this is indeed an insignificant figure to relay upon for national development. In the 1980s the livestock sector constituted 33% of gross value of agricultural production and 15% of GDP. Moreover, export of live animals mainly goats and sheep was second to coffee in the same period.[118] However, legally exported live animals declined in subsequent years but picked up again in 2001 to an impressive jump from mere 800,000 live animals to 18 million by 2004.[119] This is despite the ban imposed for the past few years on the export of live animals from Ethiopia due to the Rift Valley Fever. In contrast, from 1990 to 1999, the custom inspected livestock export was less than 50,000 annually.[120] Available data indicates that Ethiopia exported 260 million dollar worth of livestock in 2011 trailing Somalia by 100 million dollars.[121] However, this

116 Ministry of Finance and Economic Development and Ministry of Agriculture. A Review to Improve Estimation of Livestock contribution to the national GDP. Integrated Multi Disicplinary Study, (2011), 15.

117 REUTERS SUNDAY JULY, 2006 quoted from customs authority.

118 Ayele Gebre, Pastoral Development and policy orientation in Ethiopia: appraisal of policies and experience. In. *Agriculture Policy in Ethiopia. Economic development: scope, issues and prospects.* Proceedings of the 6th annual conference of Agriculture Economic Society, (2003).

119 CSA, (2005).

120 Ayele Solomon, Assgid Workalemahu, Jabar M.M. Mohamed and Belachew Hurissa, Livestock marketing in Ethiopia: *A review of structure, performance and development initiative.* Soci-economic Policy Research Working Paper 52. (Addis Ababa: ILRI, 2003).

121 Yacob Aklilu, Regional Livestock Trade, lessons for South Sudan (2012) http://www.agfairsouthsudan.org/wpcontent/uploads/2012/12/Dr.YacobAklilu_FAO_28Nov2012.pdf).

export figure is unlikely to grow faster, at least not in the near future, until more emphasis is put to develop the sector.

In the dynamic world, quality is the determinant factor for achieving excellence. Because animal related health issues are of global concern, livestock and livestock products need to maintain an acceptable quality standard to stay in par with the fierce competition around[122]. There were instances in the past that Ethiopia was prevented from exporting live animals to the traditional market, the Arabian Peninsula because of disease outbreak. In fact the ban imposed since 1997 due to the Rift Valley Fever, which is suspected to affect both human and animal health, has reduced much of the trade that the country depended upon for the outlet of its huge animal resource that it have.[123] To address the matter a grassroots initiated conference of pastoralists was convened in Ethiopia in 2006 attended by 10 livestock exporting countries. The delegates and the NGOs warned that the livelihood of people is in jeopardy and that governments should intercede on their behalf to restore normalcy in the export market.[124] The export ban imposed notwithstanding, the need to improve livestock is overdue. It is clear that, the key to lifting future ban lay not in market proximity, which Ethiopia immensely enjoys, and not in quantity alone that it certainly has an edge, but in the quality of the product that it is able to provide.

In spite of the abundant livestock population and its modest contribution to the economy, livestock resource is still one of the most underdeveloped sectors that were neglected by successive governments. In order to appreciate the status that this sector enjoys in the overall development of the country, it is useful to visit the efforts thus far made in its improvement. Earlier a survey conducted on 27 different agencies engaged in agricultural research showed that out of 691 research personnel, 18 percent devoted their time to livestock research, whereas research and development on crops took the lions share.[125] This led, Alemu Ghebrewold a noted scientist who participated in the early formulation of the livestock research program in Ethiopia in late 1960 to admit grudgingly that decades after the dice rolled, animal production research is still in its infancy especially when viewed against crop production research.[126]

According to the EIAR's stated objectives, research on livestock was expected to contribute toward framing sound development policies and

---

122 Average yields per animal slaughtered or milked are estimated at 110kg of beef, 10kg of mutton and 213kg of cow's milk.

123 IRINNews.org, UN Office for the Coordination Humanitarian Affairs (September 2006).

124 The Reporter, Pastoralists Criticized the Saudi Arabia Ban on Import of Livestock (Addis Ababa: September 9, 2006).

125 Nienke M. Beintema and Menelik Solomon, ASTI Country Brief No. 9 (October 2003).

126 Dr. Alemu Ghebrewold, IAR Newsletter Vol. 12 (1997), 13.

generating technologies that improve the livestock sector through increased profitability, income distribution, biological efficiency while maintaining and enhancing biodiversity and natural resources. More specifically research on livestock was aimed at increasing milk and beef output through the genetic manipulation of cattle breeds, prevent the spread of diseases and invent cure for diseased animals and finally develop forage crops that are likely to do well under both irrigated and rain fed conditions.[127]

Unfortunately, this lofty goal set in the 1960s was short changed and therefore far from being realized. It goes without saying that, there are numerous problems that have not been tackled yet. The first and foremost is the constant outbreak of contagious cattle diseases. The frequent outbreak of *Rinderpest, Anthrax, blackleg, CBPP*, and *Haemorrhagic septicemia* usually leave trail of carcass across the landscape.[128] Ethiopia had experienced an outbreak of Rinderpest in 2001 killing hundreds of thousands of cattle. Notwithstanding, the problem facing the livestock sector, efforts so far made by the current government to close the gap is admirable. In 2004/2005 seasons, approximately 8 million of livestock were vaccinated for *Haemorrhagic septicemia* disease and between 5 to 6 million animals for blackleg and *Anthrax* diseases. The number of livestock vaccinated in 2011 has tripled with the measure taken to prevent *Blackleg* disease alone accounting to almost 6.5 million. [129] In cooperation with multilateral organization, the government has successfully stamped out the deadliest disease of all the *Rinderpest,* introduced to the region by the Italians in late 1880s.

Vaccine production for these deadly diseases remained steady for a considerable length of time ranging between to 2 to 5 million doses annually. The task of containing those easily communicable diseases, therefore, becomes a daunting task for the widely dispersed veterinary service dispensing centers in the country. From time to time the situation becomes dire in particular if the outbreak of the disease is preceded by drought conditions. Moreover, if the scenario repeats itself too soon and at frequent interval many especially the cattle herding population may permanently remove themselves from the land of their birth in search of temporary relief.

The degradation of natural pasture in the rangelands and the grasslands and the lack of industrially constituted animal feed for dairy as well as beef production in the urban centers further exacerbated livestock production in Ethiopia. In Borona, where the annual mean rainfall is in the range of 400 to 600mm the carrying capacity of the rangelands are on the order of 14 Tropical

---

127 Institute of Agriculture Research Mission Statement.

128 FAO Media Center, Rinderpest is declared eradicated by mid-2010 and the formal announcement of its eradication has come from the UN FAO in Mid 2011(2011) http://www.fao.org/news/story/en/item/80894/icode/

129 CSA, (2010/11), Livestock and Livestock Characteristics, Agricultural Sample Survey.

Livestock Unit (TLU).[130] Similarly, in the northeast rangelands, where rainfall is somewhat unpredictable and less in intensity, the carrying capacity is by far lower when compared with the southern rangelands. Clearly, the vulnerability of livestock to vagaries of nature has remained as real as ever. Together with this, the biological mishaps that often occur also play havoc upon the livestock population. No doubt, many of the factors combined greatly undermined the safety net developed long ago.

There is no widespread irrigated pasture one can speak of. In the late 1980s attempt to establish a 1000 hectare of irrigated pasture at Ingelele in the North East Rangelands was unsuccessful. Despite the technical preparedness detected in the early part of the project phase, somehow the work stalled and remained idle for a long time until it was leased to private developers in the 1990s. Some of the physical layout invested in the mid-1980s was the installation of huge water pumps accompanied by land development works necessary for the cultivation of pasture. Similarly, in Gewane some 80 kilometers north of Ingelele another pasture development scheme fizzled out of existence.

There were four range management projects executed for the benefit of the pastoralists in the East, South and North East since 1958. Each project appeared to compliment the other. Each was designed to deal with problems related to forage development, veterinary service, epidemiology, and livestock genetic improvement, roads, water points and chemical dipping basins. These World Bank/African Development Bank financed rangelands were not only badly managed but tragically became the bone of contention for the various cattle herders inhabiting in and around them. Hardly, a time passes without noticing a clash between the Boran and other ethnic groups such as Garri. The frequent but bloody confrontation between the Afars and the Issas and the seldom skirmish occurring between the Afars and the Kereyus attests to the fact that no one, not even the local experts at the time carefully studied the implication of fencing range lands of contiguous nature favoring one over the other. The Alideghe, the center of the North East Rangeland Development Unit (NERDU), which in turn is part of the Third Livestock Development Project (TLDP), has been an intermittent fighting ground for the rival Afar and Issa vying for the watering points and pasture grounds. The Alidgegh, a 200,000 hectare of traditional grazing ground of the Afars and perhaps for many pastoral groupings in the distant past, now borders the Issa and the Keryuh lands which incidentally are one of the victims of the earlier 'land grab' for public and private use.

On the other hand, the Jigjiga Range Development Unit (JIRDU) part of the TLDP development sites had experienced frequent destruction of its

130 TLU is equivalent to 250kg live weight. Cattle 0.7 TLU, Equine 0.5 TLU, and sheep/goat 0.1 TLU, Jahnke 1982.

rangeland fences and the misuse of its livestock resting places by the very nomads on whose behalf the project is purported to exist. In the past the pastoral population openly expressed their contempt for the institution. Many are oblivious of its existence and have little to do with it or rather the institution do with them. The headquarters of JIRDU, which later was incorporated into yet another grandiose project initiated in 1990, the South Eastern Rangelands Project (SERP) is a neat campus equipped with sophisticated electronic gadgets and computers, manned by expatriate and local experts, appeared more intimidating than humble and friendly edifice ready to help. Putting appearance aside, in a concrete term, JIRDU/SERP were unable to minimize the impact of the Rift Valley Fever disease, which caused disruption on the livestock export market.[131] The failure of JIRDU to live up to its expectation suggests that the project was conceived and executed without popular participation. How else, does one explain the reckless action of the nomad who has never hesitated for a minute to dismantle and get rid of paddock fences supposed to be in place for pasture regeneration? The third sub-project, the Southern Rangelands Development Unit (SORDU) is the largest area covering close to 100,000-km^2. There are more manmade ponds in this sub project than anywhere else in the semi arid regions of Ethiopia. By all accounts SORDU is believed to have performed well than the projects located in the North and East of the country

Except, for the First Livestock Development Project (FLDP) which came into effect in 1971, to address the problem of milk production in Addis Ababa and its environs, the three projects that came after it were made to focus on the pastoral way of production as well as in developing abattoir in provincial towns. In this regard, the Second Livestock Development Project (SLDP) was designed to tackle marketing and other infrastructural issue while the TLDP was set to focus on developing and managing range lands in the Afar, Boran and Somali plains. Among the benefits expected from the TLDP activities would be roads, watering points, livestock markets, veterinary services, and facilitation of inter regional trade. It is interesting to note that the last three major livestock projects were launched amid the variety of civil wars that consumed the nation for more than three decades. It is, therefore, doubtful that many of the goals are achieved. Interestingly, the Fourth Livestock Development Project (FoLDP), like the ones before it, was launched without serious assessment and evaluation of the preceding projects. This chronic malady has been the characteristics feature of the *ancien regime* and those who came after it, especially the *Derg*. The project was launched in 1988 and has a

131 According to CDC Special Pathogen Branch, Rift Valley Fever does not only infect cattle and small ruminants alike but also causes illness among humans. Unlike Kenya, Somalia and the Great Lakes country, technically Ethiopia is free from the Rift valley fever disease but its livestock market is unfairly affected by mere association.

mixed baggage of marketing, health and forage development inputs. Moreover, this time the target was the central highlands, predominantly of mixed farming but with the largest livestock population in the country.

Many of the time-consuming experimental undertakings were thought to have little relevance to peasant/pastoral production environments. If true, the need to revise of the current research strategy becomes more urgent. The assertion that many of the research findings require an important variable, that is, adequate feed to apply at pastoral level is equally true. Conversely many of the forage species under experimental scrutiny required dependable rainfall or at least guaranteed supply of moisture to fully demonstrate their potential. Understandably, in the peasant/pastoral production settings where livestock exist under bare minimum conditions, it is next to impossible to recommend research results deduced from sanitized and heavily controlled production environment. Conceivably, growing conditions like those found in the research centers were nonexistent under peasant/pastoral production and therefore implausibly naive to expect change in the lives of peasants and agro pastoralists of the country. To circumvent those prevailing production constraints, no significant attempt were made to tap the wisdom of the peasant/nomad on water harvesting and grass regeneration techniques practiced for centuries. Together with the technical skills that institutions of research claim to posses, many down to earth pasture development and water conservation practices could have been performed at a fraction of the cost spent to establish an exotic model of range management to tackle livestock related production problems.[132]

Milk in Ethiopia is produced by the traditional sector and accounts for almost ninety five percent. The lactation period rarely exceeds 190 days with an average milk yield of 2 liters per day. By contrast the cross breed maintained by few dairy farmers around major cities today have the capacity of producing between 800 liters and 1,200 liters of milk/cow/lactation.[133] One of the early livestock activities that focused on the development of high lactating herds by crossing exotic and local breeds had produced some useful results. By late 1980s research work had conclusively determined that cattle breed with ¾ exotic genetic base are 4 times more productive than the indigenous ones.[134] It is interesting to draw a comparison between milk produced by small-scale dairy farmer in England with that of Ethiopian farmer whose daily milk output does not go exceed 2 to 3 liters a day while his counterpart reaps close to 60 liter

132 The cost of TLDP and FoLDP alone is roughly put as 130 million USD.

133 Getnet Haile, The impact of Global Economic & Financial Crises On the Ethiopian Dairy Industry. UN Industrial Development Organization (Vienna: 2009), 13.

134 Melka Werer Research Center, Annual Progress Report (Addis Ababa: 1984 – 1986).

from each cow milked per day.[135] However, the information obtained as a result of painstakingly conducted experimental trials remained inaccessible for the obvious reason that research and extension did not go hand in hand to benefit the cattle herder. The scope of the experiment that went for decades at the research centers did not limit itself to transferring useful genes alone, but went as far as including different feeding regimes as the most likely dependent variable to influence milk production. Unfortunately, the feed trials as subset of the whole breeding program was never tested under local management conditions. This further hindered the direct application of the feeding regimes worked out at the research centers to pastoral/peasant production conditions where grazing habits are opportunistic to say the least.

The same can be said of the research undertakings on small ruminant carried out at two most important sheep and goat producing regions, namely Sheno in central Shewa region and at Melka Werer in the Rift Valley. In the four decades of their existence, these experimental sites have done little to alleviate sheep and goat production problems under local management conditions. To the contrary, whenever there is disease outbreak or severe drought affecting the area where the experimental sites are located, the test animals were sequestrated inside the walls of the research centers and tightly guarded to avoid infection until the misfortune is under control. Thus, instead of widening the perimeter or the theater of operation to meet the problem head on, the centers retreat to their ivory tower enclaves leaving the larger livestock population on jeopardy. In the absence of well-established rapport between the service giver and the user, thousands of the small ruminant fall victims to the natural calamity, which under normal circumstance is controllable.

The above argument suggests that in order to tap the tremendous livestock potential that the country is endowed with, it is imperative that a thorough assessment of all the projects pertaining livestock production continues. This way it gets easier to identify some of the shortcomings embodied on past policies and avoid redundancy whenever they occur.[136] The

135 An interesting program aired at Al Jazeera America on March 4, 2014 shows that a small scale English farmer sent to the Kenyan Rift valley to observe the traditional pastoral livestock management revealed to his host that out of the 120 head of cattle managed by his family his share in the market was half a million gallon or 2 million liters per year.

136 1/ N.J. Cossins, 1983 in Boron The status of ponds in the western SORDU project area. JEPSS Research Memorandum 2, (Addis Ababa: ILCA, 1983), 10. 2/ R. J. Hodgson 1990. The southern Sidamo rangelands project 1985-88: Perspectives on development interventions and extension. Unpublished Report. Addis Ababa: CARE-Ethiopia, 1990), 207. Both indicated that the Boran herdsmen were less interested in maintaining the ponds believing that the ponds were either donated or constructed as a gift by SORDU and other development agencies and therefore not of their concern.

Ethiopian Livestock Development Master Plan developed in 2007 is the first concrete step taken by the government to making livestock an important factor in food sustainability, source of income and economic diversification. The plan has embodied all aspects of the sector including, stocks, feeds, means of distribution and the projection to capitalize on both local and foreign market. It will be a while before the new policy on livestock development bear fruit and benefit the livestock rearing population.

The low laying lands have an arid to semi arid climatic conditions with annual precipitation ranging between 400mm to 700mm. The rangelands are located almost entirely in the lowlands and support more than 35% of the livestock found in the country. Although the evapo-transpiration rates in these areas are high making moisture retention in the soil and within the plant body difficult, still the plains are good for natural pastures to thrive. Based on this premise, therefore, it is imperative to devise an alternative method of developing range lands that are generally in agreement with the universally tested, accepted and approved technical innovations but rooted on community based rational exploitation of the pasture lands. Admittedly, the attempt by FLDP, SLDP, TLDP and FoLDP to address rangeland issues, milk and meat production as well as animal health is not out of place or irrelevant. What is at issue here is the approach taken to address livestock problems in particular that of the rangelands and all that are related to livestock development that is at fault. Frankly, well-scattered and small projects with down to earth approach in their implementation would have yield better result than the abstract and shortsighted development effort pursued since the 1970s.

In the Boran rangelands machinery assisted systematic flooding of the rangelands is known to have increased grazing rate by modest margin. This has been a continuous practice in the Boran area. Assuming that indigenous wisdom is incorporated into the overall range rehabilitation strategy one is assured the success of it. In fact there could be an added benefit of blending indigenous knowledge and modern water harvesting techniques to maximize water use to the fullest. It is important to note here that even the driest part of the country receives 300 mm of rain annually, therefore not difficult to envisage rain supported natural grass regeneration in the driest rangelands. To stretch the meager resources to its rational use both the development agencies and the cattle herder could collaborate in enforcing strict grazing regimes on the regenerated plot or plots of land for selected animals that are primarily targeted as food source or for those market bound. Barefoot extension agents could be valuable tool to reaching the constantly moving herd in search of grass and water. To devise long term and sustainable development for the grasslands are complex. It requires not only a sound policy and technical readiness in the part of the government but also the reshaping of the opportunistic way of pastoral life where survival instincts tend to override long-term arrangements.[137]

As indicated earlier agricultural research has been crop biased and will likely continue to be until the importance of livestock sector to the economy is fully understood. The indictment is fair in that crop focused activities does benefit the sedentary population from where traditionally political power comes from. The insignificant number of water bore holes across the country; the few chemical-dipping tanks here and there and the lack of widespread veterinary services speak for themselves. Moreover the absence of irrigated pasture to alleviate feed shortages and consequently support animal fattening programs clearly indicate the insensitivity of the policy makers toward the agro pastoral sub-sector. In short, past policies has so far downplayed the usefulness of the livestock sector despite its potential to contributing to the agriculture GDP in the range of 40% and the total GDP between 14 – 16%.[138] In a way, thanks for the modest number of boreholes dug by the TLDP project and the cisterns constructed at SORDU managed rangelands that livestock made it thus far. However, mere survival alone does not guarantee the existence of vibrant livestock industry capable of competing with other countries having a well-developed livestock sector. With population growing unabated and the weather pattern becoming unpredictable all to the detriment of livestock well being, it is not farfetched to foresee a scenario where cattle, sheep, donkey, camel, ass, goat could be reduced to insignificant level creating a catastrophic effect on the lives of 20 – 30% of the population whose livelihood is based on livestock husbandry. By the same token the peasant could loose a vital source from which a good portion of draft animals come from.

Irrigation schemes that have little to do with this sector are spread along the banks of many of the overflowing rivers further aggravating livestock condition by depriving them their traditional pasture reserve, which served as complementary to the rangelands. Thus, livestock improvement efforts cannot be seen in isolation but in association with sinking more bore holes, building cisterns and water ponds and developing irrigated pasture. All efforts aimed at livestock development and the improvement of life of those who depend on them will be fruitless unless participatory approach is chosen as an appropriate path for meaningful development.

---

137 To be able spread the message through extension activities, the wisdom and rational of owning manageable number of animals instead of possessing multitudes of livestock heads. Although not an easy task to tackle there seem to be no alternative but to try and appeal to the pastoral sensibility to mend his old ways to meet future challenge. His risk aversion tactics notwithstanding, owning selective head of cattle are eventually less cumbersome to the family and the environment.

138 Michael Halderman, Pro-Poor_-Livestock Policy Making in Ethiopia. The Political economy, FAO PPLPI Working Paper 19 (Rome: FAO, 2004).

More than half of livestock trade in Ethiopia is done clandestinely.[139] Few experts squarely put the proliferation of cross border trade on failed government policy.[140] While this accusation may have some merit in it, the marketing policy which in the past was haphazard and certainly more activity driven than based on well planned action program, has of a recent begun to take shape in favor of the state by injecting new deals such as incentives to entice the producer. No doubt the negative effect of illegal cross border trade is more felt by the state but also the primary producer. Given the fact that for years there has been no real policy for livestock in Ethiopia, one argues that the EPRDF led government has initiated policy led programs to eliminate some of the obstacles responsible for the decline-of well-structured livestock market. Some of the most important measures taken by the government are the commissioning of rural development at ministerial level, the establishment of Livestock Marketing Authority and the formation of Livestock Meat Export Association. It has also established markets in remote areas to cater pastoral need in Afar and Somali Administrative Regions. These worthwhile efforts are an important step toward the rational exploitation of the 'resource extraordinaire', the livestock. All the good reason why livestock export went by up of quarter of a billion dollars in 2011.[141]

For a long time the witty livestock traders, against all odds were able to conduct trade across the sea sometimes leisurely but most often under perilous conditions. It is indeed to the credit of the trader who in the midst of indescribable big and small wars that had had plagued the Horn of Africa region during the past half century was able to keep livestock trade afloat. The livestock herder, who by the nature of his precarious existence is an opportunist par excellence. Whenever adverse condition looms on the horizon, the pastoral is less hesitant to make prompt decision to sell his sheep, goat, camel, and cattle at the first opportune time in exchange for money, grains, sugar or medical supplies. Thus, for the benefit of all involved, that is, the cattle herder, the trader and the state, the policy need to focus on facilitating easily accessible livestock holding places equipped with adequate feed and medical supply for animals destined to local and outside market. It also needs to give an attractive incentive for the trader to channel the export through well-defined exit points.

This imply that government agencies ought to provide valuable services in return for the orderly exit of animals from the country but refrain from

---

139 Reuter (March, 21 2009) According to Berhe Igziabher, head of the Animal and Plant Health Regulatory Body, 60 percent of livestock exported by Somalia are those taken from Ethiopia through contraband trade

140 Aklilu Yacob, An Audit of the Livestock Marketing Status in Kenya, Ethiopia and Sudan (Vol. 1) AU-IBAR (Nairobi: 2002).

141 Yacobt Aklilu, Regional Livestock Trade, lessons for South Sudan (2012) http://www.agfairsouthsudan.org/wpcontent/uploads/2012/12/Dr.YacobAklilu_FAO_28Nov2012.pdf

directly involving themselves in livestock trade as was the case during the *Derg* rule. If supplying household provisions to the primary producer is proved to be a sweet bait to streamline export, government agencies may need to gear their actions toward fulfilling the goal.[142] In any case the benefit derived from well managed and carefully tend livestock is twofold; sizeable income for the state and an excellent selling price for the primary producer.[143]

In the past few decades animal and animal products have increased their values to the extent that it has become prohibitive for ordinary people to consume lavishly even in the middle and high income countries of the world, hence for Ethiopia to capitalize on market opportunity by tapping on the huge livestock resources that is blessed with. It is not lost to the policy makers that state of art abattoirs, attractive but functional packaging materials, timely transport facilities and unobstructed exit to port of destination are but some of the requirements that the Argentineans and Australians do to win the confidence of the market. Although quiet long way to go before achieving a competitive edge in world market, comparatively speaking Ethiopia has become a major exporter of chilled carcass in the region.

Unlike the pastoral population that make use of livestock in particular cattle and small ruminants as a direct source of food and on occasion as a means to access to other necessities of life, the sedentary highlander is very much dependent on draft animals to plow fields, thresh crops and haul goods. Livestock products can also be considered as an occasional treat to many of the peasant folks.[144] The importance of livestock to the sedentary population is long established. The resilience of the farming community after a long and persistent drought had always hinged on the availability of draft animals to work on the land. Historically speaking, the effect of draught animal shortage on the overall crop cultivation was nowhere as vivid as the great famine of 1888-1892. This episode which begun with the Italian import of Rinderpest, cattle disease, to the mainland and together with the long spell of drought that followed increased the vulnerability of Ethiopian agriculture long after the victorious conclusion of the battle of Adwa in 1896. It took a decade to replenish the herd from which animal power is extracted.

In the plateau where the majority of the livestock exists there is little room left for expansion. Tragically, even those traditionally reserved grazing grounds

142 Bartering livestock for industrial goods in particular sugar was common during the *Derg* rule of 1975- 1990.

143 Michael Halderman, Pro-Poor_-Livestock Policy Making in Ethiopia. The Political economy, FAO PPLPI Working Paper 19(Rome: FAO, 2004).
Quoting government source Halderman, finds that the value of cross border trade is about USD 105 million annually. This is more than 100 times greater than the average annual value of official export of livestock over the 1993- 2000 period.

144 At present the per capita consumption meat in Ethiopia is estimated to be at 10 kg per annum.

are gradually eroded within. In their place cereal and legume crops are cultivated to further endanger livestock conditions. This situation is evident that in the northern highlands the human to oxen ratio has been widening beginning the 1950s. The latest statistics indicates that in Tigrai the total cattle population is between 3.5 to 3.7 million including the number of cows hardly used for plowing. Considering the size of the rural population in this region, therefore the ratio of livestock to man is about 0.7. Although the geographic spread of Oromia Region is larger than any still the cattle population is close to 23 million making the ratio of livestock to man the same as in Tigrai and Amhara the next largest Administrative region. Nevertheless, the ceaseless move to expand cropped lands to satisfy food requirement for the exponentially growing population of the rural areas is undermining the oxen to man ratio eerily.[145] Thus, the expansion of cropped lands at the expense of the grazing grounds within the highland areas has reached its worst limit.[146] In some of the *ensete* growing regions of southern Ethiopia in particular the Wolaita-Kambata-Hadia region the land to man ratio is so wide that crop cultivation is normally performed using hoe rather than the traditional oxen-plow.

Following this argument, therefore, it is no difficult to conclude that there is little space left to expand the grazing grounds in the highlands without jeopardizing the cultivated areas already in short supply. The alternative is to inaugurate an intensive system where few desirable pedigree draft and milking animals are kept by entrepreneurs in the rural areas or by industrious peasants who in turn are in the business of dispensing livestock related services to peasants in a given locality. However, such a proposition is meaningless unless accompanied by readily available veterinary services and above all industrially constituted feed supplements. During the wet season animals are made to graze on out-of-place plants and weeds growing in and around the cultivated fields quiet contentedly. This practice is so common in Harar region that peasants make a substantial income by selling fattened animals to major cities including Addis Ababa. Cattle raised in such manner may have a live weight double and in some cases triple of those managed under stress environment.

To relieve the pressure and therefore, restore the balance of man to oxen ratio, a sort of mechanization as practiced in southern Italy and parts of China may be in order. The mini tripod tractors common in south Asia and southern Europe are not only suited to operating in small plots of steeply and undulating

145 Look at earlier surveys reveals that in the mid-1960 the oxen-less peasant was estimated at 30%, (MoA, Vol. 1), 50. Two decades later the number increased to 38% according to (Highland Reclamation Study Working paper 11). Finally in the 1990s the gap widen to close 50%, (Desalegn Rahmato, Land Reform In Ethiopia 1990).

146 Alemayehu Mengstu, Country pasture/Forage resources Profile. (FAO Crop and Grassland Service, 2006).

lands but also can perform all sorts of agro-technical operations in limited scale. Short of this one envisions the accelerated migration of people and animals from the densely populated highlands to the eastern and western escarpments that opened in the 1950s.

Both the eastern and western flanks of the plateau receive less rain and in the past has been the natural barrier between the cereal dominated culture of the highlands and the mostly cattle dependent agro/pastoral lowlands. In truth, the escarpments were more of strategic grazing reserve for the cattle herder than it has been cereal-producing ground for the sedentary peasant. The escarpments on both flanks of the elongated chains of mountains which in earlier times acted as a buffer between the cool and wet highlands and the hot and dry lowlands are today sparsely populated by peasants unable to eke out a living high in the plateau. This type of unplanned or spontaneous expansion toward the escarpment will inevitably invite environmental disaster as well as unwarranted clashes between those yearning to secure a livelihood from the land and those who historically used the area as strategic grazing ground.

VI

PEACE AND WAR: TWO FACES OF THE SAME COIN

ግጭት ካለ ምህረት አይወርድም። የድህነት ስሩ ግጭት ነው።
—**Abekrezion**,
15th century Bishop of the Ethiopian Orthodox Church[1]

Challenge from Within and Outside

The archaic feudal structure that reined over the lives of the peasantry for centuries certainly had a fettering effect on normal progression of things or toward registering any quantifiable achievements in science and technology. The pace of development in the sphere of health, education, and shelter and food production was terribly slow. The system as I attempted to describe earlier was very insular and unresponsive to calls for change. Instead it thrived by siphoning surplus produce from the peasant and worst drawing them into conflagration of periodic wars and upheavals that were so rife for

1 'With the prevalence of Conflict no development (no mihret) is expected. The source of all poverty is conflict [ed. among peoples]' Abuna Abekrezion, *15th century Bishop of the Ethiopian Orthodox Church* pp.262 DEKIKE ESTIFANOS. Getachew Haile, DEKIKE ESTIFANOS. (Collegeville, MN: 2004), 262. (Amharic)

pretty much of the last two centuries. [2] Unfortunately, these objective conditions thwarted the process of early state formation, prevented resource accumulation, hindered local and foreign technology proliferation and solidified taboos, mores and traditions all to work against absorbing and adapting novelties in social and physical sciences. In a way many of the predatory behaviors inherent in the system acted as a break for the creation of effective institutions supported by a well-developed productive base. Arguably, Ethiopian feudalism completely failed to be the mid-wife of new productive forces that will transform the socio-economic base of the country, as was the case in Europe in the 17th century.

Although the internal factors are preponderantly responsible for Ethiopia's underdevelopment and dismal agricultural performance, the external factors further exacerbated the situation setting a retarding effect on the overall development pace of the country. No doubt, the bold and aggressive attitude displayed by Europe over the non-white population of the world beginning the Nineteenth and the greater part of the twentieth century negatively influenced the affairs in the semi-fractured Ethiopian state by creating fertile ground for an amalgam of vicious war lords to go amok at the expense of a centralized system of rule. Never in the long history of Ethiopia had firearms played such a major role in the decentralization process of the country as witnessed in the nineteenth century. This is a phenomenon contrary to the one observed on the segmented societies of West Africa during and after the end of slave trade where the availability of guns helped create consolidated power around larger ethnic groups or around the Islamic communities led by Fulbe/Hausa and Taklores. At this time the hereditary chiefs and many of the pretenders to the throne procured firearms entering through the ports of Massawa and Zeila to the determent of a unified state so much dreamed by successive Emperors. As Caulk aptly observed 'rifles and even machine-guns were widely enough spread at the turn of the century to reinforce the fragmentation of power long characteristic of the Ethiopian state'.[3] In the first quarter of the 19th century where the civil war was raging there were close to 14,000 known guns in the country.[4] Before the conclusion of the century there were hundreds of thousands firearm at the end of the provincial lords

---

2 1/The author has meticulously compiled some of the minor and big wars that occurred during the epoch using different historical texts as a source. It is possible there may have been small and large battles waged between and among warlords that escaped the search of the author 2/ Richard Pankhurst, 1968. Economic History of Ethiopia. 1800-1935 (Addis Ababa: Haile Slassie University Press 1968). The book gives detailed account on the amount of arms available in Ethiopia before the conclusion of the *zemene mesafint* and the dawning of the 20th century.

3 R.A. Caulk, Firearms and Princely Power in Ethiopia in the Nineteenth Century. Journal of African History, xiii, 4 (1972), 609 – 630.

4 Richard Pankhurst, 1968. Economic History of Ethiopia 1800 – 1935. (Addis Ababa: Haile Selassie University Press 1968), 579.

Once European mind was dead set to keeping the African continent as an appendage to the burgeoning Europe, Ethiopia was put into military and political target. To the European powers in particular Great Britain, there is no urgent matter than controlling the upper reaches of the Blue Nile that has been the life line for Egypt from time immemorial. It is known that by the 1870s Egypt was in effect under subtle control of the Europeans whose numbers in 1876 was more than 100,000 strong. Correspondingly, they had vested interest in the production of agricultural commodities, namely cotton, rice and sugar, all produced by the waters of the Nile. [5] Despite the widely accepted "temporary" but direct occupation of Egypt by England in 1882, the two major wars fought in what is today's Eritrea in 1875-76 is perhaps the first attempt, albeit indirectly, by a European power to extend its influence over Ethiopia. Moreover, it was more than coincidence that during the Ethio-Egyptian war of 1876, the Egyptian finances was placed under the control of an international consortium because of Ismail's debt to European financial institutions to the tune of 91 million pounds.[6] It is plausible therefore that Ismail was doing the bidding of the powers that financed his lavish projects including the money spent to open the Suez Canal.

After the Egyptian debacle at Gura and Guda Gudi, Great Britain continued it's behind the scene maneuver when it groomed Italy to play a deceptive role in the game of occupying Massawa and consequently in the ascension to the high plateau directly overlooking the Blue Nile Basin. Ismail's ill advised adventure to conquer Abyssinia, notwithstanding, the supper ambition of his grandfather Mohammed Ali the Great to control the source of the Nile from head to toe remained the dream of any power controlling Egypt.[7]

Although, the elements that were believed to have been responsible for the demise of Emperor Tewedros's tumultuous reign (1885-1869) was long in the making, nevertheless, the lack of support from the war lords in the face of European invasion and the direct collaboration of Kassa later Yohness the IV with the English yet brought more chaos in the north. Tewedros left Ethiopia as he found it war-like and in chaos, but this time with the variety of well-armed warlords who have no qualm of challenging each other for territorial gain. To fill the power vacuum created by Tewedros's defeat and his subsequent suicide it took three years for the strong houses from the north, namely Tigrai, Gojam, Lasta and Shoa to settle the question of succession to the throne. When this

---

5 John Iliffe, Africans: the history of a continent. (Cambridge University Press, 2007), 169.

6 Ibid., p170.

7 Burhan Ali in an article posted at Awate.com. November 2010 under the heading, the *Politics of Nile: Egypt's kiss of death* quotes a noted Egyptian historian Abdul al-Rahman al-Rafai in one of his Volumes, *the Era of Mohamed Ali* the Great, 175. '*The farsightedness and strong will which this genius have acquired fame for, must have made him perceive that Independence cannot be realized unless Egypt owns the Nile course from source to mouth*".

was settled in favor of Yohaness of Tigrai the challenge for power grab continued as modern weapons become available even to the lowest of the low in the feudal hierarchy.

In the meantime, lesser lords under the tutelage of Menelik, the Shewan prince, were busy conquering southern lands with voracious intensity paying little attention to the breach of borders by foreign invaders and power challengers in the west and north. In fact, power in the so-called Ethiopian state was so devolved that Emperor Yohaness was helpless to do anything when Menelik, the Shoan prince was entering a trade and political agreements with the Italians behind his back. As the matter of fact the Treaty of Wuchalie, the linchpin of the Adwa battle of 1896, was signed weeks after Yohanness laid his life to protect the western frontier from foreign invaders.[8]

After the fall of Yohaness at Gelabat defending the western frontier from Sudanese Mahdists raids and habitual incursions by unruly tribes deep into the Ethiopian territories, Menelik II ascended to the throne with no or little opposition from the strong provincial nobilities. His unusual ability to use the traditional Ethiopian polity of neutralizing enemies before they pose serious threat and rewarding loyal servants to lubricate the system helped bring the expansion of the empire unhindered. More significantly, there was enthusiasm and support for his grand plan of restoring ancient Ethiopia's frontier by letting the European powers know that 'While tracing today, the actual boundaries of my Empire, I shall endeavor, if God gives me life and strength, to re-establish the ancient frontiers (tributaries) of Ethiopia up to Khartoum, and as Lake Nyanza with all the Gallas'.[9]

Obviously, Italy's audacious claim of protectorate-ship over Ethiopia slowed down his dream of competing with Great Britain and France, the most potent of all. Nevertheless, such false claim infuriated Menelik to call null and void the treaty signed at Wuchalie, shortly before his coronation as Emperor of Ethiopia in 1889. In less than a decade after the abrogation of the treaty, an African army made up of peoples from almost all parts of Ethiopia including subjects from the new colony of Eritrea confronted Italy at Adwa a mere 40 kilometers from the de' jure Italian frontier, delivering a heavy blow to Europe's open design of subjugating Ethiopia.[10] It is after this decisive battle that the colonial powers led by Great Britain begun the policy of containment through other means. In their diabolic design the three powers secretly joined hands to divide the country into their sphere of influence. The Tripartite Agreement of 1906 signed between France, Italy and England to the exclusion

8 Yohaness IV died in March 9, 1898 and Menelik spent no time to sign the friendship treaty with the Italians in May 2, 1898.

9 Letter addressed to Britain, France, Germany, Italy and Russia, 10th of April 1891. *Public Records Office (London), Foreign Office 1/32 Rodd to Salisbury,* No. 15, 4th May.

10 Many of the notables from what is now *colonia* Eritrea such as Dejtach Abera of Tzazega and Ras Woldemicael Solomon of Hazega and not to mention the ex-governor of Midri Bahri, Ras Alula took part in this decisive battle.

of Menelik was one of the sinister measures taken to isolate Ethiopia and when opportunity avails itself to gulp it mercilessly. Actually, the European machination to isolate and weaken the country for later conquest started a bit earlier, that is, few years after the failed attempt by Egypt to occupy the headwaters of the Nile became apparent. Following the failed Egyptian attempt to conquer Ethiopia, the British government negotiating on behalf of the trapped Egyptian forces seeking a safe conduct from troubled Sudan through the Ethiopian territories promised Emperor Yohaness IV the return of the right to use the vital sea outlet. Perhaps he is the only sovereign after Zera Yakob and Dawit to seriously consider of regaining an outlet to the sea lost sometime in 8th or 9th century by establishing a frontier town not far from the sea, Ghinda. After the fact and to the surprise of the Emperor, England renegade its earlier promise by shamefully facilitating the occupation of Massawa by the Italians in 1885 effectively locking the country from importing/exporting opportunities as well as depriving an interaction with the outside world.[11]

It is important to note here that, the Adwa campaign was costly in that it drained the meager resources of the country at the time when the land was ravaged by years of famine and epidemics of tragic proportion. It is estimated that during the great famine of 1888-1892 probably one-third of the population and 80 percent of the livestock perished[12]. This event as an important milestone as it may appear did not usher radical structural change to meet the challenge. In other wards it was not significant or a sign vital enough to awaken the consciousness of the ruling class for rapid transformation. But following the showdown with the Europeans, Menelik began consolidating his empire with no fundamental change to the power base or in the administration of land, the valuable asset at hand. However, unlike Emperor Haileslassie after him, he was reluctant to give the nobility *cart blanch* in the administration of the land. Thus, short of *gulti* ownership, the nobility without whom the expansion to the southern, eastern and western regions of the country would have been impossible were given the right to extract tributes from peasants living in those lands and nothing more. Nevertheless, until his death in 1913, the country he so much labored to unify remained in relative peace with itself. Indeed, one can argue that after brutally crushing the resistance of lesser kings in the south and

11 The Hewitt treaty was signed on the 3 of June 1884 at Adwa between Ethiopia and Great Britain, the de facto colonial ruler of Egypt. Three of the most important articles embodied in the treaty in as far as Ethiopia is concerned were a) free transit for Ethiopia through the port of Massawa for all goods including firearms. b) Ethiopian agreement to help in the safe evacuation of the trapped Egyptian army via Massawa.c) restoration to Ethiopia of Bogos (held by the Turks and the Egyptians) and other regions with their buildings, stores and munitions.

12 Girma Kebede, States and Development in Ethiopia. (Humanity Books 1992), 158.

the Emir of the ancient walled city of Harar, not to mention his ability to mollify the most potentially contending force, the Tigrai nobility, the country was in relatively much better position to focus in the introduction and adoption of modern novelties such as, post office, railway construction, public school and the formation of cabinet ministries to conduct the affair of the state.

However, his death yet brought more challenge, typical of Ethiopia's transition to power where contending forces circle the wagon to battle for the reign of the fragile state. Among those who came in the forefront to claim the vacant post was Lij Iyasu the product of matrimonial arrangement between the Shoan ruling house and another powerful house from the north. However, Lij Iyasu's candidacy to the Ethiopian throne did not settle well with many in particular with the Shoan nobility to instigate yet another bitter struggle between the feudal war lords. Finally the young prince was chased out of his capital living as fugitive while Zewditu, the daughter of Menelik II was crowned as Empress of Ethiopia in 1916. She in turn has to contend as second to Teferi the future Emperor Haile Selassie but not without powerful lords challenging Teferi himself.[13] Thus, when Mussolini of Italy was ready to conquer Ethiopia, the country was in such miserable state, both economically and militarily it took less than 7 months for the Italians to occupy the rugged landscape in 1936.

In the mean time the attack against the Ethiopian state intensified to the extent of creating fissure between the sovereign represented by Zewditu and Teferi on one hand and the various feudal lords whose basic interest was to keep Ethiopia decentralized as much as possible.[14] As pointed out earlier, the ultimate objective of the tripartite agreement of 1906 was in effect, to strangulate the ancient nation from conducting free trade and preventing it from entering political association or conducting diplomatic agreement with whomever it wishes.[15] In today's lingo this can be equated to neo-colonialism were crucial national business and trade deals are basically decided, controlled and even managed by the metropolis. The unholy alliance of the European powers was further aimed at physically parceling Ethiopia into their sphere of influence as the Italian Ambassador Conte Colli di Felizziano bluntly remarked to his colleagues in Addis Ababa 'it is up to us to convince our respective governments to consent to the 'partition' plan.

No doubt the victory scored at Adwa was in serious jeopardy, gradually being dimmed by the implication of the treaty. Despite the positive outcome of the Adwa campaign, which saved Ethiopia from direct European rule, the country nevertheless, paid heavy price for losing access to the sea, which further confounded its predicament to outmaneuver the hurdles put by the

13 Mersie Hazen Woldekirkos, Memoire, on the reign of Menelik II. (Addis Ababa: Zamra Printing press 1935 (reprinted 2002), 120-121. (Amharic)

14 Zewdie Reta, Teferi Mekonen: and the Long Rein of Zewditu and Iyassu 1884 - 1922. (Addis Ababa: Central Printing Press, 2005), 328.

15 British Foreign Office, Peace Handbooks, no. 129: Ethiopia (London: H.M Stationary O, 1920), 97-102.

treaty. For all intent and purposes, the treaty placed Ethiopia in a sequestered state of existence throughout the reign of Zewditu and Hailesellasie to finally fall victim to open Italian aggression in 1935.[16]

The primary objective of Italy's aggression against Ethiopia is innocently assumed as an act of revenge to the humiliating defeat it suffered at the hand of the Ethiopians at Dogoli in early 1887 and a decade later at Adwa.[17] Contrary to this assumption the invasion was part of the grand design by the powerful European nations led by Great Britain and France to replace the trading posts that they have established all around Africa four centuries earlier with the direct colonization of the peoples of Africa. Certainly, such course of action will enable Europe to extract resources at will, both human and material. Thus, Italy's untiring effort to occupy Ethiopia together with the peripheries cannot be viewed otherwise. Italy more than any European country before it had an intense desire and continuous interaction with the country beginning the fourth quarter of the 19th century coveting its minerals, waters and agricultural lands.

Italy's Ambition: Rude Awakening

Italian conquest of Ethiopia did little to uplift the state of the Ethiopian agriculture and by extension benefit the peasantry. Of course no one supposes that the peasant will be an integral part of the equation in the Italian adventure of conquest to expect benefit of any sort. After all, the Abyssinian campaign was conceived and executed to ultimately benefit the Italian state and its citizens by exploiting the agricultural potential that the country was perceived to posses. In this regard their first plan of action was to repossess all the prime lands belonging to the ruling class and expropriate peasant lands in the fertile regions of central Shewa, south west and western part of the country to the benefit of large scale agricultural corporations. Moreover, at the core of Italian agricultural plan was the stemming of Italian migration to the Americas by offering opportunity of settlement in Ethiopia[18]. Unmindful of its earlier but failed attempt to resettle Italian peasants in Eritrea, Italy once again embarked upon huge plan to import hundreds of thousands of impoverished Italians to settle as farmers in Ethiopia.

According to the Chief of The Fascist Party in Addis Ababa there are nearly 50,000,000 hectares of fertile land ready to be exploited. This he rationalized, need people of Italian extraction, capital and know how to put it

---

16 Article 4 (a) of the treaty also states that "To act together... to safeguard; ... the interests of Great Britain and Egypt in the Nile Basin, more especially as regards the regulation of the waters of that river and its tributaries without prejudice to Italian interests".

17 Yebio Woldemariam, Eritrea: historical perspective. Mimeograph paper. Papyrus Bookstore. New York (1976).

18 Pankhurst, Richard. 1964. Italian Settlement Policy in Eritrea and its repercussions, 1889–1896, (Boston University Papers, 1964), 125-126.

in good use.[19] The fantasy that Ethiopia is the land of milk and honey was propagated by the Minster of the Colonies, A. Lessona who reporting to his government declared that the country is abundant with precious metals, meat, milk, wheat, hide and skins, cotton, coffee, cereals, and vegetable oil seeds. He asserted that raw materials obtained from Ethiopia would meet metropolitan and the colonies needs.[20] It is on this premise that the colonial power set out to establish settlement schemes and develop parastatal and private agricultural enterprises. However, this activity driven Italian agricultural policy in Ethiopia was plagued with numerous problems one of which was lack of enough funds to execute the ambitious projects. The continuous harassment by the patriots and the deliberate disinformation disseminated on their activities also hampered quick agriculture progress, thus, preventing the expected chain reaction to occur.

The hype that Ethiopia holds the golden opportunity for Italy's rapid economic development that was not available in the early-acquired colonies of Somalia, Libya and Eritrea were not without merit. As soon as the Italians set foot in the country they began conducting physical surveys to delineate the best lands for future Italian agriculture investment. They were also busy prospecting mineral resources to supply Italian industrial enterprises at home with raw material. Obviously, the first victims of the Italian scheme were in this case, the royal family and communities who showed active resistance to the occupation. Certainly, Italian axe fell mercilessly on the exiled nobility and the patriots who took up to the mountains. Italy did not stop there but took prime lands belonging to ordinary peasants in and around the major routes to the south, east and west of the country. Lands were either forcefully confiscated or paid for nominal price, that is, one-sixths to one sixteenth of its true value.[21]

When settlement schemes envisaged by the colonial authorities become impossible to achieve, Italian agriculture policy shifted from direct colonization by civilian farmers to enticing Italian soldiers in active duty, to pursue commercial farming. We are not sure what caused the policy shift but obviously lack of funds and instability must have compelled the colonial administration to put on hold both the induction of the army to farming and the transplantation of people from the home country. Instead the military's first objective became the pacification of the unruly countryside, which turned increasingly restive. Suffice is to say that none of the agriculture projects initiated by Italy came into fruition. As a matter of fact by the end of the occupation, the demand for wheat in Ethiopia soured to 100,000 tons of which only a third was produced in the country. Obviously, the balance was met by importing from somewhere else. To the chagrin of the *Duce,* the millions of

19 G. Cotese. Problemi dell'impero (Rome: 1938), 60 -61.

20 A. Lessona, Memorie (Florence: 1958), 356 -357.

21 Alberto Sbacchi. 1997. Legacy of Bitterness. Ethiopia and fascist Italy, 1935 – 1941. (Trenton: The Red Sea Press, 1997), 111.

Italians expected to settle in Ethiopia to work on the presumed millions of hectares did not materialize. By the end of the short lived Italian adventure, there were 1,000 Italian farms commanding about 10,000 hectares.[22] So much so for the fascist propaganda, which showered lies on its citizens that as the result of the conquest, the empire will benefit from the incorporation of fertile lands and resources useful to industrial development? The total cost incurred on agriculture during the five years of Italian intervention was close to billion Liras.[23]

When the vanquished left Ethiopia, they left little or no legacy to speak of in the agriculture sector but built major road arteries leading to the four corners of the country enabling the landowning class after the liberation an easy access to their holdings. The roads leading to Gonder, Debre Marcos, Dessie, and Mekelle were improved and stretched further to reach, Jimma, Nekemte, Gambela and Harar. Thus, facilitating the plunder of the countryside without making a dent on the lives of the toiling masses.

There was no major physical investment done on land or an enlightened policy promulgated to increase productivity. Post-Italian Ethiopia was reverted back to the system it knew best this time incorporating new elements of humble origin to the ranks by providing freeholds for his/her gallant service against the fascist or to civil servants of some significance. On his return from exile, the Emperor did not follow Menelik's suit but permitted the establishment of the *guilt* system and later freehold but also allowed the nobility to get hold on the land permanently as *rist-guilt* with the provision of transferring the title to their offspring.

Ethiopian Feudalism at Crossroad

Beginning the 1940s the country was faced with new challenges of bringing the administration in par with other nations of similar standing, example Iran and Egypt. The imperial government was also keenly aware that unless Ethiopia adopts a system that is deemed 'progressive' it would fall prey yet to another intervention forces with disastrous outcome, just as predicated by many intellectuals during the Menelik-Iyassu era.[24] It is important to note here that after his accession to the throne in 1930, Haile Selassie true to his noble pedigree abandoned the intellectuals who have been his constant companions and allies against the nobility. To their chagrin he paid lip service to their usefulness while relegating most of them except four to non-executive positions leaving the country at the hand of the traditionalist who were hell

22 ibid 116.

23 Ibid 330,334.

24 Ghebrehiwot Baykedagn. *Atse Menelik na Ityopia* (Amharic version). Asmera: *Berhan Yehun* 1912), 354. (Amharic)

bent of keeping the state of affairs to pre- Zewditu era.[25] Upon his return from exile, however, he focused on the very few foreign educated elements to run the affairs of the government with the monarch at the hub. However, he never ceased playing the nobility against the modernist who yearned for constitutional rule, respect of human rights and the land reform.

It was after the restoration of the monarchy that decrees regarding land tax, stamps, monetary and banking regulation, health, income tax, formation of municipalities, post office, cigarette and alcohol production and distribution was issued. Indeed the period between 1940 and 1950, the country appeared to be laying foundation for modernization without touching the most important issue of all, land tenure. It was only after the the abortive coup of 1960, that the Imperial government set out to reform the government by appointing commission to look into constitution, administrative, judicial and land matters.[26] Investigatory committees were entrusted to study and eventually recommend the path to modernizing the state machinery by forging alliance between the nobility and the educated, albeit many of humble background. One reason why Haile Selassie turn toward the enlightened few after his return from exile was that many of the nobility who otherwise would have made his rule difficult perished in the 1935-1940 war. Those, by some miracle or through traitorous compromise survived the Italian onslaught became a minor force to deal. In spite of their misfortunes, still they remained formidable force to influence the crafting of the 1952 constitution, which left land tenure system intact. It was after the failed coup d'état of the1960 that the issues surrounding land begun to be viewed seriously obliging the sovereign to take steps to address landlessness and serfdom, hence for the establishment of the MLRA.

The imperial decree to establish MLRA was taken as the first step toward tackling the chronic land problem. Whatever objective that the new ministry has intended to achieve, it did not stop the imperial government from distributing large tracts of land to various groups of people including those of humble background but pride themselves as the servant of the crown.[27] Of the measures taken by the government the most volatile one was the land tax because it touched peasants outside the traditional serfdom areas of South and west.

---

25 Bahru Zewde, Pioneers of Change in Ethiopia. 2002. (Addis Ababa University Press, 2002), 177.

26 Seyoum Haregot, The Bureaucratic Empire: serving Emperor Haile Selassie. (Trenton: Red Sea Press, 2013), 32.

27 Micheal Stahl, Ethiopia: political contradiction in agriculture development. (Uppsula: Publications of the Political Association No 67, 1975), 64. Using primary sources from "Harrison Dwnning. Land reform in Ethiopa; a case study in non-development UCLA law Review, Vol. 18, No2 1970. Page 290." explains that up to 1969 44,000 *gasahs* were awarded and another 62,000 *gashas* were awaiting decision or pending.

The book by Gebru Tareke, *Rural Protest in Ethiopia: a study of three rebellions* deals extensively on peasant revolt in Ethiopia since the return of the Imperial power in 1940. Invariably, the cause of peasant uprising was more of land alienation, serfdom and onerous taxes, which affected the rural population in profound ways. For example in the case of Eritrea the skirmish between government forces and peasants resisting land alienation and exploitation of wood resources caused few deaths followed by retribution in mid-1950s.[28] Unfortunately, the forces of resistance were too weak and isolated to galvanize widespread support. Nevertheless, the incidents left bad after test to be exploited by later insurrectionists in Eritrea.

The Gideo peasant armed insurrection in the south was one of the most direct challenges to the feudal system with regard to land aggrandizement. At first the peasant appealed to the consciences of the government and in fact went to court expecting justice to end the horrific land grab that begun with Menelik in the turn of the 20th century. But when all means failed they resorted to an armed confrontation culminating in their defeat in January 1959.[29]

Earlier in eastern Tigrai, peasant revolt to protest the onerous land tax levied on them were brutally crashed with the direct collusion of the British Royal air force flying from Aden to bomb rebel positions. The uprising popularly known as *woyanie* subdued the local gentry and their over lords to cause panic among the ruling class. However, when the battle was lost not only was the leaders imprisoned and exiled but more significantly peasants in Azebo were forced to relinquish two third of their land leaving them more destitute than they started.[30] It is to be recalled that the Azebos, supported by the advancing Italian army in Tigrai in the second Italo-Ethiopian war of 1935 had harassed the retreating Ethiopian forces for the monarchy to be extremely ruthless with them at a later date. Incidentally, they occupied an area where one of the major and last battles was fought to stop the Italian penetration deep into the central plateau. The TPLF, the major partner in the ruling EPRDF coalition today has admitted to be inspired by the *woyanie* revolt of 1943 to initiate its own-armed struggle later in 1974.

The most enduring peasant uprising and perhaps one of the rallying point for student radicals in the 1960s to intensify their struggle for land reform was the armed uprising at Bale which begun in 1963. The spontaneous revolt of the Oromos, the primary victims of the land grab, kept a contingent of the Imperial forces on their toe until the leaders were maneuvered to seek peaceful

28 The author in his teenage years witnessed some of the events that led to the confrontation. In two occasions peasants from Tzazega and Daro Paulos not far from Asmara resisted land alienation for airport expansion and the exploitation of woods for industrial purposes.

29 Lapiso G. Dilebo, A brief account of the Gideo people's history 1882-1966, (Addis Ababa: 1972), 12-13. (Amharic)

30 Gebru Tarekegn, Rural Protest in Ethiopia: a study of three rebellions. Doctoral Dissertation, (Syracuse University, 1977), 151-154.

settlement for their demand, namely less taxation and a just solution to the wanton land grab. Indeed, after the surrender of the rebel army in 1970, more land was handed out to outsiders who never shared the aspiration of the Oromo peasants. It was against this background that the Oromo Liberation Front (OLF) was formed to last longer than the life of the imperial regime and the *Derg* combined. Together with the violent uprising of the Gojam peasants in 1968 against excessive taxes and the chronic but unyielding armed struggle by the Eritrean liberation fronts since the early 1960s put the imperial government in defense. True enough, it did not take much time to end the system that shackled the country to serfdom and expose it to unrest, famine and extremism.

Taking Stock of Twentieth Century Ethiopia

The political extremism and the apparent confusion that followed soon after the downfall of Emperor Haile Selassie and the bloody confrontation that ensued among various political tendencies to fill the vacuum left by the monarchy can be traced to Ethiopia's past. Here it is meant to refer to the dysfunctional Abyssinian state of the past two centuries run by insular and myopic warlords whose *modus operandi* was to wage constant war in order to safeguard their territorial niche. In such circumstance one cannot expect conflict-resolving mechanisms born out of chaos. This lends to the fact that there was neither strong enough state institutions nor public advocacy groups inherited from the past to steer the mass movement of 1974. For a longtime the feudal system was regarded as the noose around the neck of the people stifling social life and neglecting material progress. The process of building up state institutions in favor of a centralized system continued after the ouster of the interventionist colonial forces in 1941. By all accounts the Emperor's 44 years rule was characterized by a stiff struggle to strengthen a centralized state where power emanates from one source at the center and not at the peripheries, as was the case before his accession to the throne. In retrospect decentralization hurt the country's economic and social progress seriously but also the autocratic and rigid system developed in the second half of the 20th century played a major role in its stagnation.

A cursory look at Ethiopia's economic and social development between the periods of the restoration of the monarchy in 1941 and its demise in 1974 reveals a pretty grim picture on where Ethiopia stood vs. avis other nations considered her contemporaries. In this case we refer Iran as an independent state but a giant in world history and Egypt a semi colony but the inheritor of the mighty Nile civilization and the center of Arab nationalism. It is important to take stock of this period so as to be able define Ethiopia's position in world affairs and also appreciate the struggle ahead to surmount the problem of chronic food shortages. Looking into the country's level of industrialization, the extent of its infrastructure network, its educational level and social services in particular health facilities between the expulsion of the Italian forces and the demise of the Imperial regime was not commendable.

In the sphere of modern education one can argue that it never had and was unable to create institutions that probe the skies for celestial bodies like that of University of Pisa in post-medieval Italy or the British Naval Academy to explore the high seas in search of wealth in far-off lands.[31] Before 1935, Ethiopia had 20 schools primarily catering for students below the secondary level, two of which are schools built by Menelik II in 1908 and the French four years later. Half of the schools were located in the capital. It is important to note, that even the agricultural training school at Holletta was established way after the liberation of the country from the fascist grip in 1941. In a point of fact a technical school with 225 students as the first entrant, did not come before 1953.

At this time there were close to 60,000 students pursuing elementary as well as post elementary schooling.[32] A decade later, the number of students jumped to 398,554 of which over a quarter of million students were enrolled in government-sponsored schools. By the year in 1965, 8 -10% of the school aged children were attending schools of which women made a minuscule part of it.[33] A UNESCO document published in 1961 put the ratio of girls to boys attending Secondary School 1:6.[34] Out of 1,667 schools opened for students during the mid-1960s about 1556 were primary school level scattered throughout the empire. In 1974/5, the year that marked the end of feudalism and the ushering of new socialist rule, total student enrolment reached a little over one million.[35] This figure corresponds to mere 3% of the total population and perhaps 6% of the school aged children and young people. Nevertheless, it is quite an achievement from the period where the student population was a mere 4,000 in 1935.[36]

The Addis Ababa University-College founded in the early 1950s had a limited capacity to accommodate all the secondary school leavers. In the first few years of its establishment it accepted about 84 percent applicants for degree

---

31 There has been for centuries traditional schools associated with churches and monasteries, though.

32 Edouard Trudeau, A survey of Higher Education in Ethiopia. With emphasis for future planning and development. Doctoral Dissertation, (Columbia University, 1967) 10, 16, 33.

33 Maaza Bekele. 1966. A study of Modern education: its foundation, its development, and its future with emphasis in primary education. Doctoral dissertation, (Columbia University, 1966), 7-8, and 11.

34 UNESCO, Ethiopia School statistics for the 1957/8 academic years. World Survey of Education vol 3 Secondary Education, (1961), 451-462.

35 Tekeste Negash, Crisis of Ethiopian Education, (Uppsala, 1990), 11

36 Richard Pankhurst, Language and Education in Ethiopia: Historical Background to the Post-war Period, (Haile Selassie University, 1969), 30.

program but precipitously fall to 24% of the candidates admitted in 1962.[37] Unfortunately, this trend continued way after the abolishing of the monarchy with tens of thousands poised to enter college but unable to be admitted to the three universities two of which had their status upgraded from college to university level in mid to late 1980s. However, with the support of the United States bilateral agreement three more technical schools were added in1955 to 1974 to mitigate the dropout rate. In addition to the two technical schools built with the aid of the US, the Soviet Union donated state of art comprehensive technical college at Bahr Dar.

Be that as it may, the monarchy by the end of its life boasted one university with four thousand-student populations. In comparison the number of students attending university training in Egypt during the same period was over three hundred thousand.[38] In mid-1960s the number of Ethiopians studying abroad was a little less than 1,500, not an impressive number at all. But these particular students and those who followed them a bit later were vocal anti-feudal and agitated for its downfall. Their activism was relentless and continued until the regime collapsed of its own weight in February 1974. Incidentally, the ringleader who masterminded the violent change of government in the December 1960 coup when the Emperor was on an official visit to South America was foreign trained Columbia University alum.

The physical relief of Ethiopia is as contrasting as the people who inhabit it. In other words the physical feature, that is, the physiography is unique in that it poses serious challenge for one to easily connect lands otherwise dissected by deep gorges, massive mountains, rivers and rivulets, steeply escarpments and torrid lowlands. With the exception of few road projects deemed vital to the invading Italian army to subdue the ancient land, many of the modest socio economic activities including rudimentary road building initiated by the Emperor before the debacle of 1935 came into abrupt halt. Interestingly, the Italians during their short stay in Ethiopia have added another 5,000 Km road to the existing 2,000 Km long mostly all unpaved weather roads that existed before the intervention.[39] Between the periods of 1951 and 1961, the Imperial government built roads totaling 3,460 Km in length of which a third was laid with asphalt.[40] Nevertheless, the main roads radiating from the capital to the four corners of the empire had a limited use in accelerating development because they lacked secondary and tertiary feeder roads

37 Edouard Trudeau, 1967. A survey of Higher Education in Ethiopia. With emphasis for future planning and development. Doctoral Dissertation (Columbia University, 1967), 33.

38 Alan Richards, 1992. Higher Education in Egypt. Policy Research Working Papers. WPS 862. (World Bank, 1992), 27.

39 Ethiopia – A Country Study. Roads. (1991) http://www.country-data.com/cgi-bin/query/r-4475.html.

40 Transport and Communication in Ethiopia. Ethiopia Ministry of information. Book IV, 1965. (Addis Ababa: Commercial Printing Press, 1965).

connecting the villages and townships nested on mountain tops or arid plains to the main road arteries. Unfortunately, this condition persisted long after the Italian defeat where remote villages and towns remained isolated to interact in the market or to receive social services such as education, health, and justice. Between the years of 1941 and 1991 the total length of roads that the country could boast was about 20,000 Km, a quarter of which is asphalted. Thus, lack of road infrastructure hampered rapid exchange of good and services locking rural Ethiopia into extreme poverty.

In contrast, the network of roads built under the EPRDF government was impressive. Since 2007, the total road length ranging from first class asphalt to dirty rural roads is about 111,000 Km[41]. It is significant to mention here that rural roads alone account 2/3 of the roadwork underway.[42] Moreover a railway system to connect hitherto remote sections of the country is under study. The significance of the 5,000 km long north-south and east-west rail network to be built within the coming five years is that it will connect about 50 regional cities and towns while traversing the countryside. Emperor Menelik built the 700 Km long Addis Ababa-Djibouti narrow track rail systems over a century ago. For a rugged country with forbidding massif mountain ranges and torrid low laying planes the current road networking is still inadequate.

Transport, in particular ground transport at the initial stage of nation building is vital. It constitutes a conduit through which the proliferation of knowledge and technology become possible. Without elaborate road system it is difficult to envision production surplus and hence the exchange of produce at will. Who will argue that transport does not change or shape the landscape in all its forms? This grim reality clearly explains of Ethiopia's inadequate road system at the time when the 20th century was folding.

Public health services were almost nonexistent before 1941. The few hospitals build between 1909 and 1935 were concentrated in Addis Ababa with one exception, the Mekonen Hospital in Harar. Briefly, by 1946, 24 hospitals and 50 clinics with 1,280 beds were either completed or under construction. A decade later hospital beds throughout the empire increased to 7,751 manned by 324 doctors and 604 nurses, this in a country with a population estimated at 22 million in the early 1960s.[43] In this case, the ratio of patient to hospital bed was about 1:2838 while patient to health personnel ratio amounted to 1:23,706.

In the manufacturing sector the picture was more dismal. The sector, which appeared after the liberation of the country, was small and mainly

41 Transformation National Journal No. 6; Ginbot – Sene 2005 (Amharic). Extracted from an interview in Amharic provided by the Executive Director of the Ethiopian Roads Authority, Zaid Wolde Gebriel.

42 Ken Ohashi, Sustain Growth: Ethiopessimism? World Bank's Country Director for Ethiopia and Sudan. Part I. (World Bank, 2008) http://www. Ethiopian reporter.com/content/view/1032/36/

43 Ministry of Information, Public Health in Ethiopia. Book III, (Addis Ababa: Commercial Printing Presses Addis), 5-6.

focused on manufacturing textile and food products. The bulk of the industrial enterprises processed few primary commodities namely coffee cleaning, timber work, leather and foot wear, sugar mill, flour meal, beverage etc. In 1954 there were 249 industrial enterprises in the whole country employing 18,000 workers. The ex-Italian colony Eritrea, alone boosted 83 small and medium industrial enterprises of which 36 engaged in non-agricultural based commodities such as chemical, paper, glass, match, cement etc.[44] A decade later, that is, 1963/64 the number of workers employed in 300 enterprises reached 45,000, thanks to the new sugar mill established at Methara employing agricultural field hands to perform the regular but repetitive farm chores that irrigated agriculture demands. The coffee cleaning enterprises account most of the industrial establishment. At the eve of the Revolution, 13,737 were employed in the small to medium scale factories food oil extraction, grain mills, sugar confectionaries, cement factories and textile mills. The number of workers who took part in the nationwide strike of March in 1974 against the Emperor's government was in fact larger than the 55,000 dues paying members under the Confederation of Ethiopian Labor Union' (CELU).[45]

The Ethiopian working class, if indeed such categorization fits to this particular social group at the time, was too small in size and more significantly still attached to its rural traditions and mores, something antithetical to the classic Marxian definition of working class. Bizarrely, this section of the workforce was the center of contention between the army eager to solicit sympathy from the socialist block and the student radicals under the spell of Marxism Leninism who championed for peoples government with the worker in the lead. Unfortunately, the uneven capitalist transformation that Ethiopia has to go through during the past century, made it impossible to even contemplate the emergence of a vanguard party composed and led by class-conscious workers to consummate the revolution.

No doubt, as Addis Hiwot explained in one of the early books published after the revolution, Ethiopia entered the 20th century unprepared to the intricacy of the modern world unfolding in front of it. It has neither the technical capability nor accumulated wealth to confront the harsh reality of the new world order, where countries in her position are but independent in name. In short, rather than progressing at steady pace to fulfill the innate ambition of its citizens for higher standard of living, the country succumbed to the struggle of achieving the bare minimum namely, food and shelter. Through the passage of time it became clearer that Ethiopia could not escape the reality of being underdeveloped and perhaps condemned to an unbreakable cycle of poverty. This, I believe is the product of its historical evolvement where the forces of

---

44 Ministry of Information, Pattern of Progress: industry in Ethiopia. Book VI, (Addis Ababa: Commercial Printing Press, 1966), 6.

45 Thomas Killion, Workers, Capital and the State in the Ethiopian region, 1919 – 1974. Vol. I (Stanford University, 1985), 550.

decentralization relentlessly clashed with each other hurting the producer in the process and robbing his fruit of labor and using him as cannon fodder in the endless campaigns involving wars. This lead to its present incapacitated state brought about by the negative forces played throughout the 19th and 20th century.

The Horn in Historical context

The Ethiopia region including Somalia has been the hotbed of conflicts arising from intersectional strife for centuries. Ever since the southward migration of the Axumites beginning early 7th century, first to the north central highlands and later to the south and east by the 14th century there has been a constant clashes with those coming in close contact namely the Somali/Afar coalition and the Cushitic and Cushitic-Semitic people of the south. In the span of 150 years after their expansion in what is today Shoa, the Christian dominance over these areas was indisputable. However, such physical contact, mostly antagonistic have not left a lasting legacy to ensure social integration of peoples into one nation state or at least adapt the Christian Amhara culture which by then seem to have an upper hand.[46] On the other hand, the *Walasmah*, a strong Muslim principality bordering Christian Abyssinia under whose authority numerous Muslim mini-states in the region fall did not succeed in converting Abyssinia into an Islamic *Dawla,* to bring integration, assimilation followed by development.

Except for Zera Yakob (1434-1468) who attempted to station troops/settlers in the eastern escarpment of Yifat, where the *Walasmah* made their home, there is little evidence to support that the 'conquerors' had similar attempts anywhere else.[47] It is clear that the encounter was temporary and certainly predatory in nature. The attempt to create a unified state by dispatching permanent representatives of the crown bolstered by armed militia turned into landowning class, the *neftegna*, didn't happen before the ascendency of Menelik to the throne in late 19th century. The decision made by the Emperor to establish military garrisons across the conquered lands and covert these military posts into towns, *ketemoch* created an opportunity for forced integration.

Strictly speaking, the northern plateau with its predominately Abyssinian population constituted the core of the mostly fluid and not well-defined boundary of Ethiopia up to the end of the 19th century. This ancient state as rightly portrayed by history was largely a "Christian island" standing tall like a watchtower overlooking the lowlands that are exclusively inhabited by Moslems and animist alike. It is therefore correct to state that throughout time

46 In contrast the Oromos of South West Tigrai assimilated fully to the Tigrigna culture with few still speaking the Oromo language.

47 Girma Awgichew Demeke, The Origin of Amharic. Centre Francais d'Etudes Ethiopiennes Etudes Ethiopiennes 5. 2009 (Addis Ababa: 2009), 51.

the country's boundary was never fixed but elastic depending on historical circumstance and events. Hence, the political boundary may have at one time shrunk to only half of the plateau and at times could have expanded to include the low laying lands bordering the Red Sea in the north and the Indian Ocean in the east. Once again it is safe to deduce that in the course of empire expansion and in periods of attrition where self-preservation was the primary goal, wars of varying degrees must have been waged to produce the unique scenario that we see today.

Among the earlier empire builders namely Amdezion, Dawit and Zera Yakob who lived in the 14th century was known to have waged a constant war against the Muslim principalities ruling over vast areas situated in the Eastern flank of the plateau.[48] The intermittent war continued well into the 15th century where the rulers of Christian Ethiopia plundered and subdued the peoples in the fringe who were militarily and organizationally weaker than the rest. As the result of the physical interactions, however, there is little to indicate that the invading forces had left behind historically verifiable permanent fixtures for the conquered people to adopt. Thus, it is safe to surmise that the relationship between the plateau and those encircling it was more predatory in nature that left little if at all lasting legacy in the likes of art, pottery, blacksmith or goldsmith widespread in Abyssinia then and the transplantation of *Geez* alphabet, a unique Ethiopian invention, among the subdued population. Conversely, little can be said of the impact that the cultures and skills of the conquered people had on the highlanders as a result of the contact. All this will change when the Abyssinians led by the Shoan Amhara took a permanent hold on all territories now defined as modern Ethiopia.

It must be noted here, the power vacuum created by the rise of Islam and the defeat of Persians and most significantly the ejection of the Byzantine, an ally of Abyssinia from South Arabia was a turning point for Axum from which it never recovered.[49] Similarly, the fall of North Africa and the Near East to the conquering Arab dynasties beginning the mid seventh century must have affected the psychic of the northern highlands with Axum at the center into believing that as adherents of the Christian faith they are shut off from interacting with their co-religionist the Syrians, Nubians and the Copts of Egypt. They were also resentful of the fact that the pilgrimage to the Holy Land was turned into perilous undertaking on the account that the Moslems controlled the land routes to Jerusalem. But more seriously, the rise of Islam coupled with the dissension of all sorts from inside induced self-hibernation on historic Abyssinia dulling its curiosity for wider interaction with neighbors in the rim and beyond.

---

48 TekleTsadek Mokria, Ye Gragn Ahmed Werera. Brhanena Selam Printing Press, (Addis Ababa, 1960) (Amharic).

49 Maxime Rodinson. 1974. Mohammed. (New York: Vintage Books, 1974), 267.

As the power shifted from Axum, a mere 100 kilometers from the sea coast to the rugged mountainous region south of Axum, the Abyssinians with the Agew at the helm lead more localized life mostly focusing on matters of religious nature which included literature and architect and perhaps on proselyzation mission to nearby peoples. It was a matter of time that the new power established in the interior has to face head on with Islam that have already made an inroads not only in the lowlands of the Northeast but also in the highlands of Shoa at Yifat and Tumuga in about 10th or most certainly in the 11th century. No sooner did the Muslim principality establish its rule in central Ethiopia they begun to challenge the status quo with devastating consequence to the established order. It is from this perspective that the war of 1528-1543 involving the Christians and Moslem communities living in the fringe of the Abyssinian highlands was and still is treated as proxy war waged on behalf of Arabia and the Ottoman Empire. This may have truth in it, considering the involvement of outsiders in the supply of war material and moral support and the presence of Arab ideologues to steer the campaign. Incidentally, a learned Arab historian who accompanied the much organized and effective military campaign led by Imam Ahmed popularly known as *gragn* the left handed, provided the only credible and somewhat complete account of the Moslem revolt during that period. Historians are much indebted to the chronicler Shihab Al-din Ahmed bin 'Abd al-Qader popularly known as Arab Fakih who left us a detailed account of life in 16th century Abyssinia in a narrative called, *Futuh Al-Habesha.* From historical context it is indeed the beginning of the opening of Abyssinia to travelers from Europe in search of all hidden things that Ethiopia represented.

Although this historical epoch occupies an important place in the history of the people of the Horn and in particular Ethiopia, as it conjures atrocities, defilement and upheavals that still resonate on sections of the society, nevertheless it is not as particularly serious as what followed after the Christian-Muslim tug-of-war ended. Because what is left after the religious inspired conflict ended was the intra-regional rivalry and constant war waged between and among the warlords that had turned the life of the ordinary people upside down for quite a sometime time. In a way the root cause of Ethiopia's underdevelopment is the constant in fighting among the regional feudal lords and the incessant raids on settlements, farmlands and even holy sanctuaries in both north and south that is to blame.

Earlier in the chapter an attempt was made to describe the causes of displacement and its resultant effect on the slow progress of the country. In a sense it was the centrifugal forces at play that caused havoc on life and property preventing a linear progression on the ground. With the advent of European adventure in the region, matters went from bad to worse. While the adventurous Europeans trekked on one guise after another and sought sympathetic ears from regional warlords, the shaky Emperors had to deal with host of contenders busy corresponding with various European sovereigns claiming legitimacy and therefore more arms to subdue a rival feudal lord.[50]

With the advent of modern armament in the Horn of Africa, the intersectional strife reached its climax in mid-19th century. In this regard the Abyssinian nobility in particular the northern chieftains in proximity to the sea were the primary beneficiaries. Thus, having modern weapons at their disposal, the nobility unceasingly battled with each other for supremacy.[51] What made the consolidation process more difficult and an arduous task to achieve were the regional based local rivalries that continued way into the 20th century. In other wards regional loyalties and the inaccessibility of the mountain top bases where a challenger most often find shelter made it difficult for the strong to neutralize the weaker in favor of bigger and centralized state. This was further exacerbated by the proliferation of guns in the north and the central plateau and the lack of transport network to ward off chaos and prevent the disintegration of the Abyssinian state. Understandably, the peasant was in the receiving end of the pillaging and destruction that continued well over a century and it did not stop until Haile Selassie ascended the throne in early 1930.

On the other hand, during the early and mid-period of the nineteenth century the lowlands stretching up to the Indian Ocean coast had its own dynamic of state building. These lands influenced by the most powerful and trade savvy Warsangali Sultanate, the Bari dynasties and the Gobroon almost lived peaceably or for lack of better term ignored by the Abyssinians until the European attempt to capture the entire Somaliland north and south opened hostility again. This was a period when the various Somali sultanates were busy trading with distant land not to mention the Yemen, Oman and other important Muslim enclaves in East Africa, presumably Mombassa/Lamu and Kilwa.

Other sultanates like the Gerads and the Bari built impressive palaces and fortresses and had close relations with many different kingdoms in the Near East. This unusually peaceful period soon came to an end with the arrival of European powers in the Somali coast. Moreover, after the defeat of the Harari forces at the decisive battle of Chelenko in 1887, the ancient walled city of Harar opened its four gates to the conquering Abyssinian forces. In a sense this

---

50 1/ Sven Rubenson, Internal Rivalries and Foreign Threats 1869-1879. Acta Ethiopica Vol. II 1989. (Addis Ababa Priniting Press, 1989) 2/ Erlich, Haggai, Ras Alula and the Scramble for Africa. (Trenton: The Red Sea Press, 1996), 143. Quoting correspondence of Dejach Debeb and Muhamed Abdel Hafiz to Baldesera undermining Alula's position to hold on against the European power. In Archivio Storico Ministero dell' Africa Italiana. Debab to Baldesera 24.5.89, D.I. 27.5.89, 28.5.89 and D.I 28.5.89).

51 Richard Pankhurst, Economic History of Ethiopia 1800 – 1935. (Haile Selassie University Press, 1968), 603. By the turn of the century over half a million shotguns were all over the country with greater portion of it under Menelik, the empire builder.

event also opened the entire eastern region of the Horn to serious Abyssinian challenge of the clan based Somali mini states. The increased hostility with the various Somali chieftains in the east prompted the Somali rebel leader Sheikh Hassan Abdella popularly known as 'mad mullah' to declare war on England and Ethiopia as major expansionist powers threatening the existence of the fragmented Somali nation, its people and its religion. He fought Italy, Britain and Ethiopia tooth and nail for over a decade until his army dissipated into the arid Ogaden in the 1920s. This rivalry will again manifest itself in more deadly form in the third quarter of the 20th century.

As the French jealously guarded their interest in their tiny enclave of Djibouti, the Italian holding of Eritrea was becoming tenuous as the World War II gained momentum. Finally, Mussolini's miscalculation in the power game and the inherent weakness of the Italian state precipitated in Eritrea becoming a protectorate of England while Ethiopia free from the fascist clutch. Unlike Libya and Somalia who were promised an outright statehood through the process of decolonization, Eritrea's disposal and the fate of its people were left to international wrangling with Ethiopia as a pivotal player. The UN was unable to give a definitive answer as to the status of Eritrea and how it should be handled. It therefore, named a commission to look into the matter and come up with the recommendation that would eventually be adopted by the General Assembly in late 1950. After all Eritrea was a colony of Italy same as Libya and Somalia requiring UN attention. This it must do by accommodating the wishes of the Eritrean people but also that of Ethiopia, which claimed it soon after the Italians evacuated the territory.

Based on the recommendation of the UN Commission, therefore, Eritrea was declared a federated state with Ethiopia in September 12, 1952, a verdict that satisfied neither the unionist forces nor those who advocated for complete independence. This unresolved issue fueled by the wanton intervention of the imperial government in the internal affairs of Eritrea created a big fissure that will soon develop into an armed conflagration lasting over three decades. The thirty years of protracted war involving more than 8 armed groups fighting against the center and at times against each other further plunged the region into abyss exposing citizens to become food aid recipients.

Even tiny Djibouti had its share in the widespread conflict arising from the rivalry between its two major ethnic groups most often influenced by events in the Republic of Somalia and Ethiopia. During the nominal handover of sovereignty to the politicians in Djibouti, the French together with Ethiopia worked hand in glove to thwart the Republic of Somali's plan to wrestle Djibouti to join the Greater Somalia scheme of things. It also fought a decade long war with a guerilla group, Front for the Restoration of Unity and Democracy further disorienting it. Today Djibouti lives at the mercy of the French and the WFP to feed its people. It is also major outlet for inbound and outbound Ethiopian goods. The relation of the two nations is organic in that it involves not only port services which Ethiopia highly values but because

Ethiopia export food items, construction materials and electric power to the country.

Needless to say, the Turku-European schemes to subjugate the Horn of Africa with Abyssinia as a primary target may not have been the root cause of the dysfunctional state apparatus existing in the region today or the pervasive poverty affecting over half of the population, but it contributed to the schisms that later played as a potential landmine to blow when trampled. Be that as it may, by the time that the Empire builder Menelik, had completed his campaign of territorial aggrandizement in the south, west and east, all the Somaliland save for the Ogaden, the tiny Djibouti and at one point the core of the Abyssinian civilization, the highlands of Eritrea with all its low laying territories was ceded to England, France and Italy respectively.[52] In short the dichotomies brought about by varying historical experiences in part explain the root cause of the wars waged within and between the states of the Horn of Africa in the past fifty years. More tragically, as the region struggles to make slow progress in the socio-economic fields, these historical elements remain potential to act as agents of open conflagration.

The Cause and Effect of the Wars

The skirmish and sometimes-open wars waged in the Horn of Africa for the past half century cannot be seen in isolation but in connection to the overall historical evolvement of the region. It is in fact the continuation of the regional conflict that have reached its climax in the last quarter of the 19th century with its spinoffs still felt during the decolonization of the Horn. The scramble to occupy this important part of the continent had also left a lingering effect on social relations among and between the peoples of the Horn Region. One of the legacies that colonial intervention left in the region is obviously underdevelopment with all its ugly manifestations of sectarianism based on religious, linguistic and mode of production. Famine is common phenomenon in the region while drought is cyclic. Moreover, instability, inter-pastoral-clan warfare, regional sentiment, and trench and *guerilla* wars are but some of the symptoms of this historical reality. No doubt, the elusive peace hankered by citizens, the prevalence of food insecurity, the millions of refugees and the undying but antagonistic relationship that exist between governments in the region can be traced to the past. They are the products that have their genesis in the historical past exacerbated by European intervention in the region.

It is known fact, that the deadly wars of the past 50 years drained the meager resources of Ethiopia exposing it to more famine and rendered it to be perpetually food aid dependent. It is my contention, therefore, that the multitudes of wars waged throughout the region for almost two centuries have a direct bearing first on the well-being of the present generation but also on

52 Menelik's challenging words were "I am not going wait folding my hand while European powers come to conquer lands".

their traditional institutions and livelihoods. No doubt, the most deadly wars waged so far within the past half century are the Eritrean war of self-determination and the trench war that followed soon after the separation, the nationalist war in Tigrai and the two conventional wars waged between Ethiopia and the Republic of Somalia in 1964 and in 1977. While these major wars were raging simultaneously there were more than a dozen of guerilla warfare and peasant uprising littering the landscape as well.

Unlike the 19th century where technical novelties made wars somewhat less obvious and infrequent, the wars of the mid to late 20th century with all its deadly war gadgets and toys can be equated to Armageddon. The sad part is that there is no full account given by the warring factions on human casualties. According to the EPLF the total human causalities of the civil war and later on the trench war of the late 1990s is about 85,000 without adding the civilians casualties caught up in the conflict. It is assumed that the size of the ordinary men and women who lost their lives and limbs in the Eritrean war is equal if not double to the number of the mobile and agile liberation fighters who died between 1960 and 2000. On the Ethiopian side nothing much is reported about human loss but it could be as much as those lives lost in Eritrea or even more. We have no reason to doubt that equal if not more victims are to be found in Somalia resulting from inter-state wars, conflict among the warlords and of outright brigand acts. Simply put the conservative estimate of the total human loss since the 1960s could be in the range of half a million. The countries have yet to quantify the opportunity time lost, industrial establishment destroyed, peasant lots and farms abandoned, potential investment shelved as a result of the long drawn wars and armed conflicts.

All the same, wars had negative consequence on the socio-economic progress of the countries involved, in particular Ethiopia who has to face challenges from within and without. The variety of wars waged under the guise of national oppression, national liberation, territorial integrity and even socialism have left a scar very hard to erase. Today the effect of it is easily noticeable on the lives of the people of the region who regard war as a regular phenomenon and not an aberration. The effect of the strife is also clearly reflected on the slow progress of these nations where their Human Development Index is the lowest.[53] It is no secret that as the result of the long drawn civil wars, the Horn of Africa was turned into a graveyard of humanity with nearly a million of stateless zombies' crisscrossing the frontier lines for safety.[54]

---

53 United Nations Development Program. Human Development Report 2010. The HDI for Ethiopia was guessed to be at 0.328, the 157th in the list of 161 countries measured. Although the HDI for Ethiopia has improved from 0.287 in 2005 it is still the lowest by any standard.

54 1/ Assefaw Bariagaber. 2006. Conflicts and the Refugee Experience, (Ashgate Publishing USA and England, 2006), 36. Bariagaber using the primary source from US Committee for Refugees came up with an astounding figure of 10,263,240

In the 30 year wars waged between the Imperial government and later the *Derg* against assorted liberation movements among them the ELF, the EPLF, TPLF, OLF, Western Somalia Liberation Front (WSLF) and the Ogaden National Liberation Front (ONLF), to mention few, the Ethiopian economy was completely left unattended. In this case investment in infrastructure was completely halted and industrial development almost forgotten. The once inconspicuous dots of food insecure areas in the map grew larger to call an international attention to save millions from scourge of famine. In the process, the effect of the war directly or indirectly consumed unimaginable number of peasants, pastorals and the young. Sadly enough, during the long war a number of educated and skilled people were lost irrevocably, some to migration and many caught up in the conflagration. While the conflict intensified, the Eritrean side targeted the young, the peasant as well as the educated in large numbers as cannon fodders while on the Ethiopian side the educated and the skilled were deliberately hunted by the socialist regime and by political/ideological rivals in a campaign proudly proclaimed as the *red terror*, *white terror*, and *free measure*.

Perhaps one of the most serious armed uprising that threatened national cohesion and fostered regional instability was the Eritrean war of independence which begun in 1960. The guerrilla war waged in Eritrea in the rugged northeastern mountains of Sahel, did not only attract the northerners but also inspired groups from the ethnic Oromo in the south and the West, the Marxists and socialist from the center, as well as separatist from the East. This is not because the Eritrean fronts were regarded as quintessential to the liberation of people from the unjust rule of the regime but were used as justification to realize sectarian ambitions based on regional/ethnic dreams. Finally, the thirty-year war was concluded with the Eritrean guerrillas succeeding in establishing an independent state of their own. No sooner did the war that claimed the lives of 65,000 Eritrean combatants and unaccounted hundreds of thousands of government soldiers and Eritrean/Tigrain civilians ended, that trench warfare between the newly formed state of Eritrea and Ethiopia was resumed to claim yet another hundred thousand lives.[55] The war of attrition that begun in May 1998 had again cost both countries valuable resources to the tune of billions of dollars, money foolishly spend on armaments.

Similarly, the guerilla war waged to diminish the power of the central government in Tigrai and if opportunity avails to establish an independent entity contingent to Eritrea begun to spread like a wild fire and threaten the very existence of the regime, which had arrogantly, dismissed the movement as

---

refugees at various times of the conflict with the majority living in the Sudan, Ethiopia and Somalia beginning 1974 up to 2002. (2)UNHCR. 2010 Regional Operations Profile - East and Horn of Africa. 2010. According to UNHCR following the showdown between and among the warlords in Somalia in May 2009, an additional 270,000 people were displaced bringing the total number of internally displaced persons (IDPs) to a staggering 1.4 million.

55 Figure officially reported by the leader of the EPLF in 1993.

narrow nationalists and an ally to a separatist movement.[56] As in Eritrea, the peasants of Tigrai were not spared from the punitive measures of the *Derg* causing havoc on their livelihood and forcing many hundreds of thousands to flee across the border to the Sudan.

According to the Ethiopian Economic Association (EEA), Ethiopia spent 25.4 billion EB (about USD 3 billion) to execute the 1998-2000 war with Eritrea. Before the start of the conflict with Eritrea the defense expenditure was in single digit. Tragically, within three years of the war the military expenditure increased to 40 percent.[57] Wuyi Omitoogun, who claimed that Ethiopia's expenditure on armament between the years of 1998 to 2000, was on the average 500 million USD annually is in agreement with those provided earlier[58]. Considering the military expenditure in 2005, which was accurately estimated by the CIA at 295.9 million USD, it is plausible that Ethiopia must have spent at least 2.1 billion USD. CIA report puts Eritrea's military budget in 2005 at 220.1 million USD.[59] Adding the one billion USD spent on the UN peacekeeping force stationed along the border with Eritrea, one genuinely expects that peace is at hand or more reasonably the conflict is resolved amicably.[60]

Tragically this is not to be so. The point in fact, the war of attrition morphed into what appears to look the demilitarized zone of the 38th parallel of the Korean peninsula where no interaction of peoples or trade exchange is possible. On the Eritrean side over 200,000 youth is under indefinite military service order not to be particularly useful for nation building. An equal number of those in military service live in refugee camps in Ethiopia, escaping the dreadful rule of the EPLF. Moreover, land mines planted by the warring parties had made crop cultivation impossible affecting the livelihood of the peasants further. In the meantime many of the abnormalities in Eritrea vivid to the naked eye go unchallenged while youth and the able bodied leave the country in mass with faint idea of where the destination is. It is amazing to realize that seemingly inconsequential boarder problem was all of sudden turned into more complexes and deadly affair that a continental mediation failed to solve. The

56 Aregawi Berhe, A Political History of the Tigray People's Liberation Front 1975 – 1991. (Teshai publishers and Distributors, 2009), 158. The most conspicuous feature of the manifesto 68 is its reformulation to reflect the inspiration of the people of Tigray for 'independent republic' of their own.

57 Alemayehu Geda, Does Conflict Explain Ethiopia's Backwardness? Yes and Significantly. Paper presented at Making Peace Work Conference at WIDER, (Helsinki, Finland 2004), 21.

58 Wuyi Omitagoun.2003. A Survey of Cameroon, Ethiopia, Ghana, Kenya, Nigeria and Uganda.SIPRI Research Report No. 17. (Oxford University Press, 2003), 41-45.

59 CIA World fact book (2005).

60 Martin Plaut and Sally Healy, Eritrea and Ethiopia: Allergic to Persuasion CHATHAM House, African program. AFB BP 07/01 January 2007.

protagonists in this case are two most needy nations one with chronically food deficit in record and the other euphemistically speaking a stillborn state, a product of violently induced birth. By all measures the Eritrea that is separated from Ethiopia in 1993 remained crippled state if not completely failed one with an annual exportable resource worth 30 million while importing 500 million USD worth merchandise from outside[61]. Thanks to the recent newfound fortune, gold and other valuable minerals in areas otherwise utilized as grazing ground by semi-pastoral population, the autocratic rulers of the state of Eritrea may have more money in their pocket than what they can do with it.[62]

No conflict in the African continent has drawn so much attention by continental and international diplomats like that of the Ethio-Eritrean conflicts of the late 1990s. All for the obvious reason that escalation of conflict, would not be in the best interest of both countries barely coming out of a devastating war that has consumed an entire generation. To that effect hundreds of diplomatic man-hour was expended mostly shuttling between Asmara and Addis Ababa to help mellow rhetoric and tamper emotions, all for the sake of reason to prevail. To the disappointment of foreign diplomats, statesmen and religious figures nothing tangible came out of the proxy discussions and back door meetings to avoid an impending calamity of dreadful proportion. Instead treaties were reneged, words twisted, statements read out of context and finally personality cult clashed leading to a duel where hundred thousand died and hundreds of thousands more temporary displaced.

The major stumbling block to resolving the conflict between Eritrea and Ethiopia was innocently attributed to the political culture that both the EPLF and the EPRDF acquired during the arduous armed struggle waged to oust the monarchy and later the *Derg*. In other wards mistrust, reluctance to open oneself and unwillingness to entertain suggestions for negotiated settlement to issues paramount to national survival is assumed to have been further 'deepened by years of guerilla struggle, in which compromise is equated with

---

61 Zenebe Nega Bashaw. Trajectories of Poverty, the Environment and Peasant Revolt in Ethiopia; a case study of Ethiopia. Research Project paper. According to Bashaw in the 1980s the Ethiopian government spent an annual average of 275 million dollars on waging war against Eritrean and Tigrai and other liberation movements. http://sedac.ciesin.org/openmeeting/downloads/1004891171_ presentation_bashawz.doc.

62 The Mail Guardian, Exposed: The Africans Named in the HSBC Swiss Leaks. (13 February 2015) Eritrea which until recently lived upon preying its diaspora population by extracting money for dubious reasons have recently been exposed as a country stashing 695.2 Million dollars in the HSBC Bank under different individual bank account. The guardian states " Eritrea ranked 182 out of 187 countries in the United Nations Human Development Index (HDI) report, topped the list of HSBC clients with whopping $695.2 million".

capitulation'.[63] This is partially true but does not explain fully the root cause of the political stubbornness consistently shown by the protagonists and by virtually all the minor and major opposition groups of past and present. In actual fact, there is no historical precedence one can allude to where political rivals in the region have come together with humility and political flexibility to resolve socio-political matters in the interest of peace and development.[64] This has been the case throughout the long history of Ethiopia where the powerful trying to physically suppress the weak and where the royal authority demands nothing less than complete submission.

Even today conflict resolution mechanisms are at their embryonic stage to ward-off any upheavals that could drag the country back to its old self.[65] In regional context compromise has always been regarded as a sign of weakness and a beginning to an end for absolute rule. It is important that one looks and acts as resolute. The fifty odd years of armed struggle waged between the different nationalist groups on one side and the center on the other together with the never ending squabble within and among the ethnically constituted political fronts are true to the *modus operandi* practiced in the old Ethiopian state. The same is true with Somalia where two decades since the inter-clan conflict begun, the warlords are on record of holding over twenty peace conference outside their disintegrated country to resolve the clan based political conflict that the warlord themselves has fostered in the first place.

Following the decolonization of the African continent, the Horn of Africa has passed through a series of violent episodes disrupting the lives of millions of people in the process.[66] As mentioned earlier, during the past fifty years

---

63 Sally Healy and Martin Plaut, Eritrea and Ethiopia: Allergic to Persuasion. (CHATHAM House, African program. AFB BP 07/01 January 2007), 8.

64 The EPRDF led government of Ethiopia has been negotiating with the Somali and Oromo nationalist movements opposing its rule and so far had succeeded of bringing few in the fold with some namely the ONLF a separatist Somali movement and OLF supported by Eritrea in 2010 still holding their ground.

65 In the 2005 Ethiopian general election, the opposition who stood to gain most by challenging the incumbent EPRDF did unthinkable. While assured about 150 seats in the parliament and the privilege to administer Addis Ababa, they refused to acknowledge the result of the election on the bases that it was rigged. True enough, few observers agreed that there was some irregularities but not to the extent of boycotting it altogether. The savviest EPRDF waited for the opposition to respond but to no avail. The group with only one or two sits in the parliament prior to the 2005 election had itself shut out from the process. Eventually the coalition of the assorted opposition parties disintegrated to make impact inside or outside the parliament. There is no word to describe the irresponsible act of the opposition who could have used parliamentary power to advance western style democracy in Ethiopia and perhaps the region.

66 Assefaw Bariagaber. 2006. Conflicts and the Refugee Experience. (Ashgate Publishing USA and England, 2006), 22. In 1988-1989 there were 1 million refugees from Ethiopia/Eritrea and Somalia.

there were two major armed conflicts between the Somali State and Ethiopia. This is on top of the low level guerrilla activities sponsored by the Somali state to entice the people of Ogaden for violent uprising. Likewise, each successive government responded to the Somali provocation with heavy hand resulting in massive human dislocation, destruction and death. The province, like Kashmir in the sub-Indian continent, is also a rallying point for Somali nationalists turned into warlords to promote self-motivated agenda under the cloak of achieving a unified Somali nation.[67] In a way the lack of cohesion currently exhibited among the Somali politicians toward the common objective of maintaining a united Somali Republic so as to assume the classic neo colonial status that is currently enjoyed by many African states is in part explained by their inability to achieve the elusive goal of Greater Somalia. The common denominator for forging a viable Somali nation appeared to have hinged first and foremost on the incorporation of the Ogaden followed by those Somali inhabited territories of Kenya and Djibouti.[68]

If history is a guide to the past, one cannot fail but observe that in Ethiopia internal and external contradictions are always resolved through coercion, intimidation followed by complete annihilation against those who choose to stand up rather than be taken by surprise. Of course this is accomplished at the expense of peace and tranquility, an important prerequisite for national development but most significantly for sustainable food production. In this regard, there are no mechanisms in place to resolve crisis much less prevent their escalation. If there were methods other than violent reaction to solving problem of complex nature certainly past and contemporary Ethiopian history would have read differently. The traditional method of conflict resolution if at all existed is always overshadowed by the powerful. Typical example is that of EPLF which has deliberately undervalued all indigenous methods and practices of conflict resolution and usurped administrative power of localities to supplant it with its own.[69] In such a circumstance the winner-take-all strategy reign supreme making negotiation among adversaries difficult. The recent move by

67 The forceful incorporation of the Somali inhabited territories, namely the Ogaden, the ex-North Frontier Districts (NFD) of Kenya and parts of the republic of Djibouti into the republic of Somalia has been the main preoccupation of successive Somali leaders for over half of century. Thus, whenever opportunity avail itself the government in Mogadishu never hesitated to take advantage toward fulfilling the goal of Greater Somalia.

68 Seyoum Haregot, The Bureaucratic Empire: serving Emperor Haile Selassie. (Trenton: Red Sea Press 2013), 185. The premier of Somali Ibrahim Egal expressed his candid view to the Ethiopian authorities during the rapprochement period of 1969 that no Somali leader risks of dropping the demand of Grater Somalia and still stand proud in front of his people.

69 Gaim Kibreab, Critical reflections on the Eritrean War of Independence: social capital, associational life, religion, ethnicity and sowing seeds of dictatorship. (Trenton: Red Sea Press, 2008).

the EPRDF government to peacefully deal with some of the Somali armed groups whose existences span four decades is an exception rather than the rule.[70]

Unfortunately, the one and only institution that one expects to play a moderating role in any social upheaval is the Ethiopian Orthodox Church. The Church which have miraculously survived for 1700 years as an odd post in a predominantly Muslim region have to operate hand in glove with the state to safeguard its interest and the interest of the sovereign by advocating respect and obedience for persons in power. From all accounts the church is credited to have played a minimal role in conflict prevention. To the contrary, the church has consistently taken sides in many internal conflicts save in some extraordinary circumstance where the people implore the church to intervene to stop an impending battle[71]. And perhaps when foreign elements are involved in subjugating the country.[72] It is indeed very common for the church to be on the side of one or the other depending on regional allegiance and with the expectation of largesse from the victor after the conclusion of the conflict. In the past it was customary for churchmen to come out in full ecclesiastical regalia with the *tabot* at the head of the procession to cheer up favored combatants or those entwined in a struggle for power at the expense of peace and security.[73] As recent as the Ethio-Eritrean war of 1998-2000, the Orthodox church of both countries played no role to tamper emotions among the flocks or used the ultimate weapon of power, that is, excommunicate the demagogues who bleed the people to death.[74] In an exceptional case some church orders may openly defy the Emperor and break the delicate alliance forged between the church and the state. An example of sort are the Saint Estefanos order of

70 Ethiopian Television (November 6, 2010).

71 In 1832 at the beginning of the end of the civil war a major battle between two powerful warlord Dejach Wbie and Dejach Khasai Sebagadis was averted by the church's impartial intervention. Perini, Ruffilo. 1905. Da qua A Mereb (መረብ ምላሽ). Chiaramonte Roberto, 1905), 167. (Tigrigna, 1997)

72 Abuna Pietros who in 1936 did not only openly opposed Italian occupation of the country but issued an edict of excommunication to persons cooperating with the Italian fascists.

73 A replica of the Arch of Covenant tablet handed to Moses by God in Mount Sinai. It is the most revered icon in the Ethiopian Orthodox Church. Suffice is to mention that according to the long standing church narration the original Arc of Convenient is currently housed at St Mary of Zion in Axum, the sacred city of the Ethiopians.

74 In 1999, that is, in the middle of the bloody war, there was a lukewarm effort by religious groups of both sides to mediate peace between the opposing Eritrean and Ethiopian forces. Unfortunately, the mediation yield no tangible result to deflect the duel that claimed 70,000 dead and over a quarter of million internally displace persons.

14th century Ethiopia.[75] One cannot expect different results in the too-often erupting wars between Ethiopia and Somalia.[76]

In the absence of civil societies and/or broad based political opposition groups to stem off potential danger that could develop into armed conflict, the Ethiopian state and by extension the mini states surrounding it periodically lock themselves into an impasse that they could not extricate from. In more precise term they lack the political wisdom that could lead to a resolution of issues that may have dire consequence on peace and stability. This in turn suggests a serious deficiency in the state apparatus upon which conflict resolving mechanism are built to defuse explosive issues that would eventually lead to war. In a region where the identity of nationalities are strong, where the three Abrahamic religion namely Judaism, Christianity, and Islam rigidly exist side by side and where European colonialism had its lingering effect on parts of the region together with the subtle influence of littoral Muslim states on the Muslim population, a policy to address these historically evolved issues need to assume a paramount importance. In summary, these are the major problematic issues that have bogged down successive governments in Ethiopia and those smaller states around it for decades. Without the political will to resolve minor disagreements and the capacity to tackling fundamentally divergent issues, there could neither be steady socio-economic progress nor an assured and abundant food resource for Ethiopians and the people in the region. There could be neither regional integration nor peace without addressing this life and death issues that have consumed past and present generations in an endless national, ethnic, tribal and religious conflicts.

If the ruling EPRDF succeeded in resolving conflicts that may arise in the future through peaceful means, then you cannot put it on the same scale with the past governments whose mottos was "Ethiopia or death". The recent move to narrow down political difference among potential adversaries by the EPRDF is the only and correct way to defuse explosive issues such as secession and religious fundamentalism.[77] Frankly, the burden of responsibility falls on Ethiopia, the inheritor of the onetime mighty Axum, a country with huge population and by far the largest economy and still growing.

---

75 A typical example is that the brutal persecution of a Christian order in mid-14th century named *Dekike Estifanos* by the church hierarchy and the state for advocating that the Emperor is simply human who does not deserve reverence reserved to divine. They also questioned the appropriateness of bowing to the image of Marry the mother of Jesus, a very blasphemous act for a Christian society whose regard to the virgin Marry is unparalleled in the Christian world.

76 There were 3 major wars fought between Ethiopia and Somali since the early 1960. One cannot exclude the 2007 Ethiopian armed incursion into Somali to battle the Union of Islamic Courts, the then de facto government.

77 The EPRDF led government of Ethiopia has been negotiating with the Somali and Oromo nationalist movements outside the government and had succeeded to bring some in the fold.

REFERENCES

Abebe Atilaw and Lijalem Korbu. Recent Development in Seed Systems of Ethiopia. Debre Zeit Research Center Ethiopian Institute of Agricultural Research, 2010.

Addis Hiwot. Ethiopia: from autocracy to revolution. London: Occasional Publication No1. Review of African Economy, 1975.

Afevork G. J. Guide du voyageur en Abyssinie. Rome: 1908.

Agriculture in Ethiopia. Yemastaweqia minster (Ministry of information). Addis Ababa: 1964.

Aklilu Yacob. An audit of the livestock Marketing Status in Kenya, Ethiopia and Sudan (Vol. 1) AU-IBAR Nairobi: 2002.

Albright, W.F. The Excavation of Tel Beit Mirsim. Bulletin of the American Schools of Oriental Research, 1926.

Aleme Asresie and Lemma Zemedu. Contribution of Livestock Sector in Ethiopian Economy: A Review. Advance in Life Science and Thecnology, 2015.

Alemneh Dejene. Towards Sustainable Agriculture and Rural Development in the Ethiopian Highlands. Proceedings of the Technical Workshop on Improving the Natural Resources Base and Rural Well-being, Bahir Dar, Ethiopia: 2003.

Alemu Ghebrewold. IAR Newsletter Vol. 12, 1997.

Alula Abate and Tesfay Teklu. Land Reform, Structural Changes and Rural Development in Ethiopia: case study of two widely different regions. Institute of Development Research. Addis Ababa: Addis Ababa University, 1978.

Alula Abate. Peasant Association and collective Agriculture in Ethiopia: Promise and Performance. Seminar paper presented at the Addis Ababa: Addis Ababa University Geography Department, 1981.

Alvarz, Francisco. The Prester John of the Indies; A true relation of the lands of the Prester John, being the narrative of the Portuguese embassy to Ethiopia 1520. The Translations of Lord Stanley of Alderley (1881). Additional material by G. F. Beckingham and G. W. B. Huntingford. Published for the Hakluyt Society, University Press, 1961.

Allen, R. C. Ethiopian agriculture in a world-historical context. Background note prepared for a study on Agriculture and Growth in Ethiopia, 2008. **************

Aregawi Berhe. A Political History of Tigray People's Liberation Front 1975-1991. Addis Abeba: Tsehai Publishers, 2009.

Asbe Hailu. 'sile mengst gudat'. Addis Abeba: Brhanina Selam Printing press, 1927. (Amharic)

ASMAI 3/5-33 Report of Governor Orero to MAE (*Ministere des Affaires Etrangeres)*

Bairu Tafla. Asmegiorgis and His Work: History of Gala and the Kingdom of Shoa. Stuttgart: Franz Steiner Varlag Weisbadan, 1987.

Assefaw Bariagaber. Conflicts and the Refugee Experience. Ashgate Publishing USA and England, 2006

Asseffa Bekele and Eshetu Chole. A Profile of the Ethiopian Economy. Addis Ababa-Oxford Press, 2009.

Ayele Gebre. Pastoral Development and policy orientation in Ethiopia: appraisal of policies and experience. *In. Agriculture Policy in Ethiopia. Economic development: scope, issues and prospects. Proceedings of the 6th annual conference of agriculture economic society,* 2002

Bahru Zewde. Pioneers of Change in Ethiopia. Addis Ababa University Printing Press, 2002.

Bahru Zewde. A history of Modern Ethiopia 1855-1974. James Curry London, Ohio University Press, Addis Ababa University Press, 1991.

Barlow, Maude. Blue Covenant: the global water crisis and the coming battle for the right to water.. New York: The New Press, 2007.

Barnett, Terttia. The Emergence of Food Production in Ethiopia. Cambridge Monographs in African Archeology 45. Published Basingstoke Press, 1999.

Berhanu Abegaz. "Escaping Ethiopia's Poverty Trap: The Case for a second Agrarian Reform"*Journal of Modern African Studies, Vol. 42, No. 3, 2004*

Berhanu Debele. The Vertisols of Ethiopia: their properties, classification and management. In: *Fifth Meeting of the Eastern African Sub-Committee for Soil Correlation and Land Evaluation. Wad Medani, Sudan, 5-10 December 1983. World Soil Resources Reports No.* 56. Rome: FAO, 1985.

Bernal, Martin. The Afro-Asiatic Roots of Classic Civilization. Vol III. The Linguistics Evidence. London: Free Associations Books, 2006.

Brhanu Nega, Berhanu Adenew and Samuel Gebreselassie. Current Land Policy Issues in Ethiopia. Land Reform, 2003/3, Special Edition. FAO Corporate Document Repository, 2003.

Bonner, Margaret. USAID/ Ethiopia Program from FY 1951 to FY 1978. Addis Ababa US Mission, 1978.

Bojo, J. and Cassells, D. Land degradation and Rehabilitation in Ethiopia: a reassessment. AFTES Working paper # 17. World Bank, Washington, 1995.

Borowski, Oded. Agriculture Iron Age Isreal. Eisenbruns, 1987.

Bowersock, G.W. The Thorne of Adulis: Red Sea Wars on the Eve of Islam. Oxford University Press, 2013.

Brhanu Nega, Berhanu Adenew and Samuel Gebreselassie. Land Tenure and Agricultural Development in Ethiopia. EEA Workshop. Addis Ababa: 2002.

British Foreign Office, Peace Handbooks, # 129: Ethiopia. London: H.M Stationary O, 1920.

Bruce, James. Travels to discover the source of the Nile, in the years 1768, 1769, 1770, 1771, 1772 and 1773. London: C.G.J. and J. Robinson, 1790.

Herbert, Blundell W. (ed., tr.), *The Royal Chronicle of Abyssinia, 1769–1840*, Cambridge: 1922.

Caulk, R.A.. Firearms and Princely Power in Ethiopia in the Nineteenth Century. Journal of African History, xiii, 4, 1972.

Chilot Yirga Tizale. The Dynamics of Soil Degradation and Incentives for Optimal Management in The Central Highlands of Ethiopia. Doctoral Dissertation. University of Pretoria, 2007.

Cohen, M. John and Weintraub, Dov. Land and Peasants in Imperial Ethiopia. Assen: Van Gorcum & Comp. B.V., 1975.

Cotese, G. Problemi dell'impero –Rome: 1938.

Cossins, N. The status of ponds in the western SORDU project area. JEPSS Research Memorandum 2. Addis Ababa: ILCA, 1983.

Castanhoso. The Portuguese Expedition to Abyssinia in 1541 – 1543. Ed. Whiteway, R.S. London: Hakluyt Society, 1902.

CSA, 2013.

Damtew Teferra and Philip G. Altbach, *African Higher Education: An International Reference Handbook*. Indiana University Press, 2003.

D'Andrea, A. Catherine, Andrea Manzo, Michael J. Harrower and Alicia L. Hawkins. The Pre-Axsumite and Axsumite Settlement of NE Tigrai, Ethiopia. *Journal of Field Archaeology. Vol.* 33, 2008

Daniel Gamachu. (1988). Environment and Development in Ethiopia. In Beyond the Famine: An Examination of the issues Behind Famine in Ethiopia. Geneva: International Institute for Relief and Development and Food for the Hungry International, 1988.

Deininger, Klaus, Daniel Ayalew Ali, Holden, Stein, and Zevenbergen, Jaap. Rural land certification in Ethiopia: Process, initial impact, and implications for other African countries. World Bank Policy Research Working Paper 4218, 2007.

Dercon, Stefan, Hill, Vargas Ruth and Andrew Zeitin. Revisiting Agriculture Led growth in Ethiopia. Study funded by the Department of International development-DFID-UK., 2009.

Deressa Amantie. Sile Giziew Yegenzeb Chigir. Addis Ababa: Brahanina Salem printing press, 1932. (Amharic).

Desalegn Rahmato. Agrarian Reform in Ethiopia. Scandinavian Institute of African Studies, Uppsula, 1984.

Desalegn Rahmato. Agrarian Change and Agrarian Crisis: state and peasantry in post-revolutionary Ethiopia. Africa: Journal of the International African Institute, 1993.

Desalegn Rahmato. Ethiopia: Agricultural Policy review In: Digest of Ethiopia's National policies, Strategies and Programs, 2008.

Dikötter, Frank. Mao's Great Famine: The History of China's Most Devastating Catastrophe, 1958-62 New York: Walker & Company, 2010.

Donahue, Donald and Wendy James. The Southern March of Imperial Ethiopia: Essays in history and social anthropology, *Letter from Ghebregzeiabher to Lij Eyassu Megabit 2, 1905 E.C. and 11 march 1913 G.C.* Cambridge University Press, 1986.

Donahue, Roy. *ETHIOPIA: Taxonomy, Cartography and Ecology of Soils*. African Studies Center, Michigan State University, 1972.

Dunning, Harrison. Land reform in Ethiopia: a case study in non-development. UCLA Law Review, 1970.

Edwards, Sue, Arefayne Asmelash, Hailu Araya and Tewoldeberhan Gebregziabher. Impact of Compost Use on Crop Yields in Tigray, Ethiopia. Rome: Food and Agriculture Organization, 2007.

Ephraim Isaac. The Ethiopian Orthodox Tawahido Church. Red Sea Press, 2012.

Eshetu Bekele and Mammo Muchie. DIIPER Research Series Working Paper No. 1. Aalborg University, 2009.

Ethiopian Agriculture Research Institute. www.eair.gov Addis Ababa: September 14, 2009.

Fiseha Giorgis Abyezghi, Debtera. Tarik Itopia. (The manuscript was published with Italian introduction by Dr. Yakob Beyene 1979). 1897. (Tigrigna)

Fitsum Hagos, J.Pender & Nega Gebreselassie. Land Degradation in the Highlands of Tigrai and strategies for sustainable land management 2nd edition, 2002.

Foster, William. The Red Sea and Adjacent Countries at the Close of the Seventeenth Century. London: Hakluyt Society, 1949.

Franklin, John Hope. From Slavery to Freedom, history of African-Americans. 9th edition. McGraw Hill, 2011.

Gaim Kibreab. Critical reflections on the Eritrean War of Independence: social capital, associational life, religion, ethnicity and sowing seeds of dictatorship. Trenton: Red Sea Press, 2008.

Cotese, G. Problemi dell'impero –Rome, 1938.

Gebr Tsadik Degefu. *The Nile: Historical, Legal and Developmental Perspective.* Canada: Trafford Publishing, 2003.

Gebru Mersha and Mwangi wa Githinji. Untying the Gordian Knot: The Question of Land Reform in Ethiopia. ISS/UNDP Land, Poverty and Public Action Policy Paper No. 9. , 2005.

Gebru Tarekegn. Rural Protest in Ethiopia: A Study of Three Rebellions. Doctoral Dissertation. Syracuse University, 1977.

Haggai, Erlich. *Ras Alula and the Scramble for Africa.* Trenton: The Red Sea Press, 1996.

Valentine, Visconte George. Analysis of New Works of Voyages and Travels; Voyages and travels to India, Ceylon, the red sea, Abyssinia and Egypt in 1802, 1803, 1804, 1805 and 1806. 1810. London: Richard Phillise, 1810.

Ghebrehiwot Baykedagn. *Mengstna Yehizb Astedader* 'Government and Public Administration', Addis Ababa: 1953 EC. (Amharic).

Ghebrehiwot Baykedagn. (1912). *Atse Menelik na Ityopia* Asmera: Berhan Yehun. (Amharic)

Ghose, Ajit Kumar. Transforming feudal agriculture: agrarian change in Ethiopia since 1974. Journal of development Studies, 1985

Gibbon, Edward. The History of the Decline and fall of the Roman Empire: Chapter XLVII: Ecclesiastical Discord. Part V London: Strahan and Cadell, 1776–1788

Girma A. Demeke. *The Origin of Amharic.* Centre Francais d'Etudes Ethiopiennes, Etudes Ethiopiennes' 5. Addis Ababa: 2009.

Girma Kebede.The State and Development in Ethiopia. Atlantic Highlands: Humanities Press, 1992.

Girmachew Teklehawariat. Ali Habeshi. Ye Adenu Janderba Addis Ababa: Rome Printing Press, 2000 EC.

Greenfield, Richard. *Ethiopia: A New Political History,* London: Pall Mall Press, 1965.

Harlan, J.R. Ethiopia: A Center of Diversity. *Economic Botany* 23:309-314, 1969.

Harris, Cornwallis. *The Highlands of Ethiopia.* 2nd edition Vol. 1 London: Longman, Brown, Green, 1844.

Hodgson, R J. The Southern Sidamo Rangelands Project 1985-88: Perspectives on Development Interventions and Extension. Addis Ababa: ILCA library accession no 87515, 1990.

Holden, T. Stein, Klaus Deininger, and Hosaena Ghebru. Impacts of low-cost land certification on investment and productivity *American Journal of Agricultural Economics*. 91:2; 359-373, 2009.

Hufnagle, H. P. Agriculture in Ethiopia. FAO of the United Nations, 1961.

Huntingford, G.W.B. The Galla of Ethiopia, The kingdoms of Kafa and Janjero. International African Institute. Northeast Africa Part II. London: Lowe Brydone printers Ltd., 1955.

Iliffe, John. Africans: the history of a continent. Cambridge University Press, 2007.

Jeronimo Lobo, *The Itineration of Jeronimo Lobo*, Donald Lockhart, trans., London: Hakluyt Society, 1984.

Jutzi, S and I. Haque, (1985). Management of deep clay soils (Vertisols) in sub-Saharan Africa. ILCA/ICRISAT, 1985.

Keller, Edmond J. Revolutionary Ethiopia. From Empire to Peoples Republic. Midland Book Edition, 1991.

Kidanemariam H. M. Project Proposal for the Development of an Ethiopian Potatoe Program. Addis Ababa, 1980. (Manuscript)

Killion, Thomas. Workers, Capital and the State in the Ethiopian region, 1919 – 1974. Stanford University, 1985.

Kirsch, Ottfried C., Fred V. Goricke, and Johannes F. G. Worz. Agricultural Revolution and Peasant Emancipation in Ethiopia: A Missed Opportunity. Saarbruken: Verlag Breitenbach, 1989.

Lapiso G. Dilebo. A brief account of the Gideo people's history 1882-1966. Addis Ababa, 1972. (Amharic)

Laszlo Saska. *Fascist Italian Brutality in Ethiopia 1935 – 1937*. Trenton: The Red Sea Press, 2015.

Lessona, A. Memorie. Florence, 1958.

MacLean, Kenneth. AGRARIAN AGE: *A Background for Wordsworth*. New Haven: Yale printing Press, 1950.

Marcus, Harold. History of Ethiopia, 2nd edition. University of California Press, 2002.

Maaza Bekele. A study of Modern education: its foundation, its development, and its future with emphasis in primary education. Doctoral dissertation. Columbia University, 1966.

McCann, James. People of the Plow: An Agricultural History of Ethiopia, 1800-1990.University of Wisconsin Press, 1995.

McCain, James. Maize and grace: History, Corn and Africa's new landscape, 1500-1999. Comparative studies in Society and History, Boston University, 2001.

Brooks, Miguel F. Kebra Nagast (**ክብረ ነገሥት**) ed. Red Sea Press, 1995.

Mekonen Getu. Socialism, participation and Agricultural Development in post-revolutionary Ethiopia: A study of Constraints. Doctoral Dissertation. Almqvist and Wiksell International, c1987.

Mengustu Woube. The Geography of hunger: some aspects of the causes and impact of hunger. University of Upsala, Department of Social Economy, 1987.

Mersie Hazen Woldekirkos. *Memoire, on the Reign of Menelik II*. Addis Ababa: Zamra Printing Press, 1935 (reprinted 2002). (Amharic)

Merrick, W. Thomas. Population and Poverty: New Views on an Old Controversy. International Family Planning Perspectives Vol.28 No.1, 2002.

Mengistu Lema. *Meshaf Tizita Ze Aleqa Lema Hailu*, 1954. (Amharic)

Merid Welde Aregai. Gonder and Adwa the tale of two Cities. Proceedings of the 8th International Conference of the Ethiopian Studies Vol.2, 1989.

Halderman, Michael. Pro-Poor_-Livestock Policy. The Political economy Pro-Poor Livestock Policy Making in Ethiopia. FAO PPLPI Working Paper 19, 2004.

Brooks, Miguel F. Kebra Nagast (**ክብረ ነገሥት**). Trenton: Red Sea Press, 1995.

Miressa, D, and RoBarge, W. P. (1999) Soil characteristics and management effects on phosphorus sorption by high land plateau soils of Ethiopia. Soil Sci Soc Am J, 1999.

MoA sample survey, 1974-75. (1975). Vol.1. Addis Ababa: 1975.

MoA/ FAO. Ethiopian Highland Reclamation Studies Final Report. Addis Abeba: 1986.

Mohammed Hassen. Futuh Al-Habaša: The Conquest of Abyssinia [16th Century] by Šihab ad-Din Ahmad bin Abdal-Qader bin Salem bin Utman; International Journal of Ethiopian Studies, Vol. 1, No. 2, 2004.

Mulat Demeke and Bekele Hunde. Population Pressure and Intensification of Agricultural Production in Ethiopia: The Will and Cost of Overcoming Malthusian Trap. First International Conference on the Ethiopian Economy, EEA. Addis Ababa, 2003.

Murphy, H.F. Fertility and Other Data on Some Ethiopian Soils. Experimental Station Bull. No. 11. Ethiopian College of Agriculture and Mechanical Arts. Dire Dawa, 1963.

Olson, Mancur Jr. Big Bills Left on the Sidewalk: why some nations are rich and others poor. Journal of Economic Perspectives. Vol. 10 No.2, 1996.

Pankhurst, Richard. Guns in Ethiopia. Transition, No. 20. Indiana University Press, 1965.

Pankhurst, Richard. Economic History of Ethiopia 1800 – 1935. Haile Sellassie I University Press, 1968.

Pankhurst, Richard. A Social History of Ethiopia. Institute of Ethiopian Studies. Addis Ababa University, 1990.

Pankhurst, Richard. The Ethiopian Boarderlands. Essays in regional History from ancient times to the end of the 18th century. Trenton: The Red Sea Press, 1997.

Pearce, Nathenial. The Life and Adventures of Nathaniel Pearce. London: Henry Colbum and Richard Bentley, 1831.

Perini, Ruffilo. Da qua A Mereb (መረብ ምላሽ), 1905. Tigrigna text. Chiaramonte Roberto Edition, 1905. (Reprinted Tigrigna, 1997

Phillpson, W. David. *Foundation of an African Civilization, Aksum and Northern Horn 1000BC – AD 1300.* James Curry Publishing, 2012.

Plowden, Walter Chicele. Travels in Abyssinia and The Galla Country with an account of the mission to Ras Ali,. London: Longmans green and Co., 1868.

Poncet, Charles Jacques *A Voyage to Æthiopia Made in the Year [sic] 1698, 1699, and 1700.* Ed. by Sir William Foster London: Hakluyt Society, 1949.

Randolph Barker and François Molle. Evolution of Irrigation in South and Southeast Asia Comprehensive Assessment of Water Management in Agriculture. Research Report No. 5. Colombo, Sri Lanka: Comprehensive Assessment Secretariat, 2004.

Richards, Alan. Higher Education in Egypt. Policy Research Working Papers.WPS 862. World Bank, 1992.

Rodinson, Maxime. *Mohammed.* New York: Vintage Books, 1974.

Sayer, A.J.; Harcourt, S.C. and Collins, M. N. (1992). The Conservation Atlas of Tropical Forest Africa. IUCN Cambridge, 1992.

Sbacchi, Alberto. Legacy of Bitterness Ethiopia and Fascist Italy 1935 – 1941. Trenton: Red Sea Press, 1997.

Schmidt, R. Peter, Curtis T. Matthew and Zelalem Teka. The Archaeology of Ancient Eritrea. Trenton: Red Sea Press, 2008.

Serageldin, Ismail. Water Wars? World Policy Journal, 2009.

Seyoum Haregot. The Bureaucratic Empire: serving Emperor Haile Selassie. Trenton: Red Sea Press, 2013.

Seyoum Gebregziabher. The Symphony of my Life. Trenton: The Red Sea Press, 2012.

Torjman, Sherri. What is Policy? Ottawa: The Caledon Institute of Social Policy, 2005. PDF at: http://www.caledoninst.org/ Publications/ PDF/544ENG.pdf.

Sileshi Bekele. Challenges and opportunities in Capacity Building for water resources, R&D. MoWR/EARO/IWMI/ILRI/ Workshop Ethiopia, 2002.

Singh, Harjinder. (1987). Agricultural Problems in Ethiopia. New Delhi: Gian Publishing House, 1987.

Stahl, Micheal. Research Contradiction in Agricultural Development: A study of three minimum package projects in Southern Ethiopia. Report # 4, 1973.

Stahl, Michael. Ethiopia Political Contradiction in agriculture development. Uppsula: Publications of the Political Association No 67, 1975. Using primary sources from "Harrison Dwnning. Land reform in Ethiopa; a case study in non-development UCLA law Review, Vol. 18, No2, 1970.

Tadesse Mecha Tikur Anbessa. Asmara: Corriera Eritrea Press, 1943. (Amharic)

Techane Adugna. Birth of the Ethiopian Seed Trade Association Seed Info, Official Newsletter of the WANA Seed Network - MoA, 2007.

Teffera Haileselassie. The Ethiopian revolution 1974-1991. From Monarchial Autocracy to Military Oligarchy, 1997.

The Travels of Marco Polo Compiled in 1298, First Edition. 1921, Barnes and Nobles, 2005.

Tekeste Negash. No Medicine for the Bite of a White Snake: Notes on Nationalism and Resistance in Eritrea, 1890-1940. Printed in Sweden by Reprocentralen, HSC Uppsala. From. "A Memoire of Blatta Gebre Egziabeher Gila Mariam from Tzzadda Kristian, Hamassien, 1889 EC." (1896/97), 1986.

Tekeste Negash. The Crisis of Ethiopian Education: Some Implications for Nation-Building Uppsala Department of education, 1990.

TekleTsadek Mekruia. (1960). Ye Gragn Ahmed Werera (የግራኝ አህመድ ወረራ). Addis Abeba: Brhanena Selam Printing Press, 1960.

Tewdaj Mogues. (2004). Must drought and poverty go hand-in-hand? College of Agriculture and Life Science. Madison: University of Wisconsin, 2004.http://news.cals.wisc.edu/departments/highlights/2004/03/19/must-drought-and-poverty-go-hand-in-hand

Tirfe Mammo. The Role of Indigenous Knowledge, Traditional Practices, and Local Institutions: the Ethiopian case. Trenton: Red Sea Press, 1999.

Trudeau Edouard. A survey of Higher Education in Ethiopia. With emphasis for future planning and development. Doctoral Dissertation. Columbia University, 1967.

US Census Bureau. International (2010).

Vavilov, N. I. Studies on the Origins of Cultivated Plants. Leningrad: Bulletin of Applied Botany and Plant Breeding, 1926.

Ward, M. and Yeshanew. World Sea Surface Temperatures and Kiremt Rains in Ethiopia. Addis Ababa, 1990.

Webb, Patrick, Joachim von Braun, and Yohannes Yisehac. Famine in Ethiopia: of Coping Failure at National and Household Levels. IFPRI, 1992.

Westphal, E. Agricultural Systems in Ethiopia. Agriculture Research Reports. Wageningen: Center for Agricultural Publishing and Documentation, 1975.

Wolska-Conus, *Christian Topography III 65*. 1968-73.

World Food Program. 1993.

Wong, John. Chinese Land Reform in Retrospect. Land Tenure Center, Madison: University of Wisconsin, 1974.

Wylde, A.B. Modern Abyssinia. London: Mathuen & Co, 1901.

Yaraswork Admassie. Twenty Years to Nowhere: Property Rights, Land Management and Conservation in Ethiopia. Trenton: Red Sea press, 2000.

Yebio Woldemariam. A Report on the Socio-economic Conditions, Production

Status and Institutional Strength of the Afar: Background Information Document to the Comprehensive 'needs Assessment Study'. UNDP, 1994.

Yebio Woldemariam. Land In Eritrea: a contentious commodity. Taking On Current Political Issues of Eritrea. London, 2010. Proceedings of CDRiE'S Conference. http://ssrn.com/abstract= 1664873.

Zewde Reta. Yeqedamawi Haileselassie Mengist. New Delhi: Laxmi Publications (and Boston USA, 2012. (Amharic).

Zewdie Reta. Teferi Mekonen and the Long Reign of Zewditu and Iyassu 1884 – 1922. Addis Ababa: Central Printing Press, 2005. (Amharic).

INDEX

A

B

C

D

E

F

G

H

I

J

K

L

M

N

O

P

R

S

T

U

V

W

Y

Z